Amphiphilic Molecules, Interfaces and Colloids

Amphiphilic Molecules, Interfaces and Colloids

Guest Editors

Khristo Khristov
Plamen Tchoukov

MDPI

Basel • Beijing • Wuhan • Barcelona • Belgrade • Novi Sad • Cluj • Manchester

Guest Editors
Khristo Khristov
Institute of Physical Chemistry
Bulgarian Academy of Sciences
Sofia
Bulgaria

Plamen Tchoukov
Institute of Physical Chemistry
Bulgarian Academy of Sciences
Sofia
Bulgaria

Editorial Office
MDPI AG
Grosspeteranlage 5
4052 Basel, Switzerland

This is a reprint of the Special Issue, published open access by the journal *Molecules* (ISSN 1420-3049), freely accessible at: https://www.mdpi.com/journal/molecules/special_issues/TL4T305L4S.

For citation purposes, cite each article independently as indicated on the article page online and as indicated below:

Lastname, A.A.; Lastname, B.B. Article Title. *Journal Name* **Year**, *Volume Number*, Page Range.

ISBN 978-3-7258-5959-7 (Hbk)
ISBN 978-3-7258-5960-3 (PDF)
https://doi.org/10.3390/books978-3-7258-5960-3

Contents

About the Editors

Khristo Khristov

Khristo Khristov received his PhD in 1991, and in 1993, he became an Associate Professor at the Institute of Physical Chemistry, Bulgarian Academy of Sciences (IPC-BAS). He spent three years as a visiting professor (1997-1999; 2005-2006) at the University of Alberta, Edmonton, Canada. In 2008, Dr. Khristov was the Secretary of the Scientific Council of IPC-BAS. He was a Member of the Organizing Committee of the IX International Conference on Surface and Colloid Science, 1997, Sofia, Bulgaria, and the Scientific Secretary of Nanoscale Phenomena in Colloid and Interface Science Workshop (NPCIS 2007) in Plovdiv, Bulgaria. Dr. Khristov's scientific areas of interest include foam and foam films, emulsions and emulsion films, and dispersions stabilized by synthetic, polymeric, biopolymeric, and natural surfactants. While at the University of Alberta, he studied petroleum emulsion properties and water/oil/water emulsion films and acted as a scientific adviser to two PhD students. He supervised three PhD students at the IPC-BAS. He was the principal investigator on several international and national scientific projects, including with Oil Company ELF France, Syncrude-Canada, MaxPlanck Institute Germany, etc. Lately, he has been involved in studying microrheological and microfluidic mechanical properties of blood cells. All these studies encompass more than 60 papers, including reviews in renowned scientific journals (Q1 and Q2) and a book chapter (*Foam Films and Foams, Fundamentals and Applications,* 2021, CRS Press) with over 1100 citations. Dr. Khristov holds an H-index of 19.

Plamen Tchoukov

Plamen Tchoukov is an Associate Professor at the Institute of Physical Chemistry, Bulgarian Academy of Sciences (IPC-BAS). His current research interests focus on self-assembly of amphiphilic molecules, surface forces and colloid stability, the application of green surfactants in mineral processing, oil-spill containment, functionalized colloidal systems for delivery of bioactive agents, as well as corrosion-related colloidal and interfacial phenomena occurring in pipelines for the transport and storage of CO_2. He holds an MSc degree in Physics from Sofia University and a PhD in Physical Chemistry from the Bulgarian Academy of Sciences. He continued his work as an NSERC Postdoctoral Visiting Fellow at Natural Resources Canada, CanmetENERGY, where he developed and applied state-of-the-art thin emulsion film instrumentation to study the stability of petroleum emulsions that occur during heavy oil extraction. This work contributed to the deeper understanding of the stability of oil-in-crude oil emulsions, particularly the role of asphaltenes and their most interfacially active subfractions. Subsequently, in the Oil Sands Group of Dr. Zhenghe Xu at the University of Alberta, in addition to the research on crude oil emulsions, he designed and built a novel Thin Film Force Apparatus (TFFA). This instrument enables simultaneous measurement of film thickness profiles, bubble or droplet shape, contact angles, and forces in thin liquid layers between approaching bubbles/droplets and a solid surface. The TFFA has proven highly valuable for studying hydrodynamic interactions and the forces governing bubbles and droplets' attachment to solid surfaces. In addition to his academic career, Dr. Tchoukov has more than ten years of industrial experience in the formulation of innovative specialty chemicals.

molecules

MDPI

Editorial

Amphiphilic Molecules, Interfaces and Colloids

Khristo Khristov * and Plamen Tchoukov *

Institute of Physical Chemistry, Bulgarian Academy of Sciences, "Acad. G. Bonchev" Str. Bl. 11, 1113 Sofia, Bulgaria
* Correspondence: khristov@ipc.bas.bg (K.K.); tchoukov@ipc.bas.bg (P.T.)

1. Introduction

Colloids, such as emulsions, foams, sols and gels, play integral roles in living organisms, the natural environment, resource extraction, pharmaceuticals, cleaning products, processing industries and almost all aspects of our everyday lives [1]. Generally, colloid systems consist of a continuous phase and a dispersed phase, which is in the form of finely distributed particles with sizes in the 1 nm–1000 nm range. Such a high degree of dispersity is inevitably related to a very large interface separating the two phases. Thus, the interfacial properties and related surface forces greatly determine the properties of the colloid system as a whole [2,3].

Amphiphilic molecules contain structural groups with affinity for either "water" or "oil", two immiscible solvents. Their amphiphilic character determines preferential adsorption at the interface and enables self-assembly into various structures, such as micelles, vesicles, monolayer, bilayers and lamellar phases in the bulk [4]. These self-assembled structures play essential roles in modifying and controlling the functionality of the colloid systems. Selecting amphiphiles with appropriate structures and properties is critical for designing functional colloids for specific applications.

In light of growing environmental awareness, there is increased interest in using "green" surfactants that are derived from sustainable sources. The aim is to replace conventional surfactants with alternatives that are biodegradable, non-toxic and produced through sustainable processes with a reduced carbon footprint [5]. Developing new green surfactants requires in depth understanding of the relationship between a molecule's chemical structure and its physicochemical properties.

This Special Issue (SI) attracted a diverse collection of research that highlights the important relationship between the molecular structure of amphiphiles, their interfacial behavior and the properties of the corresponding colloidal systems. This SI includes research topics ranging from fundamental thermodynamic investigations to the fabrication of novel materials with tailored properties.

2. An Overview of the Contributions

Peychev et al. (Contribution 1) present an improved approach for more accurate experimental determination of the free energy of transfer of a fluorocarbon -CF_2- group from oil to water. This determination is based on adsorption data of fluorosurfactants at the water/hexane interface. Knowledge of this thermodynamic parameter is essential for predicting the partitioning of fluorinated surfactants between water and oil, as well as between water and lipid membranes, and for evaluating their environmental impact.

Wei et al. (Contribution 2) report the fabrication of supramolecular self-assembled organohydrogels mixing double-tailed zwitterionic quaternary ammonium amphiphiles and phosphomolybdic acid in a binary solvent of water and dimethyl sulfoxide.

 https://doi.org/10.3390/molecules30010025

Lu et al. (Contribution 3) explored the effect of the orientations of the ester group of two series of newly synthesized methyl d-glycopyranoside-based esters on their physicochemical properties. The observed trends in solubility and adsorption, and the comparison with the commercial octyl d-glycosides, provide clear guidance on how the structure of molecules affects their properties.

Guo et al. (Contribution 4) entertained the idea of creating a protocell model in water using sodiummonododecyl phosphate (SDP), isopentenol (IPN) and pyrite (FeS_2) mineral particles. The FeS_2 particles promote the IPN/SDP vesicles' fusion and growth and influence their morphological evolution. This research is thought, in the long run, to shed light on the possible model systems for protocell membranes.

Gutiérrez-Fernández et al. (Contribution 5) report a new dispersive solid-phase extraction of melatonin using graphene mixtures with sepiolite and bentonite clays as sorbents combined with fluorescence detection. The method is applied to herbal samples containing melatonin.

In the study by Wang et al. (Contribution 6), double-chain lactobionic amide quaternary ammonium salts were synthesized via the amidation of lactobionic acid with N,N-dimethyldipropyltriamine to obtain glycosylamides, followed by quaternization with bromoalkanes of different chain lengths. The resulting product is a novel glucosamine-based cationic surfactant characterized by low foaming, antibacterial properties, antistatic properties, salt resistance, and the ability to form stable vesicular systems. It shows potential for applications in diverse fields, including drug delivery systems, biomimetic membranes, microreactors, specialty chemicals and the food industry.

Liu et al. (Contribution 7) studied liquid droplets and liquid marbles (droplets of 1M Fe_2Cl aqueous solution fully coated with hydrophobic polyvinylidene fluoride powder) at the liquid–liquid interface to investigate capillary forces and interactions at intra- and inter-cellular scales. Interfacial liquid marbles possess unique internal and external features, making them a promising model and a versatile tool for the investigation of different interfacial phenomena.

Kamburova et al. (Contribution 8) investigated the kinetics of amyloid aggregation indirectly by monitoring the changes in the polydispersity of a mixed dispersion of amyloid β peptide (1–40) and composite liposomes as a function of temperature and pH. Despite homotaurine's specific bioactivity in natural cell membranes, this study highlights its additional inhibitory effect on amyloid peptide aggregation, attributed to charge interactions and 'molecular crowding'.

Alvarez et al. (Contribution 9) studied the aggregation behaviors of four amino acid-based surfactants in the presence of five linear diamine counterions. Electrical conductimetry was used to determine the CMCs for each system, and dynamic light scattering was used to evaluate the micellar size. The obtained CMCs correlated with each surfactant's partitioning coefficients, logP (water/octanol) value. The obtained results highlight the relationship between the structure of the surfactants and their physicochemical properties.

Chunlin Xu et al. (Contribution 10) fabricated a novel single-tailed dynamic covalent surfactant 1-methyl-3-(2-(4-((tetradecylimino)methyl) phenoxy)ethyl)-3-imidazolium bromide (C_{14}PMimBr). In aqueous solutions with increasing concentrations, C_{14}PMimBr forms micelles, vesicles and hydrogels. The authors envision the application of their approach to the design of novel stimuli-responsive surfactant systems for drug delivery and targeted drug release.

The review paper "Law and Order of Colloidal Tectonics: From Molecules to Self-Assembled Colloids" authored by Leclercq (Contribution 11) discusses in depth the emerging concept of colloidal tectonics. Tectons are molecular building blocks capable of spontaneously forming supra-colloidal structures. The author presents colloidal tectonics as a phenomenon that bridges the gap between the soft self-assemblies of small amphiphilic molecules, driven by hydrophobic interactions (such as micelles or vesicles), and rigid crystalline structures with periodic molecular arrangements. The review refers to examples from biotic systems and theoretical insights as a potential guide for self-assembled systems with applications in drug delivery, catalysis, and other applications.

Author Contributions: Conceptualization, K.K. and P.T.; writing—original draft preparation, K.K. and P.T.; writing—review and editing, K.K. and P.T. All authors have read and agreed to the published version of the manuscript.

Acknowledgments: We thank all the contributors to this Special Issue, all the reviewers for evaluating the submitted articles, and the editorial staff of *Molecules* for their very helpful assistance.

Conflicts of Interest: The authors declare no conflicts of interest.

List of Contributions:

1. Peychev, B.; Arabadzhieva, D.; Minkov, I.L.; Mileva, E.; Slavchov, R.I. Quantifying the Hydrophobic Effect per CF2 Moiety from Adsorption of Fluorinated Alcohols at the Water/Oil Interface. *Molecules* **2024**, *29*, 1421. https://doi.org/10.3390/molecules29071421.
2. Wei, P.; Duan, Y.; Wang, C.; Sun, P.; Sun, N. Co-Assembled Supramolecular Organohydrogels of Amphiphilic Zwitterion and Polyoxometalate with Controlled Microstructures. *Molecules* **2024**, *29*, 2286. https://doi.org/10.3390/molecules29102286.
3. Lu, H.; Pourceau, G.; Briou, B.; Wadouachi, A.; Gaudin, T.; Pezron, I.; Drelich, A. Sugar-Based Surfactants: Effects of Structural Features on the Physicochemical Properties of Sugar Esters and Their Comparison to Commercial Octyl Glycosides. *Molecules* **2024**, *29*, 2338. https://doi.org/10.3390/molecules29102338.
4. Guo, D.; Zhang, Z.; Sun, J.; Zhao, H.; Hou, W.; Du, N. A Fusion–Growth Protocell Model Based on Vesicle Interactions with Pyrite Particles. *Molecules* **2024**, *29*, 2664. https://doi.org/10.3390/molecules29112664.
5. Gutiérrez-Fernández, L.; Díez-Pascual, A.M.; San Andrés, M.P. Dispersive Solid Phase Extraction of Melatonin with Graphene/Clay Mixtures and Fluorescence Analysis in Surfactant Aqueous Solutions. *Molecules* **2024**, *29*, 2699. https://doi.org/10.3390/molecules29112699.
6. Wang, Y.; Chen, Z.; Zhang, E.; Zhi, L.; Di Serio, M.; Wang, G.; Wang, Y.; Li, X.; Liu, X.; Huang, Y. Aggregation Behavior and Application Properties of Novel Glycosylamide Quaternary Ammonium Salts in Aqueous Solution. *Molecules* **2024**, *29*, 2749. https://doi.org/10.3390/molecules29122749.
7. Liu, Y.; Wang, Y.; Xin, J.H. Capillarity in Interfacial Liquids and Marbles: Mechanisms, Properties, and Applications. *Molecules* **2024**, *29*, 2986. https://doi.org/10.3390/molecules29132986.
8. Kamburova, K.; Dimitrov, I.L.; Hodzhaoglu, F.; Milkova, V. Investigation of the Aggregation of Aβ Peptide (1–40) in the Presence of κ-Carrageenan-Stabilised Liposomes Loaded with Homotaurine. *Molecules* **2024**, *29*, 3460. https://doi.org/10.3390/molecules29153460.
9. Alvarez, M.A.; Black, N.; Blanco, S.E.; Reid, K.R.; Billiot, E.J.; Billiot, F.H.; Morris, K.F. Influence of Linear Diamine Counterions on the Self-Assembly of Glycine-, Alanine-, Valine-, and Leucine-Based Amphiphiles. *Molecules* **2024**, *29*, 4436. https://doi.org/10.3390/molecules29184436.
10. Xu, C.; Sun, N.; Li, H.; Han, X.; Zhang, A.; Sun, P. Stimuli-Responsive Vesicles and Hydrogels Formed by a Single-Tailed Dynamic Covalent Surfactant in Aqueous Solutions. *Molecules* **2024**, *29*, 4984. https://doi.org/10.3390/molecules29214984.
11. Leclercq, L. Law and Order of Colloidal Tectonics: From Molecules to Self-Assembled Colloids. *Molecules* **2024**, *29*, 5657. https://doi.org/10.3390/molecules29235657.

References

1. Tadros, T.F. *Interfacial Phenomena and Colloid Stability: Industrial Applications*; Walter de Gruyter GmbH & Co KG: Berlin, Germany, 2015.
2. Butt, H.-J.; Graf, K.; Kappl, M. *Physics and Chemistry of Interfaces*; John Wiley & Sons: Hoboken, NJ, USA, 2023.

3. Dukhin, S.S.; Kretzschmar, G.; Miller, R. *Dynamics of Adsorption at Liquid Interfaces: Theory, Experiment, Application*; Elsevier: Amsterdam, The Netherlands, 1995.
4. Lombardo, D.; Kiselev, M.A.; Magazù, S.; Calandra, P. Amphiphiles Self-Assembly: Basic Concepts and Future Perspectives of Supramolecular Approaches. *Adv. Condens. Matter Phys.* **2015**, *2015*, 151683. [CrossRef]
5. Stubbs, S.; Yousaf, S.; Khan, I. A Review on the Synthesis of Bio-Based Surfactants Using Green Chemistry Principles. *DARU J. Pharm. Sci.* **2022**, *30*, 407–426. [CrossRef]

molecules

MDPI

Article

Stimuli-Responsive Vesicles and Hydrogels Formed by a Single-Tailed Dynamic Covalent Surfactant in Aqueous Solutions

Chunlin Xu [1,†], Na Sun [2,†], Huaixiu Li [1], Xingchen Han [1], Ailing Zhang [3,*] and Panpan Sun [1,*]

1 School of Bioscience and Technology, Shandong Second Medical University, Weifang 261053, China; xuchunlin2024@163.com (C.X.); 17753908079@163.com (H.L.); hxc18654780069@163.com (X.H.)
2 College of Pharmacy, Shandong Second Medical University, Weifang 261053, China; sunna@sdsmu.edu.cn
3 College of Chemical Engineering and Environmental Chemistry, Weifang University, Weifang 261061, China
* Correspondence: zhangal89@163.com (A.Z.); sunpanpan@sdsmu.edu.cn (P.S.)
† These authors contributed equally to this work.

Abstract: Controlling the hierarchical self-assembly of surfactants in aqueous solutions has drawn much attention due to their broad range of applications, from targeted drug release, preparation of smart material, to biocatalysis. However, the synthetic procedures for surfactants with stimuli-responsive hydrophobic chains are complicated, which restricts the development of surfactants. Herein, a novel single-tailed responsive surfactant, 1-methyl-3-(2-(4-((tetradecylimino) methyl) phenoxy) ethyl)-3-imidazolium bromides (C_{14}PMimBr), was facilely fabricated in situ by simply mixing an aldehyde-functionalized imidazolium cation (3-(2-(4-formylphenoxy) ethyl)-1-methyl imidazolium bromide, BAMimBr) and aliphatic amine (tetradecylamine, TDA) through dynamic imine bonding. With increasing concentration, micelles, vesicles, and hydrogels were spontaneously formed by the hierarchical self-assembly of C_{14}PMimBr in aqueous solutions without any additives. The morphologies of vesicles and hydrogels were characterized by cryogenic transmission electron microscopy and scanning electron microscopy. The mechanical properties and microstructure information of hydrogels were demonstrated by rheological measurement, X-ray diffraction, and density functional theory calculation. In addition, the vesicles could be disassembled and reassembled with the breakage and reformation of imine bonds by adding acid/bubbling CO_2 and adding alkali. This work provides a simple method for constructing stimuli-responsive surfactant systems and shows great potential application in targeted drug release, drug delivery, and intelligent materials.

Keywords: surfactant; dynamic covalent bond; responsive vesicle; aggregation behavior

1. Introduction

Molecular self-assembly is a transition process from disordered to ordered, offering a bottom-up approach to constructing functional materials on multiple scales [1]. Rich types of amphiphiles, such as surfactants [2], peptides [3], and phospholipids [4], have been studied extensively in the field of molecular self-assembly. A special type of surfactant with a novel structure [5–10] consisting of a versatile organic cation and an organic/inorganic anion, has attracted extensive attention in recent years. Compared with the conventional surfactants, the self-assembly properties of this novel structure can be regulated efficiently and easily through reasonable design of the cation and anion structures.

Various structurally well-defined assemblies, such as spherical micelles [11], rod-like/wormlike micelles [12–14], unilamellar/multilamellar vesicles [15–17], and gel nanofiber networks [18], can be formed via the self-assembly of surfactants in aqueous solutions. Among these assemblies, the stimuli-responsive vesicles have drawn much attention due to their promising applications in targeted drug release [19], micro-reactors [20], and biotherapy [21]. The common approaches for preparing stimuli-responsive vesicles are the use of surfactants with double tails [22], the responsive cation/anion [23], and the addition of external additives [24] in aqueous solutions. For instance, Shi et al. successfully created pH/CO_2 dual-switchable vesicles using gemini surfactants based on responsive cations (1-(3-aminopropl)-3-methylimidazolium chloride, AFIL) and sodium dodecyl benzene sulfonate (SDBS), and realized the fission and reversion of vesicles [25]. Wang et al. developed responsive vesicles in the system of [C_{14}mim]X (X = $[Imi]^-$, $[1, 2, 4\text{-}Tri]^-$, $[Ben]^-$, $[Tet]^-$, and $[1, 2, 3\text{-}Ben]^-$) with pH-sensitive anions, and realized the reversible structural transition between vesicles and micelles by alternating CO_2 and N_2 bubbling [26]. In addition, Huang et al. constructed a ternary supramolecular system, which made the reversible formation and dissociation of vesicles possible by introducing a light-responsive Azo@β-CD [27]. However, there have been few reports on the formation of stimuli-responsive vesicles by single-tailed surfactants with responsive hydrophobic chains in aqueous solutions for the sake of complicated synthetic procedures.

Dynamic covalent bonds (DCBs) [28,29], unlike traditional covalent bonds, can be cleaved and reformed reversibly under specific conditions, such as pH [30], temperature [31], and reductant–oxidant [32]. To date, an abundance of dynamic covalent bonds has been exploited to construct smart materials such as imine, acylhydrazone, disulfide, and boronic ester bonds. Combining the robustness of classic covalent bonds with the reversibility of non-covalent bonds, dynamic covalent bonds have been proven to be a powerful tool for the construction of adaptive and responsive vesicles, wormlike micelles, gels, and polymers. H. van Esch et al. reported that single-tailed [33] and double-tailed surfactants [34,35] are formed by non-surface-active precursors of bis-aldehyde and alkyl amines based on dynamic imine bonds, and found that the corresponding formed micelles, wormlike micelles, and vesicles can be switched by changing pH and temperature. In addition, the group also fabricated a series of conjugated polymers by using dynamic imine bonds and realized the controllable formation and interruption of the formed wormlike micelles [31]. Zhang et al. exploited various topological structures using dynamic covalent surfactants, such as bola-shaped [36] and H-shaped [37], and the obtained aggregates were the only responsive spherical micelles. Thus, it is of great potential to regulate the assembly and disassembly of vesicles using dynamic covalent bonds while modifying the hydrophobicity of single-tailed surfactants.

Herein, a stimuli-responsive vesicle system was formed by single-tailed surfactants with a responsive hydrophobic chain that was fabricated in situ by mixing an aldehyde-functionalized imidazolium cation (3-(2-(4-formylphenoxy) ethyl)-1-methyl imidazolium bromide, BAMimBr) and aliphatic amine (tetradecylamine, TDA) through a dynamic covalent bond. It was the first report on the spontaneous transition from micelles to vesicles and hydrogels in aqueous solutions without any additives. The self-assemblies are characterized by cryogenic transmission electron microscopy (Cryo-TEM), dynamic light scattering (DLS), scanning electron microscopy (SEM), rheology, and X-ray Diffraction (XRD). Additionally, the vesicle system can be reversed by adding alkali and acid/bubbling CO_2.

2. Results

2.1. Characterization of Single-Tailed Dynamic Covalent Surfactant C_{14}PMimBr

A novel dynamic covalent single-tailed surfactant, C_{14}PMimBr, was successfully designed and synthesized in situ by simply mixing TDA with the non-assembling building block BAMimBr at a molar ratio of 1:1 through dynamic covalent bonding at room temperature, as illustrated in Scheme 1. In consideration of the reactivity of dynamic covalent bonds, the pH of the solution was chosen to be 12 in this work. To confirm the formation of dynamic covalent bonds, ^{1}H NMR and mass spectrometry (MS) were employed. As can be seen in Figure 1, in the ^{1}H NMR spectrum of BAMimBr, a unimodal peak 'f' appeared at 9.91 ppm, which is the characteristic peak of aldehyde (-CHO proton) [38]. After the combination of TDA, the signal corresponding to the -CHO proton of BAMimBr nearly disappeared and a new peak '6' appeared at 8.18 ppm, which is generally attributed to the -CH=N- proton [39], confirming the incorporation of BAMimBr and TDA via a dynamic imine bond, and thus, the formation of the single-tailed surfactant C_{14}PMimBr was confirmed (the details of ^{1}H NMR and ^{13}C NMR are illustrated in Figure S2). Additionally, the appearance of the peak at 426.3485 ([M]$^+$) in the ESI-MS spectrum is in accordance with the theoretical value for the C_{14}PMim$^+$ cation ([M]$^+$ = 426.3479), indicating the formation of C_{14}PMimBr (Figure S3).

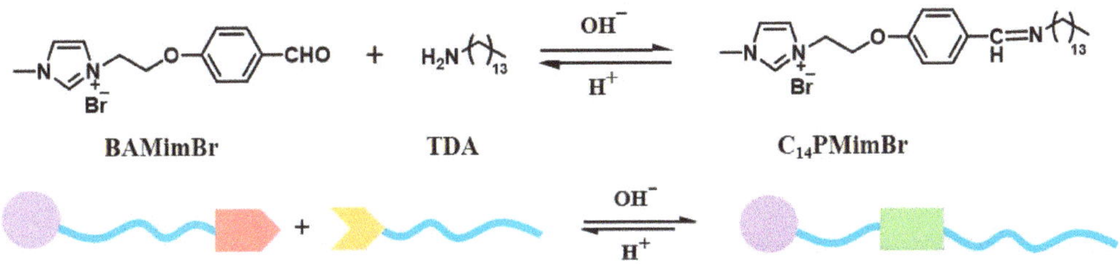

Scheme 1. Chemical structures and schematic diagrams of the surface-inactive cationic headgroup (BAMimBr), tetradecylamine (TDA), and the single-tailed dynamic covalent surfactant, C_{14}PMimBr.

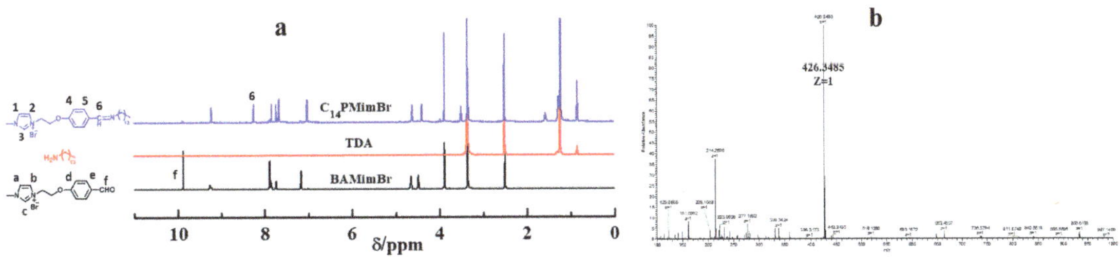

Figure 1. (**a**) ^{1}H NMR spectra of BAMimBr, TDA, and the formed single-tailed surfactant C_{14}PMimBr in DMSO-D_6; (**b**) the MS spectrum of C_{14}PMimBr.

2.2. Phase Behaviors of the Single-Tailed Dynamic Covalent Surfactant C_{14}PMimBr

It is obvious that the solubility of TDA is very poor in aqueous solutions due to its long hydrophobic alkyl chain and weak hydrophilic headgroup. The building block, ionic liquid BAMimBr, was difficult to assemble into the ordered molecule aggregates because of their short alkyl chains. Therefore, the fabricated single-tailed dynamic covalent surfactant would combine the solubility of BAMimBr with the assembly ability of TDA. We systematically investigated the phase behavior of C_{14}PMimBr in aqueous solutions, as shown in Figure 2 and Figure S4. Interestingly, the turbidity of the solutions increases gradually with the concentration increasing to 35 mM. From 35 to 60 mM, the turbidity of the C_{14}PMimBr solutions decreases, and the assemblies at concentrations of 50 and 60 mM are unstable, which are easy to aggregate to floccules standing for two weeks. Upon increasing the concentration to 75 mM, the opalescent hydrogels begin to form. The above results illustrate that the size and morphology of assemblies would change with the increase in the concentration of C_{14}PMimBr.

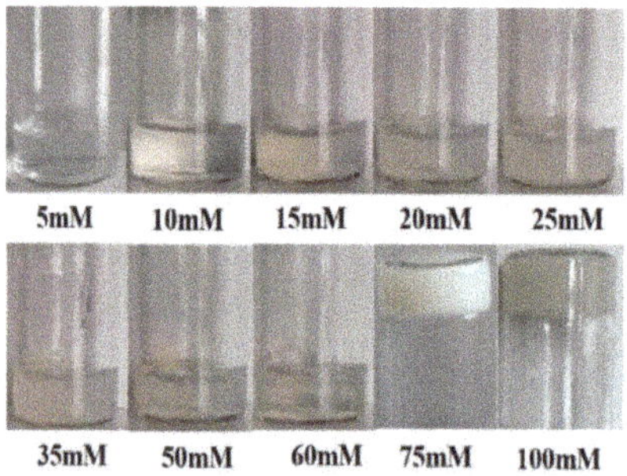

Figure 2. Optical photographs of C_{14}PMimBr (fixing the molar ratio of BAMimBr/TDA at 1:1) solutions with different concentrations.

2.3. Size Distribution and Microstructure Observations of the Single-Tailed Dynamic Covalent Surfactant $C_{14}PMimBr$

To characterize the evolution of self-assemblies with increasing concentration, two powerful tools, DLS and Cryo-TEM, were employed. The size distributions of the C_{14}PMimBr aggregates at different concentrations are illustrated in Figure 3a. At a concentration of 5 mM, the hydrodynamic radii (R) of the aggregates are about several nanometers, indicating that only micelles form at this concentration [40]. Upon increasing the concentration, the size of the aggregate increases initially but then begins to decrease when the concentration reaches 25 mM. On this basis, the morphologies of the aggregate were characterized by Cryo-TEM. As shown in Figure 3b–f, the scattered vesicles were observed at a concentration of 15 mM. With increasing concentration, the population of vesicles increased. When the concentration reaches 25 mM, the formed assemblies are densely packed unilamellar and multilamellar vesicles with diameters of 100–1500 nm, as illustrated in Figure 3e, which is in accordance with the changes in aggregate size distribution. However, increasing the concentration to 35 mM gradually reduces the size distributions of vesicles, corresponding to changes in the turbidity of solutions. As reported, the slight deformation of vesicles in hydrocarbon surfactant systems can usually be found in the Cryo-TEM picture, as shown in Figure 3f, demonstrating the flexibility of vesicles [41].

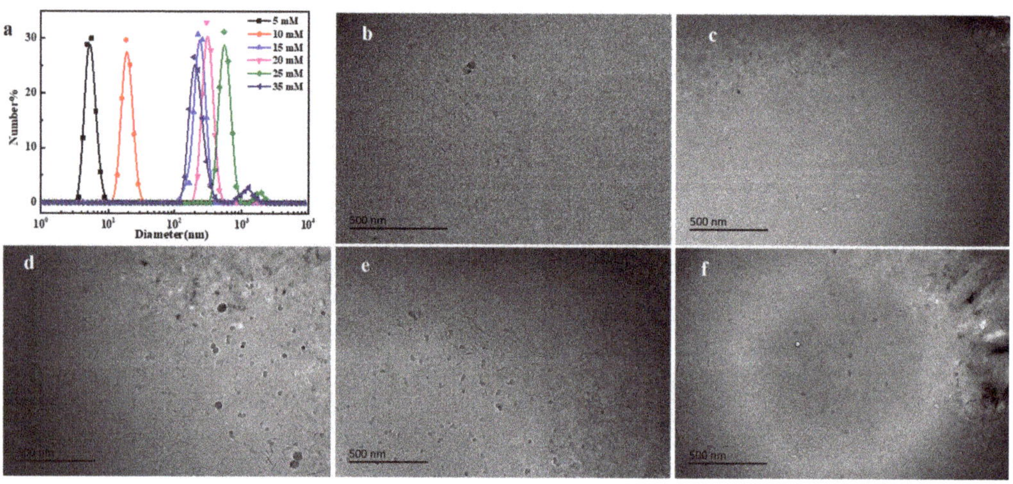

Figure 3. (**a**) Size distributions of C_{14}PMimBr solutions with different concentrations. Cryo-TEM images of C_{14}PMimBr at concentrations of (**b**) 10 mM, (**c**) 15 mM, (**d**) 20 mM, (**e**) 25 mM, and (**f**) 35 mM.

2.4. *pH—Inducing Phase Transition of Single-Tailed Dynamic Covalent Surfactant C_{14}PMimBr Solution*

The imine bond is a typical pattern of dynamic covalent bonds that can be reversibly broken and reformed by adjusting the pH of the solutions. The effect of acidity or alkalinity on the phase transition of the C_{14}PMimBr solution was explored by visual observation, with the aid of crossed polarizers. As illustrated in Figure 4, the initial state of the C_{14}PMimBr solution was homogeneous opalescent under normal light (Figure 4a) and showed typical birefringent textures under crossed polarizers (Figure 4b), indicating the formation of ordered molecule aggregates [40]. The opalescent C_{14}PMimBr solution was gradually precipitated into white solids (Figure 4c,d) by bubbling CO_2 or adding acid, and this process can be reversed by adding a base, confirming that the vesicle can be disassembled and reassembled by regulating the breakage and reformation of imine bonds.

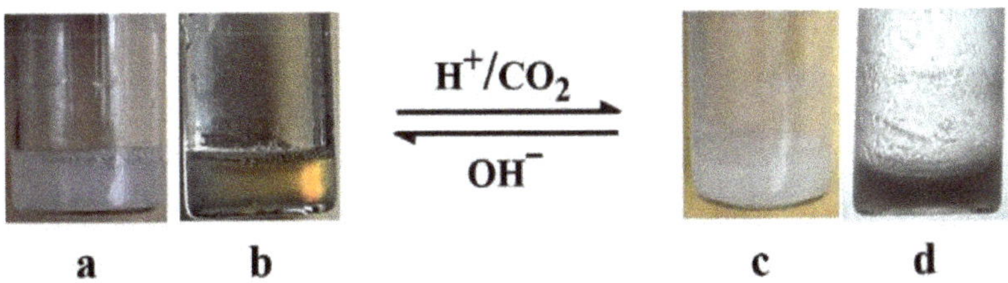

Figure 4. Optical photographs of the phase transition of C_{14}PMimBr solutions observed without (**a**,**c**) and with (**b**,**d**) crossed polarizers at a concentration of 25 mM.

2.5. *SEM Observation and Rheology Properties of the C_{14}PMimBr Hydrogels*

Rheology is a powerful method for characterizing the viscoelasticity and mechanical strength of the hydrogel system. The solid-like network structure of the hydrogels that are implanted to be sheared under increasing stress breaks suddenly at a critical shear stress, τ^*, beyond which a Newtonian-like flow occurs. The τ^*, namely "yield stress", reflects the strength of the microstructure of the hydrogels. Herein, we systematically investigated the rheological behavior of the hydrogel samples at constant temperature. As can be seen from dynamic oscillatory strain sweep tests (Figure 5a), the yield stress increases with the increase in the C_{14}PMimBr concentration from 75 mM to 100 mM, demonstrating enhancement in mechanical strength. In addition, we exploited the frequency sweep tests to detect the viscoelasticity of C_{14}PMimBr hydrogels at a constant stress of 1.0 Pa (Figure 5b). The viscoelasticity of gels can be characterized by two dynamic moduli: the storage modulus (G'), representing the elastic properties (solid-like behavior) of hydrogels, and the loss modulus (G''), estimating the viscous properties (liquid-like behavior) of hydrogels [42]. In Figure 5b, the apparent values of storage modulus (G') are higher than those of loss modulus (G'') in the whole frequency range of all hydrogels, presenting the typical rheological characteristic of solid-like materials. Furthermore, the storage modulus (G') and loss modulus (G'') are slightly dependent on the frequency, reflecting that these values increase slightly with an increase in frequency. When the concentration of C_{14}PMimBr ranges from 75 mM to 100 mM, the absolute magnitude of storage modulus (G') and loss modulus (G'') increases from 100, 20 to 1000, 200, respectively, which could be ascribed to the enhancement of the mechanical strength of the hydrogels by adding C_{14}PMimBr. Based on this, the steady shear tests are further employed to study the flow property of hydrogels. As illustrated in Figure 5c, the viscosity of all samples decreases gradually with increasing shear rate, corresponding to shear-thinning behavior and exhibiting the typical Newtonian-like flow, indicating that the structure of the hydrogels has been destroyed [43].

To further investigate the morphology of hydrogels, the SEM observation was selected. The obtained micrographs are shown in Figure 5. In all the cases, except for Figure 5i, wrinkle-like morphologies were observed; for Figure 5i, it was more like an aggregation of several stacked micrometer-long slices. However, the morphology observed herein deviated from the typically observed entangled fibrous network of supramolecular gels, which is usually observed in primary ammonium sulfonates [44] and guanidium sulfonate-derived gels [45].

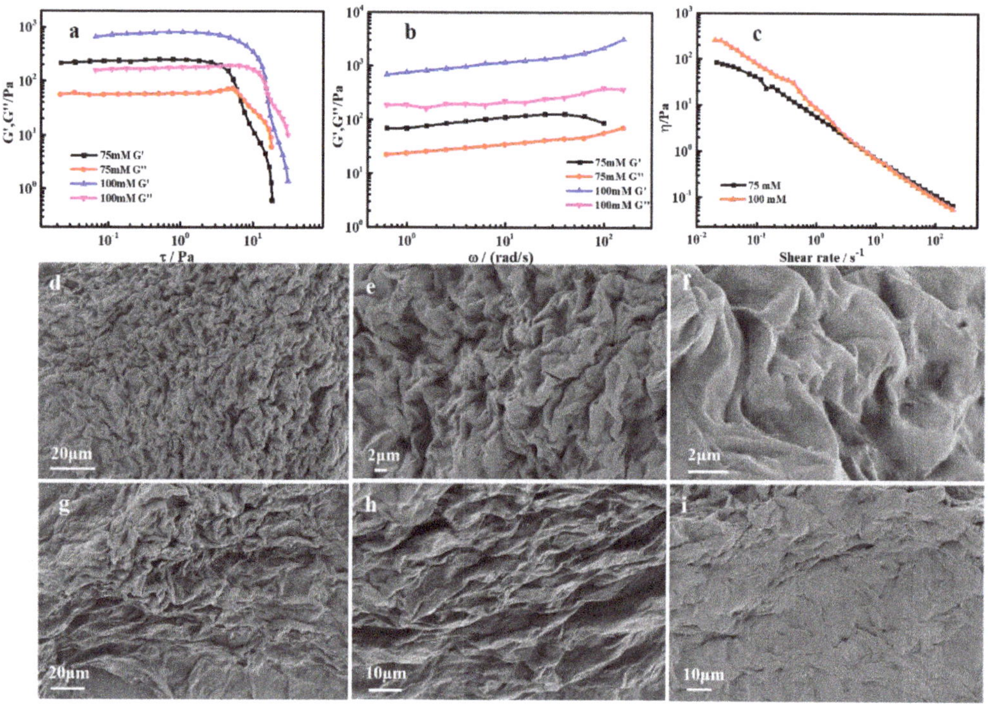

Figure 5. (**a**) Dynamic oscillatory strain sweep tests (from 0.1 Pa to 1000 Pa), fixing the frequency at 1 Hz; (**b**) frequency sweep tests (from 0.1 rad/s to 100 rad/s), fixing the shear rate at 1 Pa; (**c**) viscosity as a function of shear rate for the C_{14}PMimBr hydrogels at concentrations of 75 mM and 100 mM. SEM photographs of the C_{14}PMimBr hydrogels at concentrations of 75 mM (**d–f**) and 100 mM (**g–i**).

2.6. XRD Investigation and Density Functional Theory (DFT) Calculation

XRD measurement is a useful approach to investigating detailed information on molecular arrangements of materials. The X-ray diffractogram obtained for C_{14}PMimBr hydrogels is shown in Figure 6a. The appearance of a large number of Bragg peaks in the spectra is indicative of a highly ordered surfactant assembly [46]. The XRD pattern of the xerogel sample has five Bragg peaks (q1/q2/q3/q4/q5) with relative ratios of 1:2:3:4:5, which can be indexed 00l as the (001), (002), (003), (004) and (005), reflecting the typical feature of lamellar structure [47]. The interlayer spacing (d), consisting of the thicknesses of the amphiphile bilayer and solvent, can be calculated from the first Bragg peak ($d = \lambda/2Sin\theta$). According to the first Bragg peak $2\theta = 1.91$, the "d" is calculated to the value of 4.62 nm. To further clarify the molecular arrangement of C_{14}PMimBr in hydrogels, the density functional theory (DFT) calculation is employed. This is based on a hybrid functional B3LYP with the basis 6-31G(d, p) of the Gaussian 09 package and the obtained xyz coordinates (Figure S5 and Table S1). The fully extended length of the optimization structure of C_{14}PMimBr is 28.78 Å (Figure 6a and Figure S3). It is noted that the value of "d" is more than the length of one molecule and smaller than that of two molecules, indicating chain interdigitation of C_{14}PMimBr in hydrogels, as illustrated in Figure 6c [30].

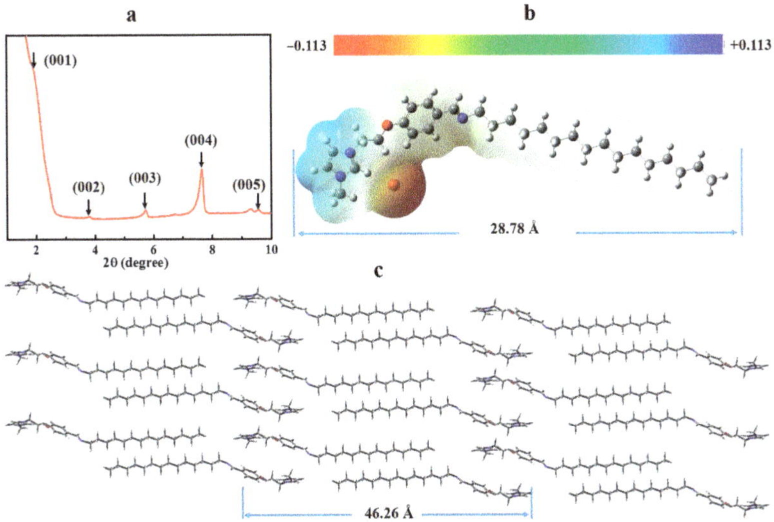

Figure 6. (**a**) XRD pattern of xerogel obtained by freeze-drying of C_{14}PMimBr hydrogel; (**b**) optimized structures and electrostatic potentials of C_{14}PMimBr by density functional theory (DFT) calculations; (**c**) schematic representation of the molecular arrangements in hydrogels.

2.7. Mechanism Analysis

Generally, when mixing BAMimBr with TDA in aqueous solutions with pH = 12, a typical dynamic imine bond is formed, and thus, a single-tailed surfactant with a responsive chain (C_{14}PMimBr) is fabricated. We systematically evaluated the aggregation behavior of surfactants in aqueous solutions by surface tension measurement. As illustrated in Figure 7, with increasing concentration, the surface tension gradually decreases until a plateau region is reached, indicating the formation of micelles. The calculated surface properties and micellization parameters of C_{14}PMimBr in an aqueous solution were illustrated in Table 1. The critical micelle concentration (CMC) of C_{14}PMimBr is about 89 $\times$ 10^{-2} mM, which corresponds to the concentration of the distinct break point.

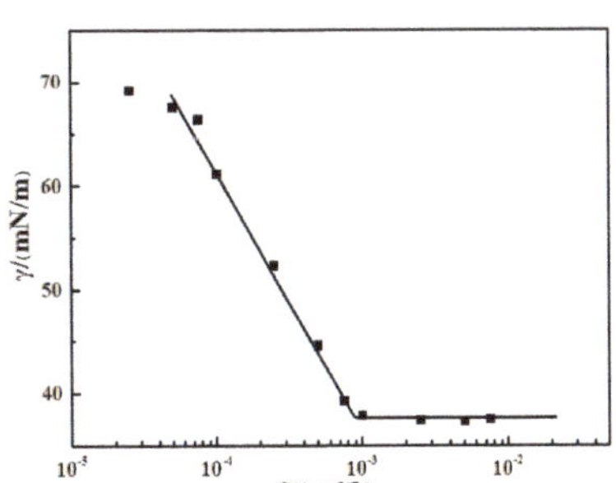

Figure 7. Surface tension of C_{14}PMimBr aqueous solution as a function of concentration at 25 °C.

Table 1. Surface properties and micellization parameters of C_{14}PMimBr in an aqueous solution at 25 °C.

Surfactant	CMC (mM)	γ_{CMC} (mN/m)	Γmax (μmol/m^2)	Amin (Å^2)
C_{14}PMimBr	0.89	37.3	2.27	73.2

Compared to the traditional surfactants, such as alkyl trimethylammonium halide, the surfactant with an imidazolium ring headgroup possessing less electron density and the π-π interaction is apt to form vesicles [7,48]. For example, Wang et al. found that the assembly structures of the single-tailed ionic liquid (IL) surfactants, 1-alkyl-3-methylimidazolium bromides $[C_n mim]Br$(n = 10, 12, 14), could transition from micelles to rodlike micelles and unilamellar vesicles in aqueous solutions without any additives [48]. The self-assembled structures of surfactants can be predicted by the critical packing parameter P, which is a classical theoretical parameter and was defined as $P = v/al$ by Israelachvili et al. [49]. Here, a is the effective headgroup area of the surfactant molecules, and l and v are the alkyl chain length and the effective hydrophobic chain volume of surfactants, respectively. According to the values of P, the different structures of assemblies can be predicted, such as, spherical ($0 < P \leq 1/3$), cylindrical ($1/3 < P \leq 1/2$), and planar ($1/2 < P \leq 1$) aggregates. As for C_{14}PMimBr, the values of l and v were calculated using Tanford equations (as listed in SI) to be 26.8 Å and 565.4 $Å^3$, respectively. Thus, the P value for C_{14}PMimBr is 0.29, indicating that C_{14}PMimBr is preferred for forming spherical micelles at relatively low concentrations. However, with increasing concentration, more Br anions bind to the imidazolium ring cations, leading to a reduction in electrostatic repulsion between head groups, which could lead to decreased a values and increased P values [48]. As a result, the change in assembly structure from micelles to vesicles and the lamellar phase is induced. Furthermore, the injection of CO_2 or the addition of acid breaks the dynamic imine bond; thus, the single-tailed dynamic covalent surfactant is disintegrated into the non-assembling building block, BAMimBr, and the insoluble tetradecylamine. According to the above analysis, the possible mechanism of phase transition and the responsive process is proposed as displayed in Scheme 2.

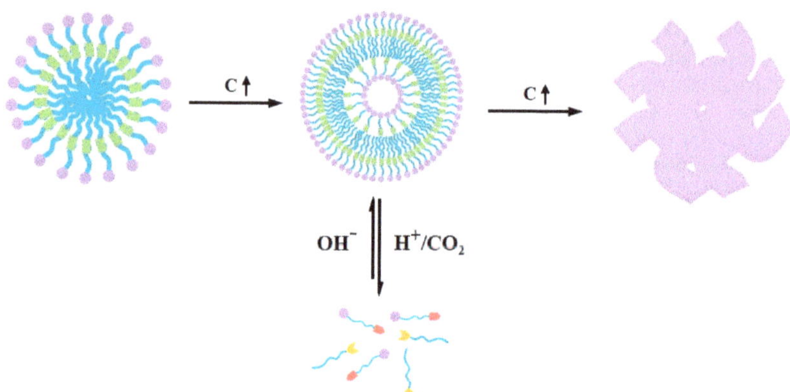

Scheme 2. Schematic representation of the responsive behaviors of the single-tailed dynamic covalent surfactant C_{14}PMimBr and its aggregates with increasing concentration and the pH/CO_2 stimuli.

3. Materials and Methods

3.1. Materials

Tetradecylamine (95%), 4-hydroxybenzaldehyde (98%), 1,2-dibromoethane (98%), N-methylimidazole (98%), $CDCl_3$ (99.96%), DMSO (99.96%), and D_2O (99.96%) were purchased from Tianjin Xiensi Biochemical Technology Co., Ltd. (Tianjin, China). Dichloromethane (CH_2Cl_2), acetonitrile (CH_3CN), petroleum ether (PE), $MgSO_4$, K_2CO_3, and KOH were obtained from Sinopharm Group Chemical reagent Co., Ltd. (Shanghai, China). All the materials were used as received without any purification. Deionized water was used in all the experiments.

3.2. Synthesis of 3-(2-(4-Formylphenoxy) ethyl)-1-methyl Imidazolium Bromide (BAMimBr) [50]

4-Hydroxybenzaldehyde (2 g, 16.4 mmol), 1, 2-dibromoethane (7.70 g, 41.0 mmol), and K_2CO_3 (6.80 g, 49.2 mmol) were dissolved in CH_3CN (120 mL) and then the mixture was refluxed by stirring for 24 h. After the reaction, the solvent was removed under reduced pressure, and thus, the wheat solid was obtained. The crude product was purified by column chromatography (silica gel, CH_2Cl_2: PE = 3:2, v/v) to give compound 4-(2-bromoethoxy)benzaldehyde as a white solid. ^{1}H NMR (300 MHz, $CDCl_3$): δH/ppm 3.67 (t, 2H), 4.38 (t, 2H), 7.04 (d, 2H), 7.89 (d, 2H), 9.90 (s, 1H). The 4-(2-bromoethoxy)benzaldehyde was abbreviated "A".

Compound "A" (3 g, 13.10 mmol) and N-methylimidazole (1.07 g, 13.10 mmol) were dissolved in CH_3CN (50 mL), and the reaction was refluxed by stirring for 24 h under a nitrogen atmosphere. After the reaction, the reaction mixture was cooled to room temperature and the resulting suspension was filtered. The obtained solid was washed with diethyl ether (3 × 50 mL) and dried to give a white solid product (BAMimBr). ^{1}H NMR (300 MHz, D_2O) (Figure S1): δH/ppm: 3.77 (s, 3H), 4.42 (t, 2H), 4.56 (t, 2H), 7.02 (d, 2H), 7.32 (s, 1H), 7.48 (s, 1H), 7.78 (d, 2H), 9.65 (s, 1H). The obtained spectrum is illustrated in Figure S2.

3.3. Sample Preparation

The details of the fabrication process for 1-methyl-3-(2-(4-((tetradecylimino) methyl) phenoxy) ethyl)-3-imidazolium bromides (abbreviation as, C_{14}PMimBr) are as follows: 0.0311 g BAMimBr and 0.0213 g TDA were mixed in 10 mL water, then the water was adjusted to pH = 12 by adding concentrated sodium hydroxide solution (10 mol/L). After aging for 12 h, the solution was used in further experiments. ^{1} HNMR (300 MHz, D_2O): δH/ppm: 0.85 (t, 3H), 1.06 (m, 2H), 1.23 (m, 22H), 1.571 (t, 2H), 3.44 (t, 2H), 3.89 (s, 3H), 4.40 (t, 2H), 4.62 (t, 2H), 7.02 (d, 2H), 7.66–7.68 (d, 2H), 7.74 (d, 1H), 7.84 (d, 1H), 8.24 (s, 1H), 9.23 (s, 1H). ^{13}C NMR (300 MHz, D_2O): δ 161.7, 160.8, 137.0, 129.8, 128.0, 123.0, 122.8, 114.5, 68.1, 61.8, 56.2, 37.1, 31.9, 31.6, 29.6, 29.3, 27.2, 22.7, 14.1. These spectra are listed in Supplementary Materials.

The vesicles were prepared by simply mixing BAMimBr and TDA; the details of the preparation process are as follows: A total of 0.0078 g of BAMimBr and 0.0053 g of TDA were added to a 1 mL aqueous solution. The pH was adjusted to 12 by adding concentrated sodium hydroxide solution with a concentration of 10 mol/L. The mixed solution was stirred for 30 min under ultrasound (KQ-520DB (10 L), Gongyi Yuhua Instrument Co., Ltd., Gongyi, China) at a frequency of 40 KHz and a temperature of 25 °C. Hydrogels were obtained with increasing concentrations of BAMimBr and TDA, which is similar to the preparation process of vesicles. All systems were aged for at least four weeks before characterization.

To regulate the breakage and reformation of dynamic covalent bonds, concentrated hydrochloric acid and sodium hydroxide solutions were added to adjust the pH values of the samples. In the bubbling CO_2 test, CO_2 was bubbled into mixed BAMimBr and TDA solutions with a fixed flow of 0.1 L/min until the solutions turned turbid.

3.4. Characterization

Dynamic light scattering (DLS): The size distributions of the vesicles were detected by DLS using a Nanotrac Particle Size Analyzer (Nanotrac NPA 250) (Hann, Germany) and the microtrac FLEX application software program. All measurements were carried out using a laser diode (780 nm wavelength, 3 mW nominal, Class IIIB at the scattering angle of 180). The temperature was controlled using a thermostat (F31C, Julabo) with an accuracy of 25 ± 0.1 °C.

Cryogenic transmission electron microscopy (Cryo-TEM) observations: The samples were prepared in a controlled environment vitrification system (CEVS) at 25 °C. A micropipette was used to load 5 µL solutions onto a TEM copper grid. Then, the solution was blotted with two pieces of filter paper, resulting in the formation of thin films suspended on the mesh holes. After waiting for about 10 s, the samples were quickly plunged into a reservoir of liquid ethane, which was cooled in nitrogen at −165 °C. The vitrified samples were then stored in liquid nitrogen until they were transferred to a cryogenic sample holder (Gatan 626) (Pleasanton, CA, USA) and examined under a JEOL JEM-1400 transmission electron microscope (120 kV) (Tokyo, Japan) at about −174 °C. The images were recorded on a GatanMultiscan CCD and processed using a Digital Micrograph.

^{1}H NMR measurements: ^{1}H NMR spectra were generated on a Bruker Advance 300 spectrometer (Billerica, MA, USA) equipped with a pulse field gradient module (Z-axis) using a 5 mm BBO probe. The instrument was operated at a frequency of 300.13 MHz at 25 ± 0.1 °C.

Rheological measurements: The rheological measurements of hydrogels were performed on an Anton-Paar Rheometer (Graz, Austria) using a cone-plate system (C35/1° Ti L07116, diameter: 35 mm, cone angle: 1°). All measurements were carried out at 25.0 °C using a cyclic water bath.

Scanning electron microscopy (SEM): A small volume of hydrogel sample was placed on a silica wafer, and most of the colloid gel was removed using small forceps to form a thin film. The wafers were freeze-dried in a vacuum extractor for 24 h using a freeze dryer. Then, the silica wafers were subjected to gold plating and observed on a JEOL JSM-6700F FE-SEM (Tokyo, Japan).

X-ray diffraction (XRD): XRD patterns of the powder of the freeze-dried hydrogels were recorded in 2θ mode (1° min^{-1}) using a DMAX-2500PC diffractometer with a Cu Ka radiation source (λ = 0.15418 nm) and a graphite monochromator.

Surface tension measurements: The surface tension measurements were carried out on a Model JYW-200B tensiometer (Chengde Dahua Instrument Co., Ltd., Hangzhou, China, accuracy ±0.1 mN/m) using the ring method. Temperature was controlled at 25 ± 0.1 °C using a thermostatic bath. Each sample was equilibrated for 10 min under the test temperature to reach equilibrium before the measurement.

Density functional theory (DFT) calculation: DFT calculation was based on hybrid functional B3LYP with the basis 6-31G(d,p) of the Gaussian 09 package.

4. Conclusions

In summary, a novel single-tailed dynamic covalent surfactant (C_{14}PMimBr) is reported for the first time. It was fabricated in situ by simply mixing the non-assembling building block BAMimBr with insoluble tetradecylamine in aqueous solutions at pH = 12. Interestingly, the distinctive molecular structure endows rich self-assembly behaviors of C_{14}PMimBr in aqueous solutions. The aggregate structures can be transformed from micelles to unilamellar and multilamellar vesicles and then to lamellar hydrogels upon an increase in C_{14}PMimBr concentrations, without any additives. The formation of various vesicles was confirmed by Cryo-TEM and the molecule arrangement of C_{14}PMimBr in hydrogels is deduced by the combination of XRD and DFT calculations. Interestingly, the dynamic covalent bond (imine bond) endows that the vesicles can be disassembled by bubbling CO_2/adding acid and reformed by adding alkali. We believe that our research would broaden the horizons for constructing novel and stimuli-responsive surfactant systems and lay the foundation for applications in drug delivery, targeted drug release, and biochemical engineering.

Supplementary Materials: The following supporting information can be downloaded at: https://www.mdpi.com/article/10.3390/molecules29214984/s1, Figure S1: ^{1}H NMR spectrum of 3-(2-(4-formylphenoxy) ethyl)-1-methyl imidazolium bromide (BAMimBr); Figure S2: ^{1}H NMR and ^{13}C NMR spectra of C_{14}PMimBr; Figure S3: Mass spectrum of C_{14}PMimBr; Figure S4: Optical photographs of C_{14}PMimBr (fixing molar ratio of BAMimBr/TDA at 1:1) solutions with different concentrations; Figure S5: Geometries of C_{14}PMimBr molecules optimized using the polarizable continuum model at the B3LYP/6-31G(d,p) level; Table S1: DFT calculation information for C_{14}PMimBr molecule. References [51,52] are cited in the supplementary materials.

Author Contributions: C.X.: Data curation, Formal analysis, Investigation, Writing—original draft; N.S.: Methodology, Project administration, Funding acquisition, Writing—original draft; H.L.: Data curation, Formal analysis, Investigation; X.H.: Data curation, Formal analysis, Investigation; A.Z.: Formal analysis, Investigation, Writing—original draft Supervision, Validation; P.S.: Conceptualization, Writing—review and editing, Visualization, Supervision, Project administration, Funding acquisition. All authors have read and agreed to the published version of the manuscript.

Funding: P. Sun was funded by the Natural Science Foundation of Shandong Province (No. ZR2023QB052). N. Sun was funded by the National Natural Science Foundation of China (No. 22102118).

Institutional Review Board Statement: Not applicable.

Informed Consent Statement: Not applicable.

Data Availability Statement: Data are contained within the article and Supplementary Materials.

Conflicts of Interest: The authors declare no conflicts of interest.

References

1. Whitesides, G.M.; Mathias, J.P.; Seto, C.T. Molecular Self-Assembly and Nanochemistry: A Chemical Strategy for the Synthesis of Nanostructures. *Science* **1991**, *254*, 1312–1319. [CrossRef] [PubMed]
2. Hao, J.; Liu, W.; Xu, G.; Zheng, L. Vesicles from Salt-Free Cationic and Anionic Surfactant Solutions. *Langmuir* **2003**, *19*, 10635–10640. [CrossRef]
3. Li, Q.; Jia, Y.; Li, J. Controlled Assembly of Chiral Structure of Diphenylalanine Peptide. *Acta Chim. Sin.* **2019**, *77*, 1173–1176. [CrossRef]
4. Seoane, A.J.; Brea, R.; Fuertes, A.; Podolsky, K.A.; Devaraj, N.K. Biomimetic Generation and Remodeling of Phospholipid Membranes by Dynamic Imine Chemistry. *J. Am. Chem. Soc.* **2018**, *140*, 8388–8391. [CrossRef]
5. Shi, L.; Zheng, L. Aggregation Behavior of Surface Active Imidazolium Ionic Liquids in Ethylammonium Nitrate: Effect of Alkyl Chain Length, Cations, and Counterions. *J. Phys. Chem. B* **2012**, *116*, 2162–2172. [CrossRef]
6. Sun, P.; Shi, L.; Lu, F.; Zheng, L. Aggregation Behavior of Zwitterionic Surface Active Ionic Liquids with Different Counterions, Cations, and Alkyl Chains. *RSC Adv.* **2016**, *6*, 27370–27377. [CrossRef]
7. Wang, H.; Zhang, L.; Wang, J.; Li, Z.; Zhang, S. The First Evidence for Unilamellar Vesicle Formation of Ionic Liquids in Aqueous Solutions. *Chem. Commun.* **2013**, *49*, 5222–5224. [CrossRef]
8. Bi, Y.; Zhao, L.; Hu, Q.; Gao, Y.; Yu, L. Aggregation Behavior of Imidazolium-Based Surface-Active Ionic Liquids with Photoresponsive Cinnamate Counterions in the Aqueous Solution. *Langmuir* **2015**, *31*, 12597–12608. [CrossRef]
9. Pankaj, O.; Kumar, S.; Kumar Mahajan, A. Impact of Aromatic Counter-Ions Charge Delocalization on the Micellization Behavior of Surface-Active Ionic Liquids. *Langmuir* **2019**, *35*, 14586–14595.
10. Cao, H.; Hu, Y.; Xu, W.; Wang, Y.; Guo, X. Recent Progress in the Assembly Behavior of Imidazolium-based Ionic Liquid Surfactants. *J. Mol. Liq.* **2020**, *319*, 114351.
11. Fan, Y.; Hou, Y.; Xiang, J.; Yu, D.; Wu, C.; Tian, M.; Han, Y.; Wang, Y. Synthesis and Aggregation Behavior of a Hexameric Quaternary Ammonium Surfactant. *Langmuir* **2011**, *27*, 10570–10579. [CrossRef] [PubMed]
12. Chu, Z.; Dreiss, A.; Feng, Y.C. Smart Wormlike Micelles. *Chem. Soc. Rev.* **2013**, *42*, 7174–7203. [CrossRef] [PubMed]
13. Zhao, W.; Dong, S.; Hao, J. Colloidal Wormlike Micelles with Highly Ferromagnetic Properties. *Langmuir* **2015**, *31*, 11243–11248. [CrossRef] [PubMed]
14. Zhai, Z.; Ye, S.; Yan, X.; Song, Z.; Shang, S.; Rao, X.; Song, J. pH-Responsive Wormlike Micelles Formed by an Anionic Surfactant Derived from Rosin. *J. Agric. Food Chem.* **2020**, *68*, 10063–10070. [CrossRef]
15. Wang, L.; Zhao, W.; Dong, R.; Hao, J. Phase Structure Transition and Properties of Salt-Free Phosphoric Acid/Non-ionic Surfactants in Water. *Langmuir* **2016**, *32*, 8366–8373. [CrossRef]
16. Sun, P.; Wu, A.; Sun, N.; Qiao, X.; Shi, L.; Zheng, L. Multiple-Responsive Hierarchical Self-Assemblies of a Smart Supramolecular Complex: Regulation of Noncovalent Interactions. *Langmuir* **2018**, *34*, 2791–2799. [CrossRef]
17. Xu, H.; Wang, S.; Li, Y.; Liang, X.; He, H.; Du, N.; Hou, W. Aggregation Behaviors of Alkyl α-keto Acids in Water. *J. Mol. Liq.* **2024**, *394*, 123700. [CrossRef]
18. Wang, Y.; Xing, P.; Li, S.; Ma, M.; Yang, M.; Zhang, Y.; Wang, B.; Hao, A. Facile Stimuli-Responsive Transformation of Vesicle to Nanofiber to Supramolecular Gel via ω-Amino Acid-Based Dynamic Covalent Chemistry. *Langmuir* **2016**, *32*, 10705–10711. [CrossRef]
19. Ma, W.; Bi, J.; Wu, H.; Zhang, G. An Amphiphilic Micromolecule Self-Assembles into Vesicles for Visualized and Targeted Drug Delivery. *ACS Med. Chem. Lett.* **2020**, *11*, 1562–1566. [CrossRef]
20. Shen, J.; Xin, X.; Liu, T.; Wang, S.; Yang, Y.; Luan, X.; Xu, G.; Yuan, S. Ionic Self-Assembly of a Giant Vesicle as a Smart Microcarrier and Microreactor. *Langmuir* **2016**, *32*, 9548–9556. [CrossRef]

21. Jie, K.; Zhou, Y.; Yao, Y.; Shi, B.; Huang, F. CO_2-Responsive Pillar [5]arene-Based Molecular Recognition in Water: Establishment and Application in Gas-Controlled Self-Assembly and Release. *J. Am. Chem. Soc.* **2015**, *137*, 10472–10475. [CrossRef] [PubMed]
22. Shi, L.; Chen, F.; Sun, N.; Zheng, L. Gemini Supra-amphiphiles with Finely-Controlled Self-assemblies. *Soft Matter* **2015**, *11*, 4075–4080. [CrossRef] [PubMed]
23. Wang, H.; Tan, B.; Wang, J.; Li, Z.; Zhang, S. Anion-Based pH Responsive Ionic Liquids: Design, Synthesis, and Reversible Self-Assembling Structural Changes in Aqueous Solution. *Langmuir* **2014**, *30*, 3971–3978. [CrossRef] [PubMed]
24. Zhu, T.; Kang, W.; Yang, H.; Li, Z.; Zhou, B.; Zhang, M.; Xie, A.; Zhang, H.; Sarsenbekuly, B. The Rheological Behavior of Sodium Dodecyl Sulfate/N-hexylamine Aqueous Solution at High Concentrations. *J. Mol. Liq.* **2021**, *335*, 116140. [CrossRef]
25. Shi, L.; Liu, F.; Liu, T.; Chen, J.; Xu, S.; Zeng, H. Reversible Fabrication and Self-assembly of a Gemini Supra-amphiphile Driven by Dynamic Covalent Bonds. *Soft Matter* **2018**, *14*, 5995–6000. [CrossRef]
26. Li, X.; Guo, Y.; Wang, H.; Li, Z.; Wang, J. Regulation of Anionic Structure and CO_2 on the Vesicle Formation of Imidazolium-based Ionic Liquids in Water. *J. Mol. Liq.* **2022**, *352*, 118699. [CrossRef]
27. Wang, X.; Gao, X.; Xiao, X.; Jiang, S.; Yan, Y.; Huang, J. Photoresponsive Supramolecular Strategy for Controlled Assembly in Light-inert Double-chain Surfactant System. *J. Colloid Interf. Sci.* **2021**, *594*, 727–736. [CrossRef]
28. Corbett, P.T.; Leclaire, J.; Vial, L.; West, K.R.; Wietor, J.-L.; K., M.; Sanders, J.; Otto, S. Dynamic Combinatorial Chemistry. *Chem. Rev.* **2006**, *106*, 3652–3711. [CrossRef]
29. Jin, Y.; Yu, C.J.; Denman, R.; Zhang, W. Recent Advances in Dynamic Covalent Chemistry. *Chem. Soc. Rev.* **2013**, *42*, 6634–6654. [CrossRef]
30. Sun, P.; Ren, S.; Wu, A.; Sun, N.; Shi, L.; Zheng, L. Chirality Transfer based on Dynamic Covalent Chemistry: From Small Chiral Molecules to Supramolecules. *Chem. Commun.* **2019**, *55*, 9861–9864. [CrossRef]
31. Janeliunas, D.; van Rijn, P.; Boekhoven, J.; Minkenberg, C.B.; van Esch, J.H.; Eelkema, R. Aggregation-Driven Reversible Formation of Conjugated Polymers in Water. *Angew. Chem. Int. Ed.* **2013**, *52*, 1998–2001. [CrossRef] [PubMed]
32. Yang, X.; Liu, G.; Peng, L.; Guo, J.; Tao, L.; Yuan, J.; Chang, C.; Wei, Y.; Zhang, L. Highly Efficient Self-Healable and Dual Responsive Cellulose-Based Hydrogels for Controlled Release and 3D Cell Culture. *Adv. Funct. Mater.* **2017**, 1703174. [CrossRef]
33. Minkenberg, C.B.; Florusse, L.; Eelkema, R.; J., M.; Koper, G.; H. van Esch, J. Triggered Self-Assembly of Simple Dynamic Covalent Surfactants. *J. Am. Chem. Soc.* **2009**, *131*, 11274–11275. [CrossRef]
34. 34. Minkenberg, C.B.; Li, F.; van Rijn, P.; Florusse, L.; Boekhoven, J.; Stuart, M.C.; Koper, G.J.; Eelkema, R.; van Esch, J.H. Responsive Vesicles from Dynamic Covalent Surfactants. *Angew. Chem. Int. Ed.* **2011**, *123*, 3483–3486. [CrossRef]
35. Minkenberg, C.B.; Hendriksen, W.E.; Li, F.; Mendes, E.; Eelkema, R.; van Esch, J.H. Dynamic Covalent Assembly of Stimuli Responsive Vesicle Gels. *Chem. Commun.* **2012**, *48*, 9837–9839. [CrossRef]
36. Wang, G.; Wu, G.; Wang, Z.; Zhang, X. Asymmetric and Symmetric Bolaform Supra-Amphiphiles: Formation of Imine Bond Influenced by Aggregation. *Langmuir* **2014**, *30*, 1531–1535. [CrossRef]
37. Wang, G.; Wang, C.; Wang, Z.; Zhang, X. H-Shaped Supra-Amphiphiles Based on a Dynamic Covalent Bond. *Langmuir* **2012**, *28*, 14567–14572. [CrossRef]
38. Xie, Y.; Xu, Y.; Xu, J. pH-Responsive Pickering Foam Created from Self-Aggregate Polymer Using Dynamic Covalent Bond. *J. Colloid Interf. Sci.* **2021**, *597*, 383–392. [CrossRef]
39. Wang, P.; Zhu, T.; Hou, X.; Zhao, Y.; Zhang, X.; Wang, T.; Yang, H.; Kang, W. Responsive Wormlike Micelle with pH-Induced Transition of Hydrotrope Based on Dynamic Covalent Bond. *J. Mol. Liq.* **2019**, *286*, 110935. [CrossRef]
40. Sun, N.; Shi, L.; Lu, F.; Xie, S.; Sun, P.; Zheng, L. Spontaneous Vesicle Phase Formation by Linear Pseudo-Oligomeric Surfactant in Aqueous Solutions. *Langmuir* **2015**, *31*, 2281–2287. [CrossRef]
41. Long, P.; Song, A.; Wang, D.; Dong, R.; Hao, J. pH-Sensitive Vesicles and Rheological Properties of PFLA/NaOH/H_2O and PFLA/LiOH/H_2O Systems. *J. Phys. Chem. B* **2011**, *115*, 9070–9076. [CrossRef] [PubMed]
42. Song, S.; Song, A.; Feng, L.; Wei, G.; Dong, S.; Hao, J. Fluorescent Hydrogels with Tunable Nanostructure and Viscoelasticity for Formaldehyde Removal. *ACS Appl. Mater. Interfaces* **2014**, *6*, 18319–18328. [CrossRef] [PubMed]
43. Sun, N.; Shi, L.; Lu, F.; Xie, S.; Zheng, L. Spontaneous Vesicle Phase Formation by Pseudogemini Surfactants in Aqueous Solutions. *Soft Matter* **2014**, *10*, 5463–5471. [CrossRef] [PubMed]
44. Chakraborty, P.; Das, B.; Pal, P.; Datta, S.; Bera, S.; Dastidar, P. A Supramolecular Hydrogel Derived from a Simple Organic Salt Capable of Proton Conduction. *Chem. Commun.* **2020**, *56*, 5251–5525. [CrossRef]
45. Martin, S.M.; Ward, M.D. Lyotropic Phases Reinforced by Hydrogen Bonding. *Langmuir* **2005**, *21*, 5324–5331. [CrossRef]
46. Belman, N.; Israelachvili, J.N.; Li, Y.; Safinya, C.R.; Bernstein, J.; Golan, Y. The Temperature-Dependent Structure of Alkylamines and Their Corresponding Alkylammonium-Alkylcarbamates. *J. Am. Chem. Soc* **2009**, *131*, 9107–9113. [CrossRef]
47. Liu, H.; Luo, J.; Shan, W.; Guo, D.; Wang, J.; Hsu, C.-H.; Huang, M.; Zhang, W.; Lotz, B.; Zhang, W.-B.; et al. Manipulation of Self-Assembled Nanostructure Dimensions in Molecular Janus Particles. *ACS Nano* **2016**, *10*, 6585–6596. [CrossRef]
48. Shi, L.; Wei, Y.; Sun, N.; Zheng, L. First Observation of Rich Lamellar Structures formed by a Single-tailed Amphiphilic Ionic Liquid in Aqueous Solutions. *Chem. Commun.* **2013**, *49*, 11388–11390. [CrossRef]
49. Israelachvili, J.; Mitchell, D.; Ninham, B. Theory of Self-Assembly of Hydrocarbon Amphiphiles into Micelles and Bilayers. *J. Chem. Soc. Faraday Trans. 2* **1976**, *72*, 1525–1568. [CrossRef]
50. Sun, P.; Ren, S.; Liu, F.; Wu, A.; Sun, N.; Shi, L.; Zheng, L. Smart Low Molecular Weight Hydrogels with Dynamic Covalent Skeletons. *Soft Matter* **2018**, *14*, 6678–6683. [CrossRef]

51. Rosen, M.J.; Cohen, A.W.; Dahanayake, M.; Hua, X. Relationship of Structure to Properties in Surfactants. 10. Surface and Thermodynamic Properties of 2-Dodecyloxypoly(ethenoxyethanol)s, $C_{12}H_{25}(OC_2H_4)_xOH$, in aqueous solution. *J. Phys. Chem.* **1982**, *86*, 541–545
52. Tanford, C. Micelle Shape and Size. *J. Phys. Chem.* **1972**, *76*, 3020–3024

molecules

MDPI

Article

Influence of Linear Diamine Counterions on the Self-Assembly of Glycine-, Alanine-, Valine-, and Leucine-Based Amphiphiles

Margarita Angel Alvarez [1], Nathan Black [1], Saylor Estelle Blanco [1], Katelyn Ruth Reid [1], Eugene J. Billiot [1], Fereshteh H. Billiot [1,*] and Kevin F. Morris [2,*]

[1] Department of Physical and Environmental Sciences, Texas A&M University-Corpus Christi, 6300 Ocean Drive, Corpus Christi, TX 78412, USA; malvarez20@islander.tamucc.edu (M.A.A.); nblack1@uw.edu (N.B.); sblanco3@islander.tamucc.edu (S.E.B.); kreid7@islander.tamucc.edu (K.R.R.)

[2] Department of Chemistry, Carthage College, 2001 Alford Park Drive, Kenosha, WI 53140, USA

* Correspondence: fereshteh.billiot@tamucc.edu (F.H.B.); kmorris@carthage.edu (K.F.M.)

Abstract: Electrical conductimetry and dynamic light scattering (DLS) were used to investigate the aggregation behaviors of four amino acid-based surfactants (AABSs; undecanoyl-glycine, undecanoyl-L-alanine, undecanoyl-L-valine, undecanoyl-L-leucine) in the presence of five linear diamine counterions (1,2-diaminoethane, 1,3-diaminopropane, 1,4-diaminobutane, 1,5-diaminopentane, 1,6-diaminohexane). Electrical conductimetry was used to measure the CMCs for each system, which ranged from 5.1 to 22.5 mM. With respect to counterions, the obtained CMCs decreased with increases in the interamine spacer length; this was attributed to the improved torsional binding flexibility in longer counterions. Strong linear correlations (mean R^2 = 0.9443) were observed between the CMCs and predicted surfactant partition coefficients (logP; water/octanol), suggesting that micellization is primarily driven by the AABS's hydrophobicity for these systems. However, significant deviations in this linear relationship were observed for systems containing 1,2-diaminoethane, 1,4-diaminobutane, and 1,6-diaminohexane (p = 0.0774), suggesting altered binding dynamics for these counterions. pH measurements during the CMC determination experiments indicated the full deprotonation of the AABSs but did not give clear insights into the counterion protonation states, thus yielding an inconclusive evaluation of their charge stabilization effects during binding. However, DLS measurements revealed that the micellar size remained largely independent of the counterion length for counterions longer than 1,2-diaminoethane, with hydrodynamic diameters ranging from 2.2 to 2.8 nm. This was explained by the formation of charge-stabilized noncovalent dimers, with each counterion bearing a full +2 charge. Conductimetry-based estimates of the degrees of counterion binding (β) and free energies of micellization (ΔG°_{M}) revealed that bulky AABSs exhibit preferential binding to counterions with an even number of methylene groups. It is proposed that when these counterions form noncovalent dimers, perturbations in their natural geometries result in the formation of a binding pocket that accommodates the AABS steric bulk. While the direct application of these systems remains to be seen, this study provides valuable insights into the structure–property relationships that govern AABS aggregation.

Keywords: amino acid-based surfactants; diamine counterions; critical micelle concentration

1. Introduction

Amino acid-based surfactants (AABSs) are typically derived from amino acid head-groups and biolipid tails. Because their composition is derived from natural products, AABSs function as green surfactants. As such, they are known to be more bioavailable, eco-friendly, and sustainable than traditional industrial surfactants [1–3]. This allows certain AABSs to function more effectively in sensitive applications, including environmental remediation [4], drug delivery [5], antimicrobial treatments [6], and cosmetic products [7]. Moreover, the customizability associated with amino acids may allow for the optimization of surfactant performance in areas that do not depend on bioavailability or eco-friendliness [7–9].

Like other surfactants, AABSs aggregate only above the CMC, which is a key indicator of their performance. Low CMCs are advantageous in applications like environmental remediation, where micelles sequester nonpolar contaminants, while higher CMCs are beneficial in antimicrobial treatment, as individual surfactants bind more effectively to pathogen membranes than micelles [10,11]. The micellar size is also crucial, particularly in drug delivery, where micelles must balance the encapsulation of therapeutic materials with the minimization of circulation disruption [12]. The counterion dynamics play a significant role in regulating both the CMC and micellar size, as the interaction between counterions and surfactant headgroups can be optimized to enhance the AABS performance across these applications [13–18]. While research on the exact relationship between the pH and CMC has yielded inconsistent results, it is widely understood that the pH directly impacts the aggregation dynamics by altering the protonation states of surfactants and counterions [19]. By fine-tuning the counterion properties, such as the size and charge density, it may be possible to tailor AABSs' behavior to maximize their effectiveness for specific applications.

This study utilized a combinatorics approach to study the aggregation behaviors of four AABSs (undecanoyl-glycine, undecanoyl-L-alanine, undecanoyl-L-valine, undecanoyl-L-leucine) in the presence of five diamine counterions (1,2-diaminoethane, 1,3-diaminopropane, 1,4-diaminobutane, 1,5-diaminopentane, 1,6-diaminohexane). These surfactants were chosen for their systematic increases in the amino acid R-group complexity, while the counterions were chosen for their systematic increases in the interamine spacer length. Furthermore, the surfactants and counterions' pH sensitivity allowed some insight into their charge-stabilizing behavior based on the protonation states. The structures of the AABSs and diamine counterions are shown in Figure 1, along with their abbreviated names (which will be used hereafter in this publication) and counterion pK_a values. The CMCs of each combination were measured by electrical conductimetry, while the micellar sizes were measured by dynamic light scattering (DLS). From the conductimetry data, the degree of counterion binding (β) and standard free energy of micellization (ΔG°_{M}) were estimated for each system. The pH was measured for each experiment to gain insights into the protonation states exhibited by all pH-sensitive functional groups.

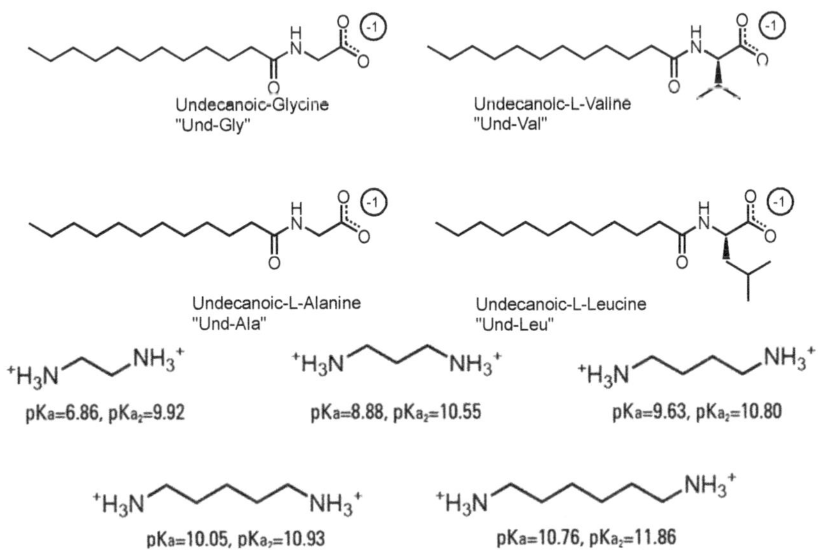

Figure 1. Chemical structures of amino acid-based surfactants (AABSs) and diamine counterions. The corresponding pKa values are noted.

2. Experimental Procedures

2.1. Production of Amino Acid-Based Surfactants and Acquisition of Diamine Counterions

The amino acid-based surfactants (AABSs) were produced by the stepwise coupling of undecanoic acid with *N*-hydroxysuccinimide (NHS) and amino acids, as reported previously [20]. Once the acidic form of each AABS was recovered gravimetrically at a low pH, the reaction completeness and purity were verified by relative peak integration on ^{1}H NMR spectra acquired by a Bruker Avance II 300 MHz spectrometer (Supplemental Information 1).

All synthetic materials were utilized as received from their respective manufacturers. The amino acids glycine, L-alanine, L-valine, and L-leucine (>97% purity) were received from Acrotein, Hoover, AL. The other reagents were undecanoic acid (>99%), *N,N′*-diisopropylcarbodiimide (>99%), NHS (98%), tetrahydrofuran (THF; ≥99.9%), sodium bicarbonate (≥99.7%), and hydrochloric acid (37%); all were acquired from Sigma-Aldrich, St. Louis, MO, USA.

MilliQ water was also utilized during AABS synthesis (Millipore, Bedford, MA, USA). The diamine counterions 1,2-diaminoethane, 1,3-diaminopropane, 1,4-diaminobutane, 1,5-diaminopentane, and 1,6-diaminohexane were utilized as received (>97%; TCI America, Portland, OR, USA).

2.2. Measurement of Critical Micelle Concentrations by Conductimetry

For each AABS/diamine counterion combination, three 10 mL solutions were prepared in MilliQ water containing 40 mM of both the surfactant and counterion at 25 °C. The initial conductivity of each solution was assessed using a Vernier potentiometric conductivity meter. Next, a series of dilutions was performed, each involving a 10% reduction where 1 mL of the solution was systematically replaced with 1 mL of water. The electrical conductivity of these solutions was recorded as a function of the concentration, and an in-house Python script was employed to identify the CMC (Supplemental Information 2). Triplicate solutions were then prepared at the calculated CMC (rather than 40 mM) and measured with a pH meter to gain insights into the protonation states exhibited by each AABS and counterion during micellization.

2.3. Measurement of Micellar Size by Dynamic Light Scattering

For each AABS/diamine counterion combination, three 3 mL solutions were prepared at 5× CMC in MilliQ water at 25 °C. The solutions were prepared at this high concentration, as opposed to the CMC, to ensure that the micellar species were predominant and readily detectable. Each solution was then filtered through a 0.020 µm Whatman syringe filter into a new 3 mL cuvette. Next, measurements were performed using a Malvern Nano Series Zetasizer at a backscattering angle of 173 degrees. A series of 12 scans was carried out with a 10-second interval between each. The largest peak present in both the by-intensity and by-volume spectra was identified, and the size corresponding to that peak in the by-intensity spectra was taken as the hydrodynamic micellar diameter. Triplicate solutions were then prepared at 5× CMC and measured with a pH meter to determine whether the pH significantly changed between the CMC and micellar size measurements.

2.4. Estimation of Degrees of Counterion Binding from Conductimetry

The conductimetry data collected in Section 2.2 were treated with analysis intended to estimate the degrees of counterion binding (β) for each studied system. This parameter was calculated by determining the relative change in slope of the conductimetry vs. concentration plot at the CMC, as shown in Equation (1) and previously reported [17]:

$$\beta = \frac{m_1 - m_2}{m_1} \quad (1)$$

In Equation (1), m_1 is defined as the conductimetry plot's slope below the CMC, and m_2 is defined as the conductimetry plot's slope above the CMC. This estimation of

β assumes that any differences between m_1 and m_2 are exclusively caused by counterion charge stabilization effects at the CMC. From this perspective, the estimated β values were interpreted as binding fractions. To illustrate this, it is emphasized that $\beta = 0$ only when $m_1 = m_2$, indicating no change in surfactant conductivity at the CMC and thus corresponding to no counterion binding. However, $\beta = 1$ only when $m_2 = 0$, indicating the complete cessation of surfactant conductivity at the CMC and thus corresponding to total counterion binding. Despite the assumptions made in this estimation (and subsequent interpretation) of β, it is a widely used standard for the comparison of counterion binding between aggregation systems [21–23].

2.5. Estimation of Free Energies of Micellization from Conductimetry

Conductimetry data were used to approximate the standard free energies of micellization (ΔG°_{M}) for each studied system. This parameter was calculated by the substitution of the CMC and other structure–property values in Equation (2), as previously reported [21]:

$$\Delta G^{0}_{M} = RT\left(\frac{1}{j} + \beta\frac{i}{j}\left|\frac{Z_s}{Z_c}\right|\right)\ln(CMC) + RT\left(\frac{i}{j}\left|\frac{Z_s}{Z_c}\right|\beta * \ln\left(\frac{i}{j}\left|\frac{Z_s}{Z_c}\right|\right) - \frac{\ln(j)}{j}\right) \quad (2)$$

In Equation (2), i is the number of ionic surfactant groups, j is the number of surfactant tails, Z_s is the expected charge per ionic surfactant group, and Z_c is the expected charge per counterion. For the studied AABSs, it is clear that $i = j = 1$. The parameters Z_s and Z_c were selected based on the expected protonation states for each system from the analysis of the pH and DLS measurements, as discussed in Section 3.5.

3. Results/Discussion

3.1. Critical Micelle Concentration Measurements

Critical micelle concentrations were determined by conductimetry for Und-Gly, Und-Ala, Und-Val, and Und-Leu in the presence of 1,2-diaminoethane, 1,3-diaminopropane, 1,4-diaminobutane, 1,5-diaminopentane, and 1,6-diaminohexane; the resulting measurements are shown in Table 1.

Table 1. CMCs in mM of the AABSs in the presence of different diamine counterions.

Counterion	CMC			
	Und-Gly	**Und-Ala**	**Und-Val**	**Und-Leu**
1,2-diaminoethane	22.5 ± 0.5	20.8 ± 0.1	12.2 ± 0.8	7.7 ± 0.4
1,3-diaminopropane	21.3 ± 0.6	15.3 ± 0.6	10.2 ± 0.1	6.4 ± 0.4
1,4-diaminobutane	13.5 ± 0.3	13.2 ± 0.2	8.1 ± 0.2	5.1 ± 0.5
1,5-diaminopentane	13.4 ± 0.1	10.4 ± 0.1	7.5 ± 0.3	5.5 ± 0.3
1,6-diaminohexane	9.9 ± 0.6	9.3 ± 0.2	7.0 ± 0.4	5.7 ± 0.4

Counterions with larger alkyl chains are generally associated with significant decreases in the CMC, as can be seen in Table 1. This effect is most dramatic for Und-Gly, whose CMC decreases from 22.5 to 9.9 mM as the counterion alkyl group length increases from 1,2-diaminoethane to 1,6-diaminohexane. These CMC values could be attributed to the rotational flexibility afforded by larger counterions, allowing for optimized binding conformations. Alternatively, these data could be explained by hydrophobic interactions between the AABS R-groups and counterion interamine spacers, which functionally depend on the counterion length. Interestingly, the studied counterions have less diverse effects on the CMC as the AABS R-groups increase in complexity from Und-Gly to Und-Leu. Contrasting with Und-Gly's steeply defined decrease from 22.5 to 9.9 mM, Und-Leu slightly decreases from 7.7 to 5.7 mM as the counterion length increases from 1,2-diaminoethane to 1,6-diaminohexane. Furthermore, Und-Leu has statistically indistinguishable CMCs in the presence of 1,4-diaminobutane, 1,5-diaminopentane, and 1,6-diaminohexane (5.1, 5.5, and 5.7 mM, respectively). This gradient in the sensitivity of the surfactant aggregation behavior with respect to the counterion structure indicates that the AABS R-groups control each counterion's role in micellization.

Table 1 also shows that the CMC decreases as the AABS R-group size increases from Und-Gly to Und-Leu. For example, the CMC decreases from 21.3 mM for Und-Gly to 6.4 mM for Und-Leu in the presence of 1,3-diaminopropane. This gradient indicates that surfactant aggregation may be largely driven by hydrophobicity in AABS R-groups and that their steric bulk does not significantly hinder the micellization process. In this context, the aforementioned gradient in counterion sensitivity with respect to the CMC from Und-Gly to Und-Leu is more explainable by the relative hydrophobicity of AABSs rather than their steric bulk. In other words, it is unlikely that sterically hindered binding interactions are solely responsible for the relative independence of Und-Leu's CMC from the counterion length. Instead, Und-Leu may be so hydrophobic compared to Und-Gly that the former's aggregation is highly favorable regardless of which counterion binds to it.

To further investigate the role of the AABS's hydrophobicity in these systems, their CMCs were correlated with the AABS partition coefficients (logP) between water and octanol. This was estimated by a consensus calculation performed by ChemAxon's MarvinSketch software 23.4, in which the output was influenced by several predictive models [24,25] and the total hydrophobic contribution of all AABS molecular fragments was numerically estimated and summed under standard conditions by two datasets. When these correlations were grouped by counterion, as shown in Figure 2, it was observed that the general relationship between the CMC and surfactant logP values was strongly linear, with R^2 values ranging from 0.8833 for 1,4-diaminobutane systems to 0.9870 for 1,5-diaminopentane systems. This indicates that most of the variance in the observed CMC values is explainable solely by differences in the predicted AABS hydrophobicity. However, this correlation was significantly stronger for 1,3-diaminopropane and 1,5-diaminopentane (Figure 2A, mean $R^2 = 0.9854$) than for 1,2-diaminoethane, 1,4-diaminobutane, and 1,6-diaminohexane (Figure 2B, mean $R^2 = 0.9169$). This was determined at 90% confidence by collecting the residuals from all linear regressions in both datasets and performing a two-sample t-test between them ($p = 0.0774$). This deviation from linearity for 1,2-diaminoethane, 1,4-diaminobutane, and 1,6-diaminohexane is expected to be due to the statistically indistinguishable CMCs between Und-Gly and Und-Ala. For example, the CMC decreases from 13.5 mM for Und-Gly to 13.2 mM for Und-Ala in the presence of 1,4-diaminobutane. However, in 1,5-diaminopentane systems, this same comparison corresponds to a more significant decrease from 13.4 mM to 10.4 mM. It is interesting that these anomalous behaviors were exclusively observed for AABSs with the least complex R-groups (Und-Gly and Und-Ala) in the presence of counterions with an even number of methylene groups (1,2-diaminoethane, 1,4-diaminobutane, 1,6-diaminohexane). These structural motifs imply that while aggregation appears to be primarily driven by the surfactant hydrophobicity, as shown in Figure 2, it is still influenced by cooperative binding, which depends on the structure of both the surfactant and counterion.

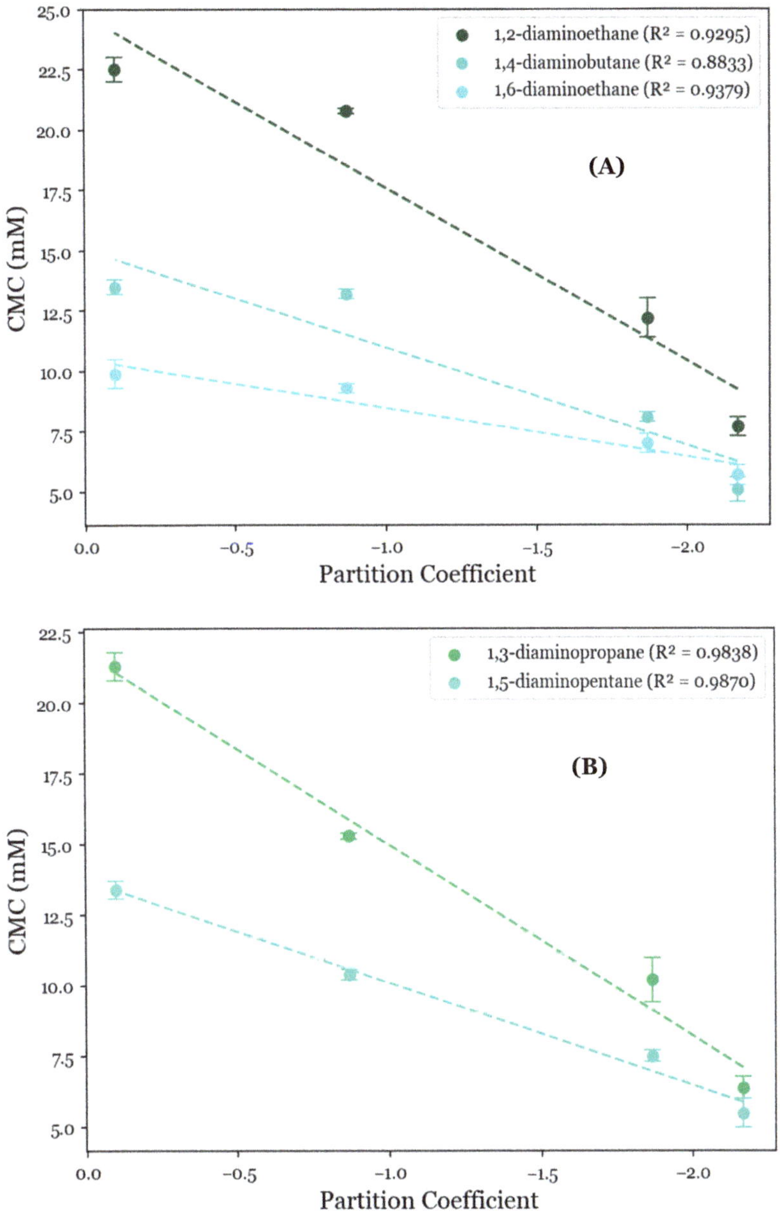

Figure 2. CMC values for the studied systems as a function of AABSs' partition coefficients between water and octanol (logP) for diamine counterions with (**A**) an even number of methylene groups (1,2-diaminoethane, 1,4-diaminobutane, 1,6-diaminohexane) and (**B**) an odd number of methylene groups (1,3-diaminopropane, 1,5-diaminopentane).

3.2. pH Measurements at the Critical Micelle Concentration

To determine the role of charge stabilization in the cooperative binding between AABSs and diamine counterions during the micellization process, the pH was recorded at each system's CMC. The observed pH values presented in Table 2 are basic, ranging from

9.8 to 11.7. This indicates that the acidic surfactants are overwhelmingly deprotonated and possess negatively charged carboxylate groups. However, because the measured pH values generally lie in the range of the counterion pK_a values reported in Figure 1, the protonation states experienced by each diamine counterion are less obvious. As such, numeric methods were used to estimate their average charge at each system's CMC. This was conducted using Equation (3), which is a weighted average built from fractional compositions for each protonation state; each fraction is estimated by Equations (4)–(6) (Supplemental Information 3). The results of these calculations are also presented in Table 2.

$$z_{avg} = (+2)\left(\alpha_{NH_3^+-R-NH_3^+}\right) + (+1)\left(\alpha_{NH_3^+-R-NH_2}\right) + (0)\left(\alpha_{NH_2-R-NH_2}\right) \quad (3)$$

$$\alpha_{NH_3^+-R-NH_3^+} \sim \frac{[H^+]}{[H^+]^2 + K_{a1}[H^+] + K_{a1}K_{a2}} \quad (4)$$

$$\alpha_{NH_3^+-R-NH_2} \sim \frac{K_{a1}[H^+]}{[H^+]^2 + K_{a1}[H^+] + K_{a1}K_{a2}} \quad (5)$$

$$\alpha_{NH_2-R-NH_2} \sim \frac{K_{a1}K_{a2}}{[H^+]^2 + K_{a1}[H^+] + K_{a1}K_{a2}} \quad (6)$$

Table 2. The average counterion charge along with the recorded pH for each system.

Counterion	Surfactant							
	Und-Gly		Und-Ala		Und-Val		Und-Leu	
	Avg. Counterion Charge	pH	Avg. Counterion Charge	pH	Avg. Counterion Charge	pH	Avg. Counterion Charge	pH
1,2-diaminoethane	0.52 ± 0.0	9.90 ± 0.0	0.04 ± 0.0	11.3 ± 0.2	0.53 ± 0.0	9.90 ± 0.0	0.59 ± 0.0	9.78 ± 0.1
1,3-diaminopropane	0.48 ± 0.0	10.6 ± 0.1	0.50 ± 0.0	10.6 ± 0.1	0.49 ± 0.0	10.6 ± 0.1	0.53 ± 0.0	10.5 ± 0.1
1,4-diaminobutane	0.29 ± 0.0	11.2 ± 0.1	0.42 ± 0.1	11.0 ± 0.1	0.25 ± 0.0	11.3 ± 0.1	0.33 ± 0.0	11.1 ± 0.0
1,5-diaminopentane	0.19 ± 0.0	11.6 ± 0.1	0.19 ± 0.0	11.9 ± 0.1	0.16 ± 0.0	11.7 ± 0.0	0.17 ± 0.1	11.7 ± 0.2
1,6-diaminohexane	1.59 ± 0.0	10.5 ± 0.1	1.61 ± 0.1	10.5 ± 0.1	1.70 ± 0.2	10.3 ± 0.4	1.60 ± 0.1	10.5 ± 0.1

Despite the numeric protonation states produced by these calculations, they are interpreted qualitatively. This is partially due to the use of kinetic pre-equilibrium approximations in Equations (4)–(6), which are not necessarily appropriate when pK_{a1} and pK_{a2} have similar values. This is true for 1,5-diaminopentane, for which the pK_a values differ by less than 1 pH unit, with pK_{a1} = 10.05 and pK_{a2} = 10.93. Furthermore, the reported counterion pK_a values do not account for binding interactions with AABSs, nor do they account for subsequent perturbations in acidity caused by the surfactant assembly; a very recent study published during the preparation of this manuscript reported the steep dependence of an AABS's pK_a values on the micellization process [18].

Even from this qualitative perspective, there are significant discrepancies between the estimated counterion charges and experimental CMC data. Firstly, there is a general reduction in the predicted charge as the counterion length increases from 1,2-diaminoethane to 1,5-diaminopentane. For example, in Und-Gly systems, 1,2-diaminoethane is predicted to exhibit an average charge of +0.52, while 1,5-diaminopentane is predicted to exhibit an average charge of +0.19. This indicates that the counterion strength should also decrease along this gradient, but, instead, significant decreases in the CMC are observed: Und-Gly's CMC drops from 22.5 to 13.4 mM as the counterion length increases from 1,2-diaminoethane to 1,5-diaminopentane. A single deviation from this trend is seen in Und-Ala systems, for which 1,2-diaminoethane has an abnormally low predicted charge of +0.04. Moreover, the high charges predicted for 1,6-diaminohexane indicate that it should be by far the most effective counterion due to charge stabilization. For instance, in Und-Val systems, 1,6-diaminohexane exhibits an estimated average charge of +1.70; the next-highest charge is +0.53, exhibited by 1,2-diaminoethane. However, this steep difference is not reflected in the CMCs of systems containing 1,6-diaminohexane, which are only slightly lower than those containing structurally similar counterions such as 1,5-diaminopentane. For example, the CMC of Und-Val is observed to be 7.5 mM in the presence of 1,5-diaminopentane but only decreases to 7.0 mM in the presence of 1,6-diaminohexane.

Overall, no significant correlation was found between the pH-based predictions of the counterion charge and the experimental CMC values. Despite the initial impression that the pH might not influence these systems, previous but limited variable-pH studies of diamine counterions have already established that the CMC is heavily dependent upon the pH [18]. Because the pH data do not provide consistent insights into the effect of the protonation state on the CMCs of these systems, it is expected that the input counterion pK_a values were flawed in describing these systems, as discussed previously. If true, this reinforces the aforementioned study, which documented changes in the AABS pK_a values due to micellization [19]. Because this research evaluated four AABSs in the presence of five diamine counterions, it comprises a more extensive set than those analyzed in the previous study. From this perspective, the effect of micellization on the surfactant/counterion pK_a values appears to be far more significant and ubiquitous than indicated by previous research. Therefore, this effect may be significant to the point that it should be considered when evaluating any counterion charge-stabilizing effects associated with micellization.

3.3. Micellar Hydrodynamic Diameter Measurements

Approximate micellar sizes were determined from hydrodynamic diameter measurements collected by dynamic light scattering (DLS). Beyond evaluating the effects of aggregate structures on the micellar size, this was performed to gain further insights into any cooperative binding processes between the AABSs and diamine counterions. The resulting measurements are shown in Table 3.

Table 3. The hydrodynamic diameters of AABS micelles in the presence of different diamine counterions.

Counterion	Surfactant							
	Und-Gly		Und-Ala		Und-Val		Und-Leu	
	Size (nm)	pH	Size (nm)	pH	Size (nm)	pH	Size (nm)	pH
1,2-diaminoethane	1.7 ± 0.2	10.2 ± 0.1	1.7 ± 0.3	11.3 ± 0.2	2.3 ± 0.3	10.1 ± 0.0	2.2 ± 0.2	10.0 ± 0.0
1,3-diaminopropane	2.3 ± 0.3	10.6 ± 0.1	2.3 ± 0.2	10.7 ± 0.0	2.3 ± 0.3	10.8 ± 0.1	2.7 ± 0.2	10.7 ± 0.0
1,4-diaminbutane	2.5 ± 0.2	11.6 ± 0.1	2.5 ± 0.4	11.4 ± 0.1	2.6 ± 0.2	11.7 ± 0.1	2.8 ± 0.2	11.5 ± 0.1
1,5-diaminopentane	2.7 ± 0.2	11.8 ± 0.1	2.6 ± 0.2	11.6 ± 0.1	2.8 ± 0.2	11.8 ± 0.1	2.7 ± 0.2	11.9 ± 0.1
1,6 diaminohexane	2.7 ± 0.2	10.8 ± 0.1	2.7 ± 0.2	10.8 ± 0.2	2.8 ± 0.2	10.6 ± 0.4	2.8 ± 0.2	10.0 ± 0.1

The hydrodynamic micellar diameters of all AABSs were statistically indistinguishable as the counterion length increased from 1,3-diaminopropane to 1,6-diaminohexane, as shown in Table 3. For example, Und-Gly's hydrodynamic diameters vaguely increased from 2.3 to 2.7 nm along this counterion gradient. Interestingly, 1,2-diaminoethane did not conform to this trend, as it induced significantly lower micellar sizes than other counterions. For example, the same surfactant (Und-Gly) yielded a hydrodynamic diameter of 1.7 nm in the presence of 1,2-diaminoethane. This trend is not unique to Und-Gly and was observed with all tested AABSs, which supports the conclusion that 1,2-diaminoethane exhibits anomalous behaviors in comparison to the other diamine counterions. Because 1,2-diaminoethane has a low number of methylene groups and is relatively small, it is possible that its constrained torsional flexibility forces it to behave like a monoatomic ion rather than a divalent counterion with a flexible spacer. This abnormal behavior would explain the size discrepancies observed in systems containing 1,2-diaminoethane, which is thus scrutinized in subsequent systematic comparisons with the other diamine counterions.

The statistical indistinguishability in the DLS measurements for systems containing counterions longer than 1,2-diaminoethane implies that the counterions' structural variations have a minimal effect on the micellar size. As such, it is likely that these counterions bind parallel to the AABSs' micellar interfaces, as shown in Figure 3. This binding orientation ensures that the counterions do not protrude significantly from each micellar surface, thus explaining the lack of correlation between the counterion length and micelle size. Furthermore, the proposed binding conformation is especially favorable because it would likely result in full counterion protonation, allowing for more effective charge stabilization through noncovalent dimerization. Therefore, despite the inconsistent results yielded by the pH data, it is expected that each diamine counterion has a +2 charge. Because the recorded pH values for each size measurement (Table 3) were extremely similar to those recorded for the CMC measurements (Table 1), it is not likely that the measured binding conformations were significantly altered between the two experiments.

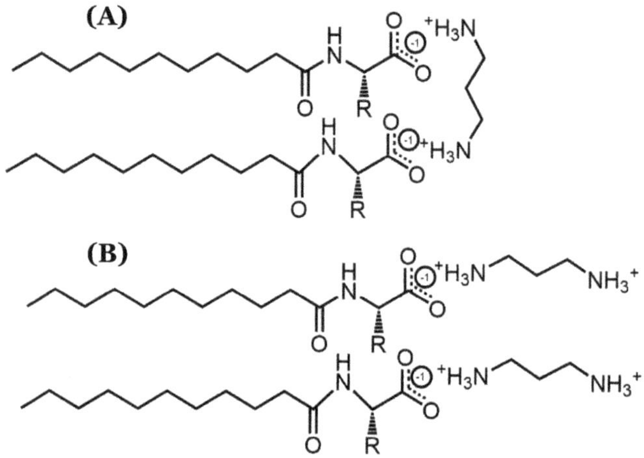

Figure 3. Visualization of possible diamine counterion binding conformations, particularly focusing on (**A**) parallel and (**B**) perpendicular orientations with respect to AABS micellar interfaces.

3.4. *Calculated Degrees of Counterion Binding*

To provide quantitative comparisons between the proposed noncovalent dimers formed by the diamine counterions, the degrees of counterion binding (β) were calculated from the conductimetry data according to Equation (1). These values are tabulated in Table 4; they are also visualized in Figure 4 with respect to (A) each counterion and (B) each AABS.

Table 4. The degree of counterion binding (β) and the free energies of micellization (ΔG°_{M}) for the AABSs in the presence of diamine counterions.

Counterion	Surfactant							
	Und-Gly		Und-Ala		Und-Val		Und-Leu	
	β	ΔG (kJ/mol)	β	ΔG (kJ/mol)	β	ΔG (kJ/mol)	β	ΔG (kJ/mol)
1,2-diaminoethane	0.44 ± 0.0	−12.1 ± 0.1	0.05 ± 0.0	−12.8 ± 0.2	0.38 ± 0.0	−13.3 ± 0.3	0.38 ± 0.0	−14.8 ± 0.5
1,3-diaminopropane	0.44 ± 0.0	−12.1 ± 0.1	0.61 ± 0.0	−14.4 ± 0.2	0.47 ± 0.0	−15.1 ± 0.5	0.51 ± 0.0	−16.4 ± 0.2
1,4-diaminobutane	0.59 ± 0.0	−15.2 ± 1.1	0.64 ± 0.0	−14.8 ± 0.1	0.61 ± 0.0	−16.3 ± 0.2	0.65 ± 0.0	−17.7 ± 0.1
1,5-diaminopentane	0.67 ± 0.1	−15.4 ± 0.5	0.69 ± 0.0	−15.9 ± 0.3	0.56 ± 0.0	−16.0 ± 0.2	0.62 ± 0.0	−17.2 ± 0.3
1,6-diaminohexane	0.73 ± 0.0	−16.2 ± 0.3	0.72 ± 0.0	−16.3 ± 0.1	0.56 ± 0.0	−16.2 ± 0.3	0.69 ± 0.0	−17.8 ± 0.2

β generally increases as the counterion length increases from 1,2-diaminoethane to 1,6-diaminohexane, as can be seen in Figure 4A. For example, the inspection of Und-Gly yielded β = 0.44 in the presence of 1,2-diaminoethane and β = 0.73 in the presence of 1,6-diaminohexane, corresponding to a ramp from 44 to 73% counterion binding. This aligns with the earlier proposal that longer counterions generally facilitate better binding, whether due to increased rotational flexibility or hydrophobic interactions with the counterions' interamine spacers. Interestingly, all AABSs appeared to be equally affected by the counterion gradient from 1,2-diaminoethane to 1,6-diaminohexane regardless of the R-group complexity, with even Und-Leu exhibiting a ramp from 38 to 69% counterion binding. This observation supports the earlier proposal in which Und-Leu's significantly reduced sensitivity to the counterion (with respect to the CMC) compared to Und-Gly was considered to be independent of counterion binding interactions, instead being primarily driven by hydrophobicity.

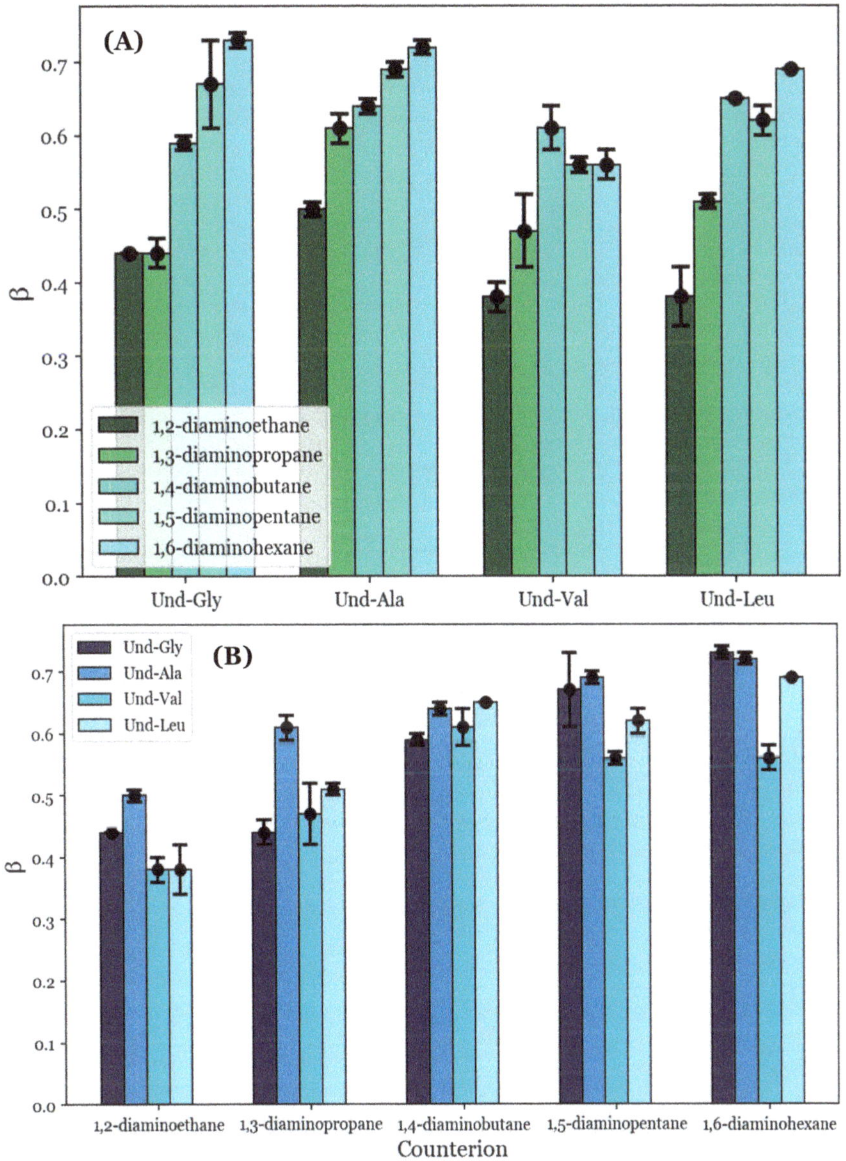

Figure 4. The degrees of counterion binding (β) with respect to (**A**) each AABS and (**B**) each diamine counterion.

It is interesting to note that Und-Leu does not exhibit a consistent increase in β along the gradient from 1,2-diaminoethane to 1,6-diaminohexane; Und-Leu binds significantly better to 1,4-diaminobutane and 1,6-diaminohexane (65% and 69% binding) than to 1,3-diaminopropane and 1,5-diaminopentane (51% and 62% binding). Furthermore, Und-Val exhibits 61% binding with 1,4-diaminobutane, which is significantly better than with 1,3-diaminopropane (47% binding) and 1,5-diaminopentane (56% binding). While the relative binding of Und-Val to 1,5-diaminopentane and 1,6-diaminohexane could not be statistically resolved due to the similar β values, it appears that the binding behaviors of Und-Leu and Und-Val exhibit a general dependence on whether the number of methylene groups in the counterion interamine spacer is even or odd. This is very interesting given that it mirrors previously discussed deviations from linearity in the relationship between the CMC and AABS hydrophobicity in the presence of 1,2-diaminoethane, 1,4-diaminobutane, and 1,6-diaminohexane, as shown in Figure 2B. However, it is interesting that this dependence in the β values occurred in every AABS except for Und-Gly and Und-Ala, as these surfactants were seemingly responsible for this linear deviation. 1,2-Diaminoethane was notably excluded from this analysis, as the β values for its systems did not adhere to the observed trend; this is attributed to its abnormal binding behaviors as proposed in Section 3.3.

To further understand the anomalous binding behaviors of Und-Val and Und-Leu, the β values were inspected with respect to each surfactant. Figure 4B shows that the steepest change in β between two consecutive AABSs (with respect to their R-group complexity) occurs between Und-Ala and Und-Val. More specifically, Und-Val exhibits significantly worse counterion binding than Und-Ala in the presence of most counterions. For example, in 1,5-diaminopentane systems, Und-Ala exhibits 69% counterion binding, while Und-Val exhibits 56% counterion binding. By comparison, Und-Gly's binding to 1,5-diaminopentane (67%) only differs from that of Und-Ala by two percentage points. This could be due to Und-Val's relative steric bulk compared to Und-Ala, which would severely limit the allowable binding conformations of each counterion to Und-Val. However, it should be noted that sterics cannot completely explain this trend, as there are significant increases in counterion binding from Und-Val to Und-Leu with 1,5-diaminopentane and 1,6-diaminohexane. The magnitude of this effect can be illustrated by comparing the β values for Und-Ala, Und-Val, and Und-Leu in 1,6-diaminohexane systems: Und-Ala exhibits 72% counterion binding, Und-Val exhibits 56% counterion binding, and Und-Leu (despite having the most steric bulk) exhibits 69% counterion binding—a value very similar to that of Und-Ala. Because this effect only occurs for the longest counterions (1,5-diaminopentane and 1,6-diaminohexane), it is proposed that Und-Leu's increased steric bulk promotes significant repulsion between Und-Leu's headgroups, thus creating extended binding distances, which may be optimized for noncovalent dimerization by these counterions.

In sum, the analysis of the data shown in Figure 4A demonstrates that counterion binding is generally improved with longer counterions. An exception seems to occur for Und-Val and Und-Leu, for which counterions with even numbers of methylene groups in their interamine spacers exhibit generally improved binding. The analysis of the data shown in Figure 4B complements these observations by suggesting that the AABS steric bulk generally inhibits counterion binding to Und-Val and Und-Leu. With this information, along with the proposed theory of noncovalent dimerization by the diamine counterions discussed in Section 3.3, it may be possible to explain the dependence of the β values for Und-Val and Und-Leu on the evenness/oddness of the number of methylene groups in the counterion interamine spacers. In order to orient both amines in the same direction (i.e., to bind both amines to the AABS micellar interfaces), the methylene groups of 1,2-diaminoethane, 1,4-diaminobutane, and 1,6-diaminohexane would require torsional strain that deviates from their typical sp^3 geometries. It is proposed that this torsional strain introduces binding pockets in their molecular geometry (visualized in Figure 5), which can at least partially accommodate the bulky R-groups associated with Und-Val and Und-Leu during binding. Because this torsional strain and subsequent binding pocket formation would not be necessary for the dimeric binding of 1,3-diaminopropane and

1,5-diaminopentane, the steric hindrances would be expected to intensify in their binding with Und-Val and Und-Leu. This suggests that the increased β values and decreased CMCs generally induced by longer counterions are not driven by hydrophobic interactions between their interamine spacers and AABS R-groups, but instead by the improved torsional flexibility of these counterions. Furthermore, because 1,2-diaminoethane likely does not have the torsional flexibility or size to produce a significant binding pocket, its abnormal binding behaviors are further explained.

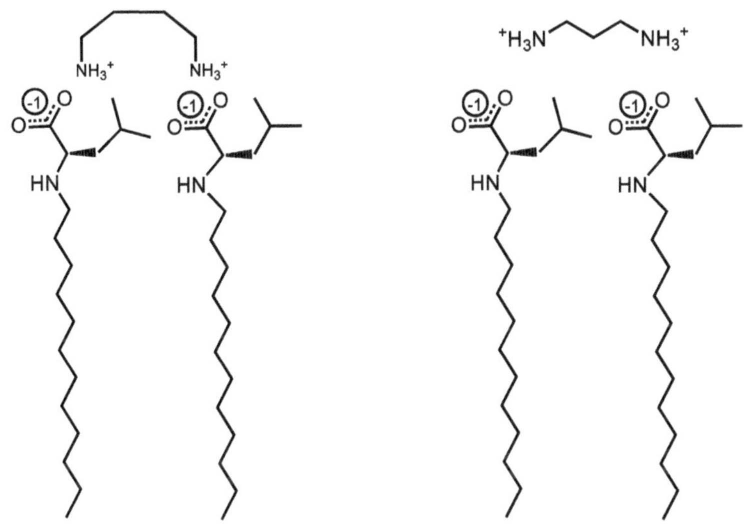

Figure 5. Illustration of torsional strain creating a binding pocket for the bulky R-groups of Und-Leu in the presence of 1,4-diaminobutane and 1,3-dimainopropane.

In sum, this proposed binding conformation explains why Und-Val and Und-Leu exhibit generally improved binding with counterions with an even number of methylene groups. It also explains why the β values for Und-Gly and Und-Ala do not show this trend: because these AABSs do not have significantly bulky R-groups, their binding is not expected to depend on the presence of a binding pocket in a counterion's molecular geometry. Finally, this similarity in behavior between Und-Gly and Und-Ala explains why they exhibited similar CMCs in the presence of counterions with an even number of methylene groups.

3.5. Estimated Free Energies of Micellization

As an extension to the previous results, the free energies of micellization (ΔG°_M) were estimated from the conductimetry data for the selected systems using Equation (2). This was performed in order to approximate the thermodynamic favorability of each micellization process. As discussed in Section 2.5, the condition $i = j = 1$ was applied for all studied systems. Based on the pH data reviewed in Section 3.2, the AABSs were deemed to be fully deprotonated, so Z_s was set to -1; however, the counterion charges were not as obvious from these data. A review of the micellar sizes obtained by DLS in Section 3.3 led to the proposal that the diamine counterions form fully charge-stabilized noncovalent dimers, so +2 was substituted for Z_c. The derived ΔG°_M values are depicted in Table 4 and graphically represented in Figure 6.

The ΔG°_M values appear to generally decrease with the counterion length, as shown in Figure 6A. For example, Und-Gly micellizes with a free energy of -12.1 kJ/mol in the presence of 1,2-diaminoethane but does so with a free energy of -16.2 kJ/mol with 1,6-diaminohexane. This indicates that the favorability of AABS aggregation generally increases with the counterion length, in agreement with previous observations. However, Und-Val and Und-Leu exhibit reduced ΔG°_{M} values for counterions with an even number of methylene groups. This effect is most pronounced for Und-Leu, which has lower free energies for 1,4-diaminobutane and 1,6-diaminohexane (−17.7 and −17.8 kJ/mol) than for 1,3-diaminopropane and 1,5-diaminopentane (−16.4 and −17.2 kJ/mol). While Und-Val exhibits statistically indistinguishable ΔG°_{M} values for counterions longer than 1,3-diaminopropane, this effect can still be observed qualitatively. Again, 1,2-diaminoethane was excluded from this analysis due to its abnormal binding behaviors, as discussed in Sections 3.3 and 3.4.

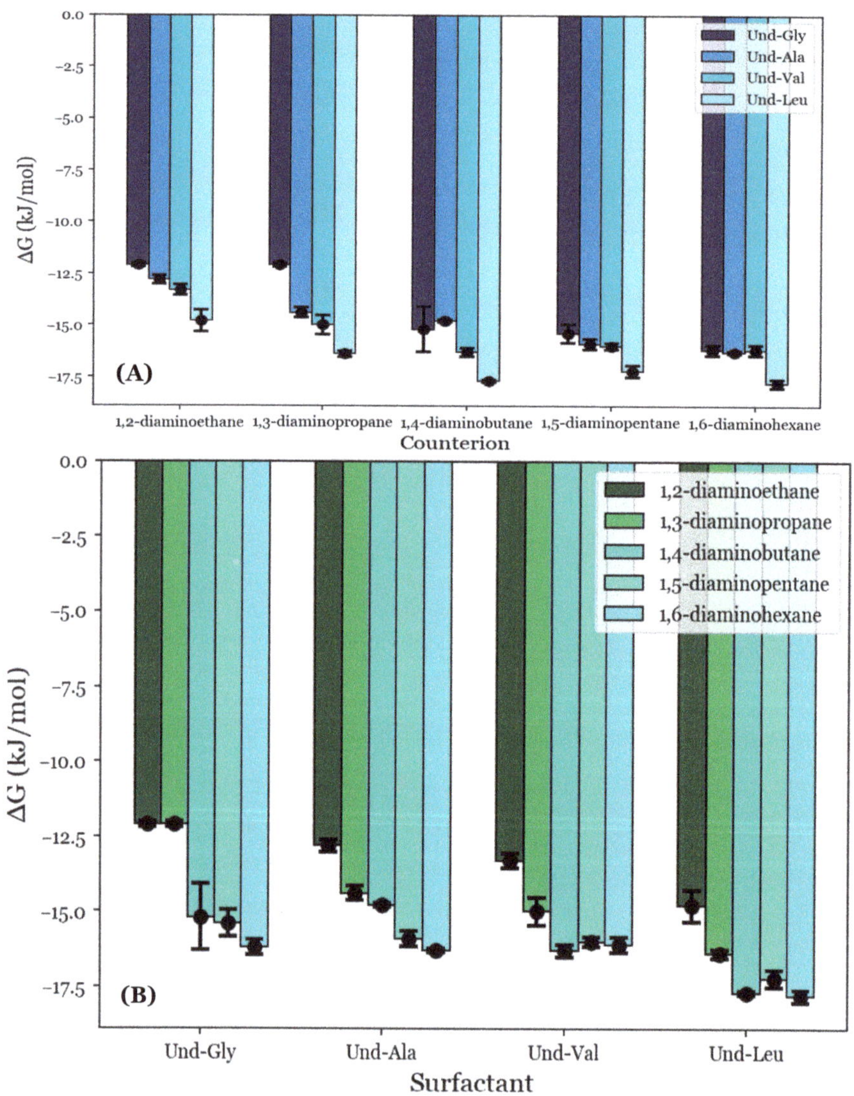

Figure 6. The free energies of micellization (ΔG°_{M}) given with respect to (**A**) each AABSs and (**B**) each diamine counterion.

The trend of reduced ΔG°_{M} values for Und-Val and Und-Leu with these counterions agrees with previous trends in β for the same systems as illustrated in Figure 4A and further supports the binding conformation proposed in Figure 5. This is because a lower ΔG°_{M} indicates that micellization is more favorable, while a higher β indicates stronger binding. However, unlike the reported β values, the free energies of each AABS were not equally affected by the counterion gradient from 1,2-diaminoethane to 1,6-diaminohexane. For example, while Und-Gly's free energies of micellization decrease from −12.1 to −16.2 kJ/mol along this gradient, Und-Leu shows a less significant decrease from −14.8 to −17.8 kJ/mol. This change in counterion sensitivity appears to be correlated with the AABS R-group complexity, as it decreases along the gradient from Und-Gly to Und-Leu. Interestingly, this aligns with earlier trends that indicated that AABSs with less complex R-groups exhibit CMCs that are more sensitive to the counterion length, as discussed in Section 3.1. Moreover, in further agreement with the CMC trends, the ΔG°_{M} values generally decrease with the AABS R-group complexity, as shown in Figure 6B. For example, in 1,5-diaminopentane systems, the free energy of micellization is −15.4 kJ/mol for Und-Gly and −17.2 kJ/mol for Und-Leu.

While the reported CMC values were correlated almost exclusively with the AABS hydrophobicity, as discussed in Section 3.1, the reported ΔG°_{M} values appear to account for the merging of the trends observed from the CMC and β values. This indicates that the overall thermodynamic favorability of micellization depends heavily on both the AABS hydrophobicity and counterion binding interactions, rather than hydrophobicity alone, as would be implied by the CMC data.

4. Conclusions

Electrical conductimetry and DLS were used to determine the CMCs and approximate micellar diameters for four AABSs in the presence of five linear diamine counterions. The obtained CMCs correlated extremely well with each surfactant's logP (water/octanol) value, suggesting that aggregation is primarily driven by AABS hydrophobicity. The recorded pH values indicated that each AABS was deprotonated but did not yield consistent insights into the counterion protonation states, suggesting that the counterion pK_a values were significantly perturbed by micellization. The micellar sizes obtained by DLS were independent of the diamine counterion length, indicating the formation of fully charge-stabilized noncovalent dimers. Estimates of β obtained from the conductimetry data indicate that counterion binding generally improves with the counterion length and reduced AABS sterics, although Und-Val and Und-Leu exhibit generally preferential binding for counterions with an even number of methylene groups. It is proposed that these counterions form a binding pocket during the formation of noncovalent dimers, which accommodates the steric bulk of Und-Val and Und-Leu. The relatively small size and lack of expected torsional flexibility from 1,2-diaminoethane explain its deviation from this trend, as well as its induction of abnormally small micelles. The estimates of ΔG°_{M} obtained from the conductimetry data and other structure–property parameters further support this theory, indicating that the trends in counterion binding interactions predicted by β greatly influence the overall thermodynamic favorability of AABS micellization, in addition to the R-group hydrophobicity.

Supplementary Materials: The following supporting information can be downloaded at: https://www.mdpi.com/article/10.3390/molecules29184436/s1, Supplemental Information 1- HNMR spectra of AABSs. Supplemental Information 2- Python script to calculate CMCC, Supplemental Information 3- Explanation of how the equations are derived.

Author Contributions: Conceptualization, N.B., E.J.B., F.H.B. and K.F.M.; methodology, N.B., E.J.B., F.H.B. and K.F.M.; software, N.B. and K.F.M.; validation, M.A.A. and F.H.B.; formal analysis, M.A.A., N.B., F.H.B., S.E.B., K.R.R. and K.F.M.; investigation, M.A.A., N.B., F.H.B., S.E.B., K.R.R. and E.J.B.; resources, E.J.B., F.H.B. and K.F.M.; data curation, N.B., E.J.B., F.H.B. and K.F.M.; writing—original draft preparation, M.A.A. and N.B.; writing—review and editing, M.A.A., N.B., S.E.B., K.R.R., F.H.B., S.E.B., K.R.R., E.J.B. and K.F.M.; visualization, M.A.A. and N.B.; supervision, E.J.B., F.H.B. and K.F.M.;

project administration, E.J.B., F.H.B. and K.F.M.; funding acquisition, E.J.B., F.H.B. and K.F.M. All authors have read and agreed to the published version of the manuscript.

Funding: This work was supported by the National Science Foundation under Grant Nos. 2203506 and 2203652, as well as by The Welch Foundation under Grant No. BT0041.

Institutional Review Board Statement: The study was conducted in the laboratory of Eugene and Fereshteh Billiot at Texas A&M University Corpus Christi, Texas, USA.

Informed Consent Statement: Not applicable.

Data Availability Statement: The data presented in this study are available on request from F. Billiot at fereshteh.billiot@tamucc.edu.

Acknowledgments: F.H. and E.J. Billiot acknowledge the Welch Chemistry Departmental Grant #BT-0041 awarded to the Texas A&M University Corpus Christi Chemistry Program. Morris acknowledges the generosity and support of the Ralph E. Klingenmeyer family.

Conflicts of Interest: The authors declare no conflicts of interest.

References

1. Rebello, S.; Asok, A.K.; Mundayoor, S. Surfactants: Toxicity, remediation, and green surfactants. *Environ. Chem. Lett.* **2014**, *12*, 275–287. [CrossRef]
2. De, S.; Malik, S.; Ghosh, A.; Saha, R.; Saha, B. A Review on Natural Surfactants. *RCS Adv.* **2015**, *5*, 65757–65767. [CrossRef]
3. Urata, K.; Takaishi, N. A perspective on the Contribution of Surfactants and Lipids toward "Green Chemistry": Present States and Future Potential. *J. Surfactants Deterg.* **2001**, *4*, 191–200. [CrossRef]
4. Mosler, R.; Hatton, T.A. Surfactants and Polymers for Environmental Remediation and Control. *Curr. Opin. Colloid Interface Sci.* **1996**, *1*, 540–547. [CrossRef]
5. Lu, Y.; Yue, Z.; Xie, J. Micelles with Ultralow Critical Micelle Concentration as Carriers for Drug Delivery. *Nat. Biomed. Eng.* **2018**, *2*, 318–325. [CrossRef]
6. Cheng, M.; Zeng, G.; Huang, D. Advantages and challenges of Tween 80 surfactant enhanced technologies for the remediation of soil contaminated with hydrophobic organic compounds. *Chem. Eng. J.* **2017**, *314*, 98–113. [CrossRef]
7. Rieger, M.N. Surfactant chemistry and classification. In *Surfactants in Cosmetics*, 2nd ed.; Rieger, M.N., Rhein, L.D., Eds.; Mercel Dekker: New York, NY, USA, 1997; p. 2. [CrossRef]
8. Al Ghatta, A.; Aravenas, R.C.; Wu, Y.; Perry, J.M.; Lemus, J.; Hallett, J.P. New Biobased Sulfonated Anionic Surfactants Based on the Esterification of Furoic Acid and Fatty Alcohols: A Green Solution for the Replacement of Oil Derivative Surfactants with Superior Proprieties. *ACS Sustain. Chem. Eng.* **2022**, *10*, 8846–8855. [CrossRef]
9. Lal, H.; Akram, M.; Kabir-ud-din. Deciphering the Mechanism of Interactions of an Ester Functionalized Cationic Gemini Surfactant with Bovine Serum Albumin: A Biophysical and Molecular Modeling Study. *Colloids Surf. A Physiochemical Eng. Asp.* **2022**, *646*, 128944. [CrossRef]
10. Ahmad, Z.; Shahg, A.; Siffiq, M. Polymeric Micelles as Drug Delivery Vehicles. *RSC Adv.* **2014**, *4*, 17028–17038. [CrossRef]
11. Moon, S.; Yoon, B.K.; Jackman, J.A. Effect of Membrane Curvature Nanoarchitectonics on Membrane-Disruptive Interactions of Antimicrobial Lipids and Surfactants. *Langmuir* **2022**, *38*, 4606–4616. [CrossRef]
12. McFadzean, B.; Dicks, P.; Groenmeyer, G.; Groenmeyer, G.; Harris, P.; O'Connor, C. The Effect of Molecular Weight on the Adsorption and Efficacy of Polysaccharide Depressants. *Miner. Eng.* **2011**, *24*, 463–469. [CrossRef]
13. Maynard-Benson, A.; Alekisch, M.; Wall, A.; Billiot, E.; Billiot, F.; Morris, K. Characterization of Single Amino Acid Based Surfactant Undecanoic L-Isoleucine and Undecanoic L-Norleucine in the Presence of Diamine Counterions with Varying Chain Lengths. *J. Surfactants Deterg.* **2023**, *7*, 28. [CrossRef]
14. Fletcher, J.; Mahant, G.; Witzleb, T.; Busche, R.; Garcia, M.; Fang, Y.; Billiot, E.; Billiot, F.; Morris, K. NMR investigation of counterion binding to undecyl LL-leucinevalanate micelles. *J. Dispers. Sci. Technol.* **2022**, *45*, 284–295. [CrossRef]
15. Rothbauer, G.; Rutter, E.; Reuter, C.; Billiot, E.; Fang, Y.; Billiot, F.; Morris, K. Nuclear Magnetic Resonance Investigation of the Effect of pH on Micelle Formation by the Amino Acid-Based Surfactant Undecyl L-Phenylalaninate. *J. Dispers. Sci. Technol.* **2018**, *21*, 139–153. [CrossRef]
16. Lewis, C.; Hughes, B.; Vasquez, M.; Wall, A.; Northrup, V.; Morris, K.; Billiot, F.; Billiot, E.; Fang, Y. Effect of pH on Binding of Sodium, Lysine, and Arginine Counterions to an L-Undecyl Leucinate. *J. Surfactants Deterg.* **2016**, *19*, 1175–1188. [CrossRef]
17. Sharker, K.K.; Islam, M.N.; Das, S. Interactions of Some Hofmeister Cations with Sodium Dodecyl Sulfate in Aqueous Solution. *J. Surfactants Deterg.* **2018**, *22*, 249–258. [CrossRef]
18. Gao, A.-T.; Xing, H.; Zhou, H.-T.; Cao, A.-Q.; Wu, B.-W.; Yu, H.-Q.; Gou, Z.-M.; Xiao, J.-X. Effects of Counterion Structure on the Surface Activities of Anionic Fluorinated Surfactants Whose Counterions Are Organic Ammonium Ions. *Colloids Surf. A Physicochem. Eng. Asp.* **2014**, *459*, 31–38. [CrossRef]
19. Mayer, J.D.; Rauscher, R.M.; Fritz, S.R.; Fang, Y.; Billiot, E.J.; Billiot, F.H.; Morris, K.R. An Investigation of the Effect of pH on Micelle Formation by a Glutamic Acid-Based Biosurfactant. *Colloids Interfaces* **2024**, *8*, 38. [CrossRef]

20. Lapidot, Y.; Rappoport, S.; Wolman, Y. Use of esters of N-hydroxysuccinimide in the synthesis of N-acylamino acids. *J. Lipid Res.* **1967**, *8*, 142–145. [CrossRef]
21. Zana, R. Critical Micellization Concentration of Surfactants in Aqueous Solution and Free Energy of Micellization. *Langmuir* **1996**, *12*, 1208–1211. [CrossRef]
22. Ono, Y.; Kawasaki, H.; Annaka, M.; Maeda, H. Effects of Micelle-To-Vesicle Transitions on the Degree of Counterion Binding. *J. Colloid Interface Sci.* **2005**, *287*, 685–693. [CrossRef] [PubMed]
23. Baskshi, M.S.; Singh, J.; Kaur, J. Estimation of Degree of Counterion Binding and Thermodynamic Parameters of Ionic Surfactants from Cloud Point Measurements by Using Triblock Polymer as Probe. *J. Colloid Interface Sci.* **2005**, *287*, 704–711. [CrossRef] [PubMed]
24. Laxmi, K.L. Characterization of Ligand N′–[(1E)-1-Phenylethylidene]-1, 3-Benzothiazole-2-Carbohydrazide by Using Marvin Sketch 20.8 Software. *Orient. J. Chem.* **2022**, *38*, 77–84. [CrossRef]
25. Kaushik, M. A Review of Innovative Chemical Drawing and Spectra Prediction Computer Software. *Mediterr. J. Chem.* **2014**, *3*, 759–766. [CrossRef]

molecules

MDPI

Article

Investigation of the Aggregation of Aβ Peptide (1-40) in the Presence of κ-Carrageenan-Stabilised Liposomes Loaded with Homotaurine

Kamelia Kamburova, Ivaylo L. Dimitrov , Feyzim Hodzhaoglu and Viktoria Milkova *

Institute of Physical Chemistry 'Acad. R. Kaischew', Bulgarian Academy of Sciences, 1113 Sofia, Bulgaria; kamelia@ipc.bas.bg (K.K.); idimitrov@ipc.bas.bg (I.L.D.); feyzim@ipc.bas.bg (F.H.)
* Correspondence: vmilkova@ipc.bas.bg

Abstract: The kinetics of amyloid aggregation was studied indirectly by monitoring the changes in the polydispersity of mixed dispersion of amyloid β peptide (1-40) and composite liposomes. The liposomes were prepared from the 1,2-dioleoyl-sn-glicero-3-phoshocholine (DOPC) phospholipid and stabilised by the electrostatic adsorption of κ-carrageenan. The produced homotaurine-loaded and unloaded liposomes had a highly negative electrokinetic potential and remarkable stability in phosphate buffer (pH 4 and 7.4). For the first time, the appearance and evolution of the aggregation of Aβ were presented through the variation in the standard percentile readings (D10, D50, and D90) obtained from the particle size distribution analysis. The kinetic experiments indicated the appearance of the first aggregates almost 30 min after mixing the liposomes and peptide solution. It was observed that by adding unloaded liposomes, the size of 90% of the particles in the dispersion (D90) increased. In contrast, the addition of homotaurine-loaded liposomes had almost minimal impact on the size of the fractions of larger particles during the kinetic experiments. Despite the specific bioactivity of homotaurine in the presence of natural cell membranes, this study reported an additional inhibitory effect of the compound on the amyloid peptide aggregation due to the charge effects and 'molecular crowding'.

Keywords: amyloid β peptide; homotaurine; κ-carrageenan; liposomes; aggregation; marine drugs

1. Introduction

The neurodegenerative brain disorder (NBD) is a common term for the processes of progressive loss of the structure or function or the death of neurons resulting in the progressive loss of the ability to learn, mental, behavioural, or functional decline, or the gradual loss of emotions and movements of the body. Alzheimer's disease (AD) is a more representative NBD. The World Health Organization (WHO) estimates that approximately 55 million people (2019) aged 65 years or older have AD diagnosis, and additionally, about 4.6 million new cases are registered every year. The number of cases is expected to increase to 66 million in 2030 and to 139 million in 2050 [1].

There are few theories (hypotheses) for the presumable molecular and pathophysiological mechanisms for the explanation of the manifestation and the progression of the disease [2]. Among them, the most well-supported is 'the amyloid cascade hypothesis' (ACH). The Aβ peptide (Aβ) is a short segment of the transmembrane amyloid precursor protein (APP) released after abnormal cleavage by the enzymes β- and γ-secretase. The Aβ molecules can aggregate to form low-molecular weight flexible and soluble oligomers, which may coalesce to form high-molecular weight oligomers, with β-sheet conformation, protofibrils, and insoluble fibrils deposited in diffuse senile plaque, which disturbs the normal function of neurons and is a characteristic feature of AD.

The presence of the soluble aggregates of amphiphilic proteins that can undergo evolution in ordered fibrils is considered a main hypothesis and common feature in the pathology of diseases related to NBD. Concerning AD, many postmortem observations indicate that the disease is characterised by the extracellular deposition of fibrous aggregates or intracellular inclusions containing Aβ peptide in the brain parenchyma and cerebral blood vessels of the patients.

Amyloid aggregation and fibrillation have been theoretically described using a nucleation-dependent polymerisation process [3–5]. The model was presented with two phases—nucleation and extension and secondary nucleation steps. The formation of the nucleus results from a series of association steps of Aβ monomers. These interactions are thermodynamically unfavourable, and this stage was determined as the rate-limiting step in fibril formation. The transition of monomer to fibril is a multi-step process. The energy barrier of the primary nucleation step is higher compared to the energy barrier for the incorporation of the new monomers in the existing protofibrils or fibrils [6]. Fibril growth has been described using a 'dock-lock' model. According to that model, the monomers interact rapidly with the existing fibril ends ('dock'), followed by a conformational change to find the more thermodynamically favourable and stable state of the peptide (β-sheet conformation) ('lock') [7]. The variation in the pH, temperature, ionic strength, or presence of cell membrane can affect the aggregation process.

Currently, there is no effective therapy for AD. The existing therapeutic strategies are based on the presumable pathophysiological mechanisms of the disease, and they can only ameliorate the suffering of the patients, but they do not cure. At present, there are several active molecules approved by the US Food and Drug Administration (FDA) for the treatment of AD—tacrine, donepezil, rivastigmine, galantamine, and memantine.

Concerning the neuroprotection activity of the natural compounds, the substances extracted from marine organisms possess a promising biotechnology potential. More than 10,000 natural components have been isolated from bacteria, fungi, algae, and other marine organisms. There are a significant number of extracted bioactive molecules (chitosan, fucoidan, homotaurine, gracillin, caniferolides, bryostatin-1, dictyostatin, anabasine, rifampicin, etc.) with potential application in the therapy of NBDs [8,9].

Homotaurine (HT, 3-amino-1-propanesulfonic acid or Tramiprosate) is a natural amino acid extracted from red, green, or brown algae. It has been also found in marine roseobacters. The therapeutic properties of HT have been tested in a few phase II and phase III clinical studies, and it has been confirmed that HT decreases the concentration of soluble Aβ in the cerebrospinal fluid (SCF) and affects the deposition of amyloid fibrils in a plaque in the brain [10]. According to a presumable mechanism of therapeutic action, the molecules of HT are BBB-permeable and can interact with the peptide to prevent (limit) the formation of neurotoxic Aβ oligomers and prevent the misfolding processes that lead to amyloid aggregation and cause neurotoxicity and progression in AD [11–14].

Homotaurine is classified as a compound that mimics glycosaminoglycan (GAG), which binds to beta-amyloid peptide molecules and thus facilitates their fibrillation. In the competitive scenario, the compound has an affinity for the GAG-binding region of Aβ molecules and thereby displaces GAGs, reduces fibrillation, and reverses the conformational changes in Aβ molecules that lead to amyloid plaque deposition [10,12].

κ-carrageenan (CAR) is a polysaccharide, a product of extraction from certain species of red seaweeds. It is a linear, water-soluble polymer that typically forms highly viscous aqueous solutions. Moreover, recently, it has been reported that the oligosaccharides of κ-carrageenan postpone the progression of AD by preventing damage to the neurons [15].

The present study is focused on the experimental investigation of the effect of the presence of homotaurine-loaded liposomes on the kinetics of Aβ aggregation.

The liposomes were chosen for this study because they are drug delivery platforms with many advantages. They are non-toxic and have a high drug-loading capacity. Concerning their application in neurological studies, they are extremely promising delivery platforms because nano-sized liposomes could cross the blood–brain barrier (BBB) [16].

In this study, the appearance and evolution of the amyloid aggregation in the presence and absence of liposomes were presented through the variation in the standard percentile readings (D10, D50, and D90) obtained from the particle size distribution analysis from kinetic experiments. The percentiles are statistical terms that correspond to the fraction of particles with a size equal to or less than a certain value. For intensity-weighted particle size distributions, the parameters are reported based on the maximum particle size for a given percentage intensity of the sample. The standard percentile readings from the analysis are denoted as D10, D50, and D90, where D stands for diameter, and 10, 50, and 90 represent the percentage of the sample below a certain particle size (e.g., 10%, 50%, and 90%). D50 represents the median of the intensity distribution and is defined as the particle size at which half of the population lies below this value. By monitoring these three parameters in the kinetics experiments, it is possible to observe any significant changes in the main particle size and changes at the extremes of the distribution, which could be due to the presence of fines or oversized particles/agglomerates.

The novelty of the present work is addressed to the investigation of amyloid aggregation in the mixture dispersion of peptide molecules and composite liposomes loaded by anti-amyloid agents. Moreover, all components in the system are completely natural compounds extracted from marine organisms (excluding lipid and amyloid peptides). The appearance of amyloid aggregation in conditions close to the physiological ones was indirectly investigated by monitoring the variation in the polydispersity of the species in dispersion.

2. Results

2.1. Characterisation of the HT-Loaded Liposomes

Table 1 presents the experimental data for the hydrodynamic size and ζ-potential of the produced unloaded and HT-loaded liposomes.

Table 1. Hydrodynamic diameter (D), polydispersity index (PDI), and electrokinetic potential of the liposomes. The concentration of CAR in stabilised dispersion was 0.1 mg/mL.

Sample	D * [nm] (PDI)	ζ-Potential [mV]
unloaded liposomes	37.1 ± 1.2 (0.12)	−55.1 ± 3.0
unloaded liposomes stabilised with CAR	52.2 ± 1.6 (0.11)	−62.7 ± 1.3
HT-loaded liposomes	55.6 ± 1.2 (0.22)	−59.8 ± 2.1
HT-loaded liposomes stabilised with CAR	92.1 ± 9.2 (0.19)	−65.0 ± 0.7

* Evaluated mean size by intensity.

According to the presented results, the values of size and charge of the HT-loaded liposomes are higher than those of the unloaded ones. One possible explanation is a variation in the liposome structure and the incorporation of some HT molecules in the lipid bilayer during the encapsulation process. In support of this assumption is the registered very high encapsulation efficiency of the compound in the liposomes (ca. 99%). The HT molecules have a slightly negative charge, and their incorporation in the bilayer leads to an increase in the net negative charge of the liposomes. When negatively charged carrageenan molecules are adsorbed onto liposomes, they stabilise them, forming a thick polymer layer (about 18 nm) and increasing their net negative charge. A visualisation of the produced liposomes (unloaded and HT-loaded) is presented in Figure A1 (Appendix A).

2.2. Estimation of the Amount of HT Encapsulated into the Liposomes

A spectrophotometric assay is applied to evaluate the quantity of loaded HT in the liposomes. In the applied methodology, methylene blue is used as an oxidant. Methylene blue exhibits redox properties, appearing blue in its oxidised form and colourless in its reduced form. Under acidic conditions, the methyl groups of the compound preferentially dissociate due to the protonation of the dimethylamino group [17,18]. When methylene blue reacts with HT in an acidic medium, its absorbance value gradually decreases. By using a suitable calibration curve, the entrapped amount of HT in the liposomes is extremely high (ca. 0.099 mg/mL), and the estimated encapsulation efficiency is ca. 99%.

2.3. Kinetics of Aggregation of Aβ at Different pH and Temperatures

The changes in the size of the peptide molecules in solution at pH 7.4 and pH 4 with an increase in temperature are illustrated by standard percentiles in Figures 1 and 2 (the experimental results for the system at pH 5 are presented in Figure A2, Appendix A). The figures present results from three samples measured at three different temperatures.

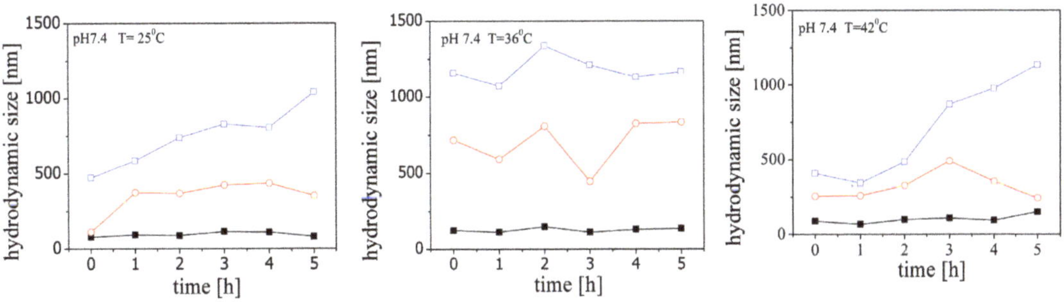

(a) (b) (c)

Figure 1. A variation in the hydrodynamic size (diameter) of Aβ peptide at pH 7.4 in the kinetic experiments at 25 °C (**a**), 36 °C (**b**), and 42 °C (**c**). The size is presented through the percentiles in the size distribution D10 (■), D50 (○), and D90 (□).

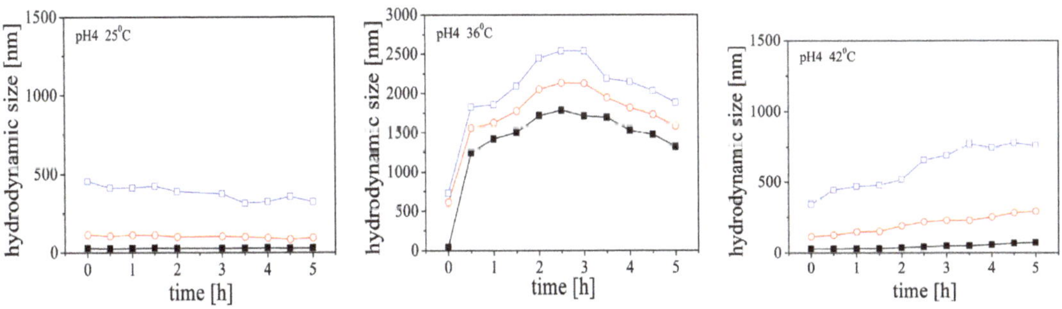

(a) (b) (c)

Figure 2. A variation in the hydrodynamic size of Aβ peptide at pH 4 in the kinetic experiments at 25 °C (**a**), 36 °C (**b**), and 42 °C (**c**). The size is presented through the percentiles in the size distribution D10 (■), D50 (○), and D90 (□).

The findings of the kinetic experiment indicated that a notable aggregation process was observed when the temperature was 36 °C and the pH was 4. Interestingly, the emergence of structures with a size of approximately 2000 nm was observed within the first 30 min (Figure 2b). On the other hand, when the sample is subjected to a gradual temperature increase in the cell (Figure 3), the size values and their dependencies take on a different shape. The structures in the samples were observed to gradually increase in size. This suggests distinct aggregation mechanisms induced by temperature, which may explain the results obtained.

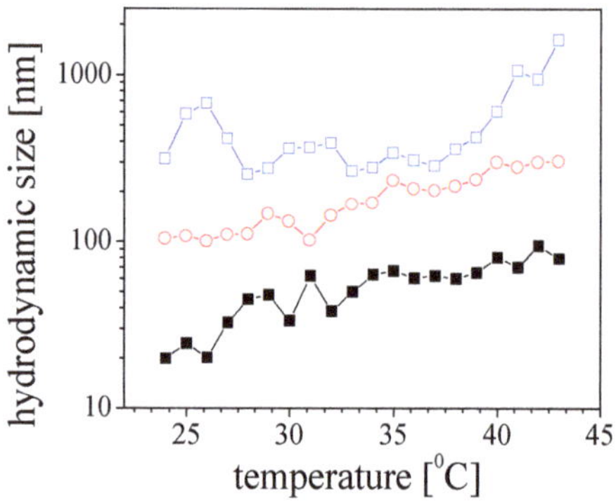

Figure 3. A variation in the hydrodynamic diameter of Aβ peptide at pH 4 in a stepwise regime of increasing the temperature. The size is presented through the percentiles in the size distribution D10 (■), D50 (○), and D90 (□).

2.4. *Kinetics of Aggregation of Aβ in the Presence of Homotaurine*

The aggregation behaviour of peptides in the presence of homotaurine is presented in Figure 4. The experimental results indicate significantly lower values of the size of the fraction of particles D10 and D90 in the presence of homotaurine compared to the distribution in a pure solution of Aβ.

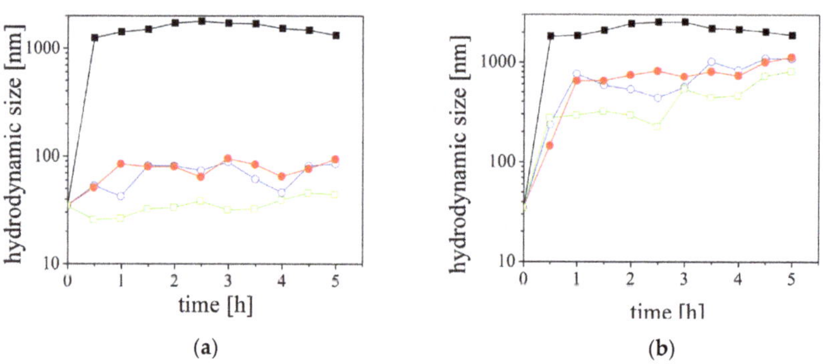

Figure 4. A variation in the hydrodynamic diameter of pure Aβ peptide (■) and mixtures of peptide and the following different concentrations of homotaurine: 0.05 mg/mL (○), 0.1 mg/mL (•), and 0.5 mg/mL (□). The size is presented through the percentiles in the size distribution D10 (**a**) and D90 (**b**). The pH of the dispersions is 4.

2.5. *Kinetics of Aggregation of Aβ in the Presence of Liposomes*

Figure 5 illustrates the percentiles (D10, D50, and D90) in the size distribution of particles in a solution of Aβ during kinetic experiments in the presence of unloaded or HT-loaded liposomes. The results suggest that larger particles or aggregates dominate, and their size increases with reaction time. The decrease in the size of fraction D90 at the last measurement (almost 5 h after mixing the peptide solution and liposomes) is likely due to the sedimentation of the largest particles in the experimental cell. Additionally, the values of the size of the fraction D50 decrease, indicating that disruption of the lipid bilayer is negligible.

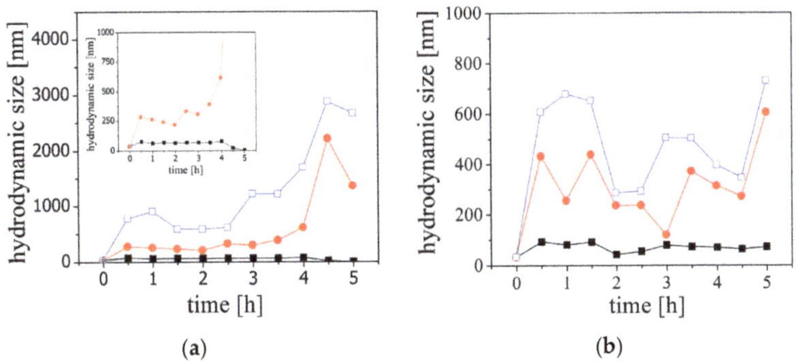

Figure 5. A variation in the hydrodynamic size of the Aβ in the presence of unloaded (**a**) and HT-loaded liposomes (**b**). The size is presented through the percentiles in the size distribution D10 (■), D50 (○), and D90 (□). The inset presents the expanded dependence of the percentiles D10 and D50.

The addition of the HT-loaded liposomes does not indicate significant aggregation during the kinetic experiments. The first change in the size distribution was observed about 30 min after mixing the peptide solution and liposomes, and the sizes corresponding to percentiles D10, D50, and D90 were almost constant until the end of the experiment.

The raw data from the DLS measurements are more representative. In Figure 6, the shift of the peaks during the kinetics experiment can be observed.

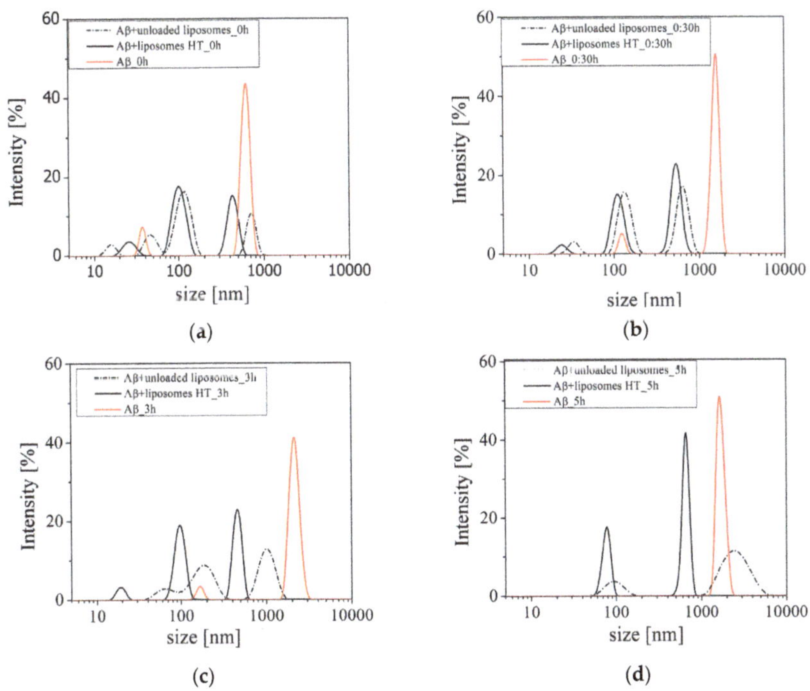

Figure 6. The size distribution during the kinetic measurements at pH 4 in pure Aβ peptide solution (red line), solution of the peptide in the presence of unloaded (dash line), and HT-loaded liposomes (solid black line) immediately after mixing the solutions (**a**), 30 min (**b**), 3 h (**c**), and 5 h (**d**).

3. Discussion

The liposomes as delivery platforms suggest many possibilities and advantages for application in successful therapies for NBDs. Some studies have reported that small liposomes (up to 100 nm) can pass through the BBB and deliver drugs or bioactive molecules, with potential activity against amyloidosis in the brain. However, the presence of liposomes, as a model of biological membranes, can also induce the aggregation of amyloid peptides.

According to ACH, the critical micromolar concentration of free peptide monomers is required to form the first oligomers. However, postmortem observations have shown that the amyloid concentration found in the brain is exceptionally lower, and the probability for the observation of a spontaneous oligomerisation of Aβ at the estimated pico- or nanomolar peptide concentration is very low. The observed limitation in the ACH has been explained with the participation of additional effects on the aggregation kinetics as Aβ–membrane interactions.

It has been previously reported that the interaction of peptides with the cell membrane (or the surface of a liposome in the model studies) is governed not only by the physicochemical characteristics of the membrane (size or the curvature, composition, elasticity, charge, degree of hydration, conformation, and dynamics of lipid headgroups) but also by the thickness of the hydrophobic part of the bilayer or surface modification of the membrane, the pH of the medium, or the presence of metal ions in the solution [19,20].

In the present study, in order to elucidate the influence of the liposomes and HT on the aggregation, kinetic experiments were performed with the peptide solution mixed with a solution of homotaurine at acid conditions (pH 4). The experimental results indicate that the Aβ peptide aggregation is increasingly suppressed with the increasing HT concentration in the system (Figure 5). The results indicate significantly lower values of the size of 90% of the particles in the solution (D90) in the presence of the homotaurine compared to the distribution in a pure solution of Aβ at the same experimental conditions. The peptide molecules are positively charged at the experimental conditions (ζ-potential is ca. + 10 mV). The addition of HT (weak sulfonic acid) does not affect the ionic strength or pH of the solution. Therefore, it was expected that attractive electrostatic interactions or specific attractive interactions between the peptide and homotaurine are governed by the very complicated structure of Aβ (hydrogen bonding or hydrophobic interactions). Additionally, as the concentration of HT increases, the total concentrations of species in solutions increase, leading to 'molecular crowding'. This phenomenon can also inhibit the aggregation process.

Postmortem observations have shown that the pH of the brain parenchyma is lower for patients with an Alzheimer's diagnosis (<pH 4). Furthermore, experimental studies have confirmed that the variation in pH is a key factor for amyloidosis [21,22]. Therefore, a detailed screening of the conditions in model experiments can ensure optimal conditions for investigations of the mechanisms of the Aβ peptide aggregation.

The present study reports preliminary investigations on the kinetics of aggregation of Aβ peptide in the presence of unloaded and HT-loaded liposomes. The electrostatic interactions between the positively charged peptide monomers and zwitterionic liposomes are additionally affected by the adsorption of carrageenan. It was presumed that the behaviour of Aβ peptide in the presence of highly negatively charged polymer-coated liposomes would be fundamentally different compared to the properties of pure peptide solution or a dispersion of carrageenan-free liposomes. Experimental data for the time dependence of the sizes of 90% of the particles in the dispersion (D90) indicate that a significant part of the peptide molecules is involved in aggregate formation (Figures 1 and 2).

According to the results from the kinetic experiments, the size of the fractions of species D10, D50, and D90 in the mixtures of peptide and liposomes suggest that the larger particles or aggregates dominate, and their size increases with reaction time (Figure 5).

The time dependence of the particle size in the solution of Aβ peptide in the presence of unloaded liposomes can be divided into two stages (Figure 5a). An increase in the size is registered in the first stage, and the beginning of the aggregation is observed 30 min after mixing the peptide solution and the liposomes. After that, the size of 90% of the particles in the dispersion (D90) gradually increases up to 3 h. Meanwhile, the fractions of aggregates are significantly larger compared to the fraction of smaller particles and increase with time. For example, two hours after mixing, the size of 90% of the particles (D90) is ca. 600 nm, the size of 50% of the particles (D50) is ca. 200 nm, and the smallest particles are ca. 70 nm. In the second stage, sharp increases in the size of larger particles are observed. The fraction of the smaller particles is almost constant during the measurement (D10).

The time dependence of Aβ aggregation in the presence of HT-loaded liposomes shows different kinetic behaviour (Figure 5b). The aggregation in the dispersion is also registered 30 min after the mixing, and the size of 90% of the particles (D90) is ca. 700 nm. However, in the presence of HT-loaded liposomes, the size of the fractions D50 and D90 almost do not change during the kinetic experiment.

The analysis of the peaks in the size distribution displayed in Figure 6 indicates that the peaks have shifted throughout the experiment. Notably, the positioning of the peaks in the most recent measurements is particularly intriguing (Figure 6d). There are two peaks registered from the dispersion of peptides in the presence of unloaded liposomes. One peak is at the position of Aβ, similar to the pure peptide solution but with lower intensity, while the other peak corresponds to the smallest particles in the system. In the dispersion containing HT-loaded liposomes, there are also two peaks, but they correspond to particles with smaller sizes than the amyloid peptide aggregates. This experimental result clearly illustrates the effect of the HT-loaded liposomes on amyloid aggregation.

The observed kinetic behaviour is in line with previously reported mechanisms of amyloid aggregation in the presence of a membrane. According to the presented models, the aggregation process can be described as a sigmoidal function of time on the fraction of aggregates. The initial nucleation phase (lag phase) is a relatively slow process of the formation of first aggregates, whereas the sequential elongation phase is rapid [23]. However, aggregation is a very sensitive process, and variations in the conditions can promote or hinder the aggregation.

Previously, it has been reported that the presence of a membrane may induce conformational changes in the amyloid molecules, and the formation of a partially unfolding molecular structure has been suggested as a critical step in fibrillation [24]. The conformational changes have been interpreted as a result of the local accumulation of protons in the vicinity of the negatively charged lipid membrane. This local decrease in pH promotes an increase in the peptide charge density and repulsive electrostatic interactions with the membrane. The protein molecule becomes more open, and the hydrophobic parts are more favourable for aggregation. Therefore, the charge of the membrane is a key factor in the structural transition of the peptide molecules. It has been shown that in the case of Aβ peptide, the fibrillation in the presence of anionic and zwitterionic membranes is governed predominantly by charge effects instead of conformation changes. These membranes promote aggregation at low peptide concentrations [25].

At a low pH, Aβ peptide is negatively charged, and attractive electrostatic interactions with positively charged domains on the zwitterionic membranes will occur. As a result, the increase in a local peptide concentration on the membrane is observed. Wilson and Binder [26] have proposed that the high local peptide concentration overcomes the energy barrier for nucleation and promotes aggregation.

Following these hypotheses, we supposed that the surface modification and variation in the electrical properties of the lipid membrane could enhance, promote, or hinder peptide aggregation. At the experimental conditions of a pH of 4, the DOPC liposomes have an almost neutral net charge [27]. The sequential adsorption of negatively charged carrageenan on the positively charged domains leads to an increase in the net negative charge of the liposomes. At the same time, the peptide molecules are positively charged at a pH of 4 (the estimated pI of Aβ (1-40) peptide is 5.4 [28]). Thus, it is expected that the adsorption of CAR would promote the electrostatic interaction between the liposomes and the peptides and peptide aggregation, respectively. However, the comparison between the aggregation behaviour in the solutions of pure Aβ peptide and the aggregation in the dispersions of unloaded liposomes and Aβ peptide (Figures 2b and 5a) indicates that the aggregation is suppressed in the presence of the liposomes. While the maximal size of 90% of the particles (D90) is approximately 2600 nm for the solution of Aβ and 3000 nm for the mixture, the processes of aggregation kinetic are different. The retardation time (ca. 3 h) was observed for the solution containing liposomes.

Under the same experimental conditions, the negative charge density of the HT-loaded particles is higher than that of the unloaded ones. We supposed that this results from the incorporation of a compound in the lipid layer. The sequential polymer adsorption leads to an additional increase in the negative charge of the liposomes, and contrary to the expectation, the peptide aggregation is hampered.

To explain the obtained results, we supposed that there were additional factors that influenced the process. Firstly, the ionic strength of the solution determined by the buffer is extremely high (ca. 10^{-1} M). Therefore, a significant screening of the electrostatic interactions between the peptide molecules and the liposomes can be expected. Secondly, CAR molecules are fully charged at the experimental conditions (pH 4). Along with the expected charge renormalisation in the CAR polyion vicinity ('counterion condensation'), the additional charge effects related to the presence of high salt concentration in the solution must be considered in the screening of the electrostatic interactions [29,30]. Thirdly, the polymer layer is formed at low ionic strength (10^{-4} M). According to the theory of strong polyelectrolytes, the adsorption layer has to be thin due to the strong electrostatic repulsion between the charges along the polyion chain. However, it is well known that carrageenans are high-molecular weight polysaccharides, and the formation of loops and tails is expected in the adsorption layer (the thickness of the CAR layer is ca. 18 nm, Table 1). Hence, the participation of additional steric interactions could be responsible for hampering the aggregation process of Aβ peptides.

Many questions remain open in this study. As far as we know, this is the first complex study of the aggregation of Aβ peptide in the presence of polymer-modified liposomes and co-encapsulated bioactive molecules. The experimental results indicate that the adsorption of polymers on the liposomes influences the peptide aggregation, but the effect of the charge density is still unclear. For this purpose, a zwitterionic membrane would be a useful model surface. Homotaurine suppresses the aggregation of the Aβ 1-40 peptide on the liposomes, but the nature of the mechanism of interaction is also unknown. Additional experimental methods have to be applied for the characterisation of the charge effects in the presence of proteins or other polyelectrolytes. A detailed visualisation of the aggregates (from oligomers to protofibrils or fibrils) under the used experimental conditions is also needed and will be the subject of another study. The distinction of the fraction of aggregates from peptide molecules or peptide–liposome complexes in the mixtures is complicated using only electrokinetic methods.

4. Materials and Methods

4.1. Materials

The phospholipid 1,2-dioleoyl-sn-glicero-3-phoshocholine (DOPC, chloroform solution, 25 mg/mL) product of Avanti Polar Lipids Inc. (Alabaster, AL, USA) was used for the preparation of unilamellar liposomes.

Amyloid beta, Aβ (1-40), was purchased from Sigma Aldrich (St. Louis, MO, USA, product number SCP0037-0.5 mg). A stock peptide solution was prepared as follows: to the lyophilised commercial peptide sample, NH_4OH was added to obtain a 12.5 mg/mL peptide concentration (the solution of 1% NH_4OH was filtered before use through 0.22 μm filters (Minisart®, Sartorius, Gottingen, Germany) to eliminate eventual dust particles). The peptide solution was then gently diluted (without vortex) to a concentration of 0.8 mg/mL with 150 mM PBS buffer with a pH of 7.4. The buffer solution was also filtered before use. It was prepared using a standard recipe with NaCl (137 mM), KCl (2.7 mM), Na_2HPO_4 (10 mM), and KH_2PO_4 (1.8 mM). The dissolved protein solution was separated into

40 aliquots of 15 μL and stored at −20 °C until use. The final concentration of the peptide in the experiments was 0.4 mg/mL, corresponding to 10 μM. For the electrokinetic measurements, the samples were further diluted with buffer to a final peptide concentration of 0.016 mg/mL.

k-carrageenan (CAR) and homotaurine (HT) were also purchased from Sigma Aldrich. k-carrageenan (product number 22048) was used in this study without any purification. The stock solutions with a concentration of 2 mg/mL were prepared in double-distilled water. The prepared polymer solutions were filtered through a 5 μm filter (Minisart®, Sartorius) to remove the possible aggregates.

4.2. Methods

4.2.1. Liposome Preparation

The liposomes were prepared by using the thin film hydration method. The procedure was described in detail in a previous study [31]. Briefly, an appropriate volume (200 μL) from the solution of lipid in chloroform (25 mg/mL) was dried under a stream of nitrogen by rotating the glass flask to form a thin lipid film on its wall.

To produce unloaded or HT-loaded liposomes, the lipid was rehydrated in 2 mL of double-distilled water or a solution of HT (0.1 mg/mL) to a final lipid concentration of 2.5 mg/mL. After 4 freeze–thaw cycles of the tube with dispersion in a bath of liquid nitrogen and tap water, the stock solution of liposomes was sonicated in an ultrasonic ice bath for 15 min.

In order to prevent a possible aggregation during the subsequent steps in the experimental procedure, the work dispersion was prepared by mixing 80 μL from the stock dispersion of liposomes and 10 mL of double-distilled water to a final lipid concentration of 0.02 mg/mL. The dispersion was filtered before measurements by extrusion through a 0.20 μm filter (Minisart®, Sartorius).

To improve the stability of the loaded liposomes, a carrageenan monolayer was adsorbed on their surfaces. The polymer layer was formed by adding a diluted dispersion of liposomes (9.5 mL) to the solution of negatively charged CAR (0.5 mL, 2 mg/mL) and stirring for 20 min.

4.2.2. Kinetic Experiments

The aggregation kinetics of Aβ was investigated in the following four different experiments:

- Investigation of the effect of pH.

The samples for the measurement were prepared by adding 15 μL of phosphate buffer with a pH of 4, 5, or 7.4 (adjusted with HCl) to the Eppendorf tube containing 15 μL of peptide solution. The final concentration of peptide was 0.4 mg/mL (10 μM). The measurements were performed by adding 25 μL of the solution to a micro cell for DLS. The variation of the size distribution of the peptide aggregates was evaluated at 24 °C, 36 °C, and 42 °C.

- Effect of the temperature at a pH of 4.

The samples were prepared similarly, as follows: 15 μL of phosphate buffer with a pH of 4 (adjusted with HCl) was added to the Eppendorf tube containing 15 μL of peptide solution. The final concentration of peptide was 0.4 mg/mL. The measurements were performed over a large temperature interval (24–43 °C) in a stepwise manner. The interval between the different values of the temperature was 10 min. Three minutes of this time interval were used for the sample equilibration, and seven minutes were used for the measurement cycle.

- Aggregation of peptide in the presence of homotaurine at a pH of 4 and 36 °C.

The samples for these experiments were prepared similarly to those in the other experiments. Phosphate buffer (pH 4 adjusted with HCl) was mixed with homotaurine (powder). The final concentrations of HT in the solutions were 0.05, 0.1, and 0.5 mg/mL.

- Aggregation of peptide in the presence of unloaded or HT-loaded liposomes.

The dispersions of liposomes stabilised by the adsorption of carrageenan (lipid concentration of 6×10^{-4} mg/mL in distilled water) were centrifuged (15,000 rpm at 19 °C for 90 min) using a laboratory centrifuge (PW-352R, Warsaw, Poland). After centrifugation, the liposomes were settled, and the supernatant was completely removed. The liposomes were redispersed in a peptide solution using sonication (5 min) in an ultrasonic ice bath. Then, 25 μL from the dispersion was added to a microcell for DLS, and 15 μL of HCl (0.1 M) was added to the dispersion. The measurements were performed at 36 °C and a pH of 4 (the peptide samples were prepared by adding 15 μL of phosphate buffer (pH 7.4) to the Eppendorf tube containing 15 μL of peptide solution).

4.2.3. Determination of the Electrokinetic Charge and Particle Size Distribution

During the experimental procedure, the size and charge of the produced liposomes were assessed using dynamic light scattering with non-invasive backscattering (DLS-NIBS) at a measuring angle of 173°. The Zetasizer Pro (Malvern, UK), equipped with a He-Ne laser with a maximum power of 10 mW and operating at a wavelength of 633 nm, was used to carry out the measurements. The liposomes were measured five times, and the average value was recorded as the final size and surface charge.

The kinetic measurements were performed using Zatasizer Pro with a low-volume quartz batch cuvette (Ultra-Micro Cell for Nano Series, with a minimal volume of 12 μL and a maximum volume of 45 μL, the width of the measurable window is 3 mm.). In the experiments, the percentiles in the particle size distributions of D10, D50, and D90 were evaluated.

4.2.4. Determination of the Encapsulated Amount of Homotaurine in the Liposomes

The concentration of HT loaded in the liposomes was estimated using a spectrophotometric assay according to the procedure proposed by Zhao et al. [17]. Briefly, the samples were prepared as follows: 1.5 mL (1 M) of sulfuric acid solution was mixed in two test tubes with 0.35 mL (0.25 mg/mL) of methylene blue solution. An HT solution with a certain concentration was added to one of the tubes, and the tubes were diluted with distilled water to a final total volume of 10 mL. The reaction was carried out at room temperature for 20 min. The absorbances of the blank solution AU0 and AU of the solutions were measured at 664 nm, and the value of ΔAU = AU0 − AU was calculated. To obtain the calibration curve from HT, the differences ΔAU were calculated from the absorbance of 0 mg/mL HT (blank solution) and HT solutions with concentrations ranging from 10^{-1} to 10^{-6} mg/mL. The same procedure was applied to estimate the concentration of free HT in the dispersion of the liposomes. For this purpose, the stock dispersion was centrifuged at 15,000 rpm and 19 °C for 90 min, and the supernatant was extracted. The calculated difference ΔAU between the maximum absorption of the blank solution and supernatant corresponds to a free concentration of HT in the dispersion—ca. 5×10^{-5} mg/mL (evaluated from the calibration curve).

5. Conclusions

This study presents an investigation of the kinetics of amyloid peptide (1-40) aggregation by monitoring the changes in the polydispersity of mixed dispersions of the peptide and composite liposomes. The influence of the experimental conditions (pH, temperature) and the addition of different concentrations of HT and loaded and HT-loaded liposomes on the peptide aggregation was studied.

The aggregation of peptides was observed when the unloaded liposomes were added to the peptide solution, but the process was delayed. The estimated retardation time is approximately 3 h. Moreover, it was registered that in the presence of homotaurine-loaded liposomes, the aggregation process was significantly suppressed. To explain this phenomenon, it was presumed that there are additional factors that influence the process, as follows: screening of the electrostatic interactions between the peptide molecules and the liposomes due to the high ionic strength of the dispersions; charge renormalisation in the vicinity of the adsorbed fully charged carrageenan and the presence of counterions with lower mobility; and the participation of additional steric interactions due to the presence of a thick polymer layer (ca. 18 nm) on the liposomal surface.

Author Contributions: Conceptualisation, V.M.; methodology, V.M. and K.K.; validation, V.M. and K.K.; formal analysis, V.M. and K.K.; investigation, K.K., I.L.D., F.H. and V.M.; resources, V.M., data curation, V.M. and K.K.; writing—original draft preparation, V.M.; writing—review and editing, V.M., I.L.D. and K.K.; visualisation, V.M. and K.K.; supervision, V.M.; project administration, V.M.; funding acquisition, V.M. All authors have read and agreed to the published version of the manuscript.

Funding: This research was funded by the Bulgarian National Science Fund, contract No. КП-06-КОСТ/8. The authors acknowledge the support from the Centre of Excellence 'National center for mechatronics and clean technologies' (Project BG05M2OP001-1.001-0008-C010), supported by the European Regional Development Fund within the Bulgarian OP 'Science and Education for Smart Growth 2014–2020'.

Institutional Review Board Statement: Not applicable.

Informed Consent Statement: Not applicable.

Data Availability Statement: Data are contained within the article.

Acknowledgments: This publication is based upon work from COST Action CA18238 (Ocean4Biotech), supported by the European Cooperation in Science and Technology (COST) programme. Research equipment of the Distributed Research Infrastructure INFRAMAT, part of the Bulgarian National Roadmap for Research Infrastructures, supported by the Bulgarian Ministry of Education and Science, was used for some investigations in the present study.

Conflicts of Interest: The authors declare no conflicts of interest.

Appendix A

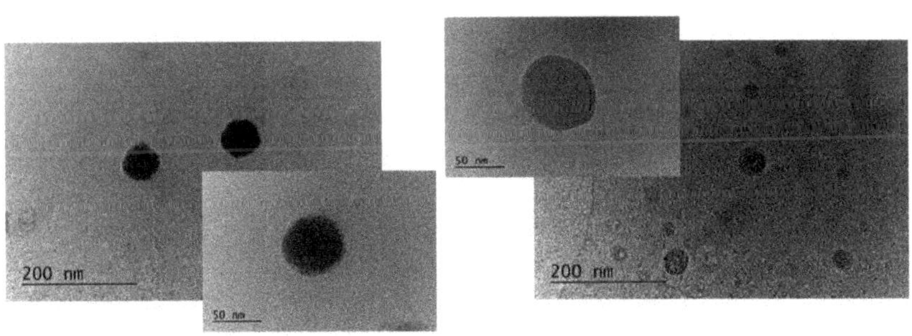

(**a**) (**b**)

Figure A1. Visualisation of the produced liposomes: (**a**) unloaded and (**b**) HT-loaded.

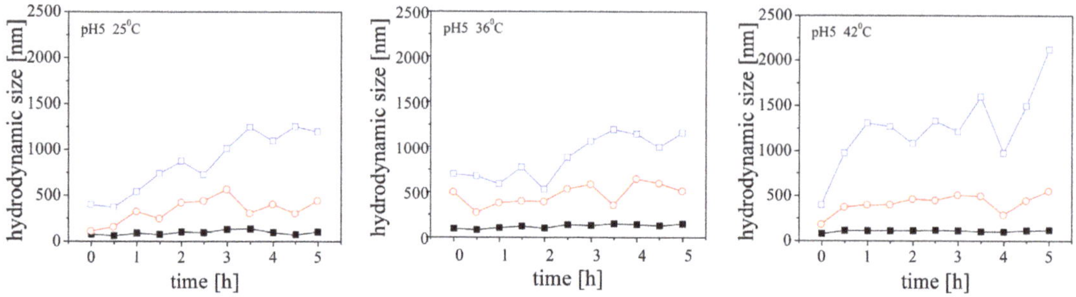

(**a**) (**b**) (**c**)

Figure A2. A variation in the hydrodynamic size (diameter) of Aβ peptide at pH 4 in the kinetic experiments: (**a**) at 25 °C, (**b**) 36 °C, and (**c**) 42 °C. The size is presented through the percentiles in the size distribution: D10 (■), D50 (○), and D90 (□).

Appendix B

(**a**) (**b**)

Figure A3. Image of the measuring cell for DLS with the sample after finishing the kinetic experiment: (**a**) observation of the aggregation of Aβ peptide in the cell at pH 4 and 36 °C. For DLS measurement, we used Ultra-Micro Cell for Nano Series. The width of the measurable window is 3 mm. (**b**) The same photo represented at a larger scale.

References

1. World Alzheimer Report 2023. Available online: www.alzint.org/resource/world-alzheimer-report-2023/ (accessed on 8 December 2023).
2. Russo, O.; Kisialiou, A.; Lamonaca, P.; Moroni, R.; Prinzi, G.; Fimi, M. New drugs from marine organisms in Alzheimer's disease. *Mar. Drugs* **2016**, *14*, 5. [CrossRef] [PubMed]
3. Ono, K.; Watanabe-Nakayama, T. Aggregation and structure of amyloid β-protein. *Neurochem. J.* **2021**, *151*, 105208. [CrossRef] [PubMed]
4. Ono, K. Alzheimer's disease as oligomeropathy. *Neurochem. Int.* **2018**, *119*, 57–70. [CrossRef] [PubMed]
5. Jarrett, J.T.; Lansbury, J.R., Jr. Seeding 'one-dimensional crystallization' of amyloid: A pathogenic mechanism in Alzheimer's disease and scrapie? *Cell* **1993**, *73*, 1055–1058. [CrossRef]
6. Buell, A.K. The growth of amyloid fibrils; rates and mechanisms. *Biochem. J.* **2019**, *476*, 2677–2703. [CrossRef] [PubMed]
7. Esler, W.P.; Strimson, E.R.; Jennings, J.M.; Vinters, H.V.; Ghilardi, J.R.; Lee, J.P.; Mantyh, P.W.; Maggio, J.E. Alzheimer's disease amyloid propagation by a template-dependent dock-lock mechanism. *Biochemistry* **2000**, *39*, 6288–6295. [CrossRef] [PubMed]
8. Silva, M.; Seijas, P.; Otero, P. Exploitation of marine molecules to manage Alzheimer's disease. *Mar. Drug* **2021**, *19*, 373. [CrossRef] [PubMed]
9. Huang, Y.; Chang, Y.; Liu, L.; Wang, J. Nanomaterials for modulating the aggregation of β-amyloid peptides. *Molecules* **2021**, *26*, 4301. [CrossRef] [PubMed]

10. Manzano, S.; Agüera, L.; Aguilar, M.; Olazaran, J.A. Review on tramiprosate (homotaurine) in Alzheimer's disease and other neurocognitive disorders. *Front. Neurol.* **2020**, *11*, 614. [CrossRef]
11. Kabir, M.T.; Sahab Uddin, M.; Jeandet, P.; Emran, T.B.; Mitra, S.; Albadrani, G.M.; Sayed, A.A.; Abdel-Daim, M.M.; Simal-Gandara, J. Anti-Alzheimer's molecules derived from marine life: Understanding molecular mechanisms and therapeutic potential. *Mar. Drugs* **2021**, *19*, 251. [CrossRef]
12. Iannuzzi, C.; Irace, G.; Sirangelo, I. The Effect of Glycosaminoglycans (GAGs) on Amyloid Aggregation and Toxicity. *Molecules* **2015**, *20*, 2510–2528. [CrossRef]
13. Tian, J.; Dang, H.; Wallner, M.; Olsen, R.; Kaufman, D.L. Homotaurine, a safe blood-brain barrier permeable $GABA_A$-R-specific agonist ameliorates diseases in mouse models of multiple sclerosis. *Sci. Rep.* **2018**, *8*, 16555. [CrossRef] [PubMed]
14. Gervais, F.; Paquette, J.; Morissette, C.; Krzywkowski, P.; Yu, M.; Azzi, M.; Lacombe, D.; Kong, X.; Aman, A.; Laurin, J. Targeting soluble Aβ peptide with Tramiposate for the treatment of brain amyloidosis. *Neurobiolol. Aging* **2007**, *28*, 537–547. [CrossRef] [PubMed]
15. Yao, Z.; Jin, L.; Li, J.; Wang, Y.; Lu, R.; Bai, Y.; Wu, H. κ-carrageenan oligosaccharides postpone the progression of Alzheimer's disease by inhibiting the neuronal damage caused by microglial overactivation. *Res. Sq.* **2022**. [CrossRef]
16. Cao, Y.; Dong, X.; Chen, X. Polymer-modified liposomes for drug delivery: From fundamentals to applications. *Pharmaceutics* **2022**, *14*, 778. [CrossRef]
17. Zhao, X.; Xu, M.; Ma, C.; Ma, Y.; Ma, X. Spectrophotometric Determination of Thiosulfate in Desulfurization Solutions by Decoloration of Methylene Blue. *Spectroscopy* **2023**, *38*, 19–22. [CrossRef]
18. Minamoto, C.; Fujiwara, N.; Shigekawa, Y.; Tada, K.; Yano, J.; Yokoyama, T.; Minamoto, Y.; Nakayama, S. Effect of acidic conditions on decomposition of methylene blue in aqueous solution by air microbubbles. *Chemosphere* **2021**, *263*, 128141. [CrossRef] [PubMed]
19. Zerovnik, E. Amyloid-fibril formation. Proposed mechanisms and relevance to conformation disease. *Eur. J. Biochem.* **2002**, *269*, 3362–3371.
20. Gorbenko, G.P.; Kinnunen, P.K.J. The role of lipid-protein interactions in amyloid-type protein fibril formation. *Chem. Phys.* **2006**, *141*, 72–82. [CrossRef]
21. Yao Tian, Y.; Viles, J.H. pH Dependence of Amyloid-β Fibril Assembly Kinetics: Unravelling the Microscopic Molecular Processes. *Angew. Chem. Int. Ed.* **2022**, *61*, e202210675. [CrossRef]
22. Assarsson, A.; Hellstrand, E.; Cabaleiro-Lago, C.; Linse, S. Charge Dependent Retardation of Amyloid β Aggregation by Hydrophilic Proteins. *ACS Chem. Neurosci.* **2014**, *5*, 266–274. [CrossRef]
23. Knight, J.D.; Miranker, A.D. Phospholipid catalysis of diabetic amyloid assembly. *J. Mol. Biol.* **2004**, *341*, 1175–1187. [CrossRef] [PubMed]
24. Uversky, V.N.; Fink, A.L. Conformation constraints for amyloid fibrillation: The importance of being unfolded. *Biochim. Biophys. Acta* **2004**, *1698*, 131–153. [CrossRef] [PubMed]
25. King, M.E.; Ahuaja, V.; Binder, L.I.; Kuret, J. Ligand-dependent tau filament formation: Implications for Alzheimer's disease progression. *Biochemistry* **1999**, *38*, 14851–14859. [CrossRef]
26. Wilson, D.M.; Binder, L.I. Free fatty acids simulate the polymerization of tau and amyloid beta peptides. In vitro evidence for a common effector of pathogenesis of Alzheimer's disease. *Am. J. Pathol.* **1997**, *150*, 2181–2195. [PubMed]
27. Zimmermann, R.; Küttner, D.; Renner, L.; Kaufmann, M.; Zitzmann, J.; Müller, M.; Werner, C. Charging and structure of zwitterionic supported bilayer lipid membranes studied by streaming current measurements, fluorescence microscopy, and attenuated total reflection Fourier transform infrared spectroscopy. *Biointerphases* **2009**, *4*, 1–6. [CrossRef]
28. Kobayashi, S.; Tanaka, Y.; Kiyono, M.; Chino, M.; Chikuma, T.; Hoshi, K.; Ikeshima, H. Dependence pH and proposed mechanism for aggregation of Alzheimer's disease-related amyloid-β(1–42) protein. *J. Mol. Struct.* **2015**, *1094*, 109–117. [CrossRef]
29. Manning, G.S. Limiting Laws and Counterion Condensation in Polyelectrolyte Solutions. *J. Chem. Phys.* **1969**, *51*, 924–933. [CrossRef]
30. Cheng, H.; Olvera de la Cruz, M. Rod-like Polyelectrolyte Adsorption onto Charged Surfaces in Monovalent and Divalent Salt Solution. *J. Polym. Sci. Part B Polym. Phys.* **2004**, *42*, 3642–3653. [CrossRef]
31. Milkova, V.; Vilhelmova-Ilieva, N.; Gyurova, A.; Kamburova, K.; Dimitrov, I.; Tsvetanova, E.; Georgieva, A.; Mileva, M. Remdesivir-Loaded Nanoliposomes Stabilized by Chitosan/Hyaluronic Acid Film with a Potential Application in the Treatment of Coronavirus Infection. *Neurol Int.* **2023**, *15*, 1320–1338. [CrossRef]

molecules

MDPI

Article

Capillarity in Interfacial Liquids and Marbles: Mechanisms, Properties, and Applications

Yang Liu [1,2,3,*], Yuanfeng Wang [2] and John H. Xin [2]

[1] Department of Biomedical Engineering, Sun Yat-sen University, Shenzhen 518107, China
[2] School of Fashion and Textiles, The Hong Kong Polytechnic University, Hong Kong 999077, China; yf.wang@connect.polyu.hk (Y.W.); john.xin@polyu.edu.hk (J.H.X.)
[3] School of Physics and Electronic Engineering, Sichuan University of Science and Engineering, Zigong 643000, China
* Correspondence: liuyang56@mail.sysu.edu.cn

Abstract: The mechanics of capillary force in biological systems have critical roles in the formation of the intra- and inter-cellular structures, which may mediate the organization, morphogenesis, and homeostasis of biomolecular condensates. Current techniques may not allow direct and precise measurements of the capillary forces at the intra- and inter-cellular scales. By preserving liquid droplets at the liquid–liquid interface, we have discovered and studied ideal models, i.e., interfacial liquids and marbles, for understanding general capillary mechanics that existed in liquid-in-liquid systems, e.g., biomolecular condensates. The unexpectedly long coalescence time of the interfacial liquids revealed that the Stokes equation does not hold as the radius of the liquid bridge approaches zero, evidencing the existence of a third inertially limited viscous regime. Moreover, liquid transport from a liquid droplet to a liquid reservoir can be prohibited by coating the droplet surface with hydrophobic or amphiphilic particles, forming interfacial liquid marbles. Unique characteristics, including high stability, transparency, gas permeability, and self-assembly, are observed for the interfacial liquid marbles. Phase transition and separation induced by the formation of nanostructured materials can be directly observed within the interfacial liquid marbles without the need for surfactants and agitation, making them useful tools to research the interfacial mechanics.

Keywords: liquid marble; liquid–liquid interface; capillarity; coalescence cascade; mass transport; microreactor

1. Introduction

Capillarity, which originated from the difference in surface tension between two immiscible phases, plays important roles in both the living and non-living processes in nature. In decades, the effects of capillary force have been widely observed in the formation of cell patterns [1], the morphogenesis of tissue [2], the formation of embryonic axes [3], and homeostasis [4], which can tune the size, morphology, structure, and function of cells and tissues through surface mechanics. Moreover, capillary force can be generated at the interfaces formed by immiscible intracellular biomolecular condensates and membranes, mediating the size, morphology, structure, and function of the intracellular organelles [5]. On the other hand, capillary forces participate in a wide range of non-living processes, such as the formation of fog [6], raindrops [7], capillary waves [8], and vortexes [9], regulating the size, shape, motion, and stability of the fluids through surface mechanics. However, the organization and morphogenesis of inter- and intra-cellular structures are mechanically complex, and involve mechanical forces with disparate origins, e.g., motor proteins [10], gravitational field [11], and hydrostatic pressure [12], other than the capillary force. Much effort has been devoted to solving the morphogenesis homologies in different species, which can be described by using the coarse-grained approach under the theoretical frames of fluid mechanics and phase transitions [13]. For example, the rigidity phase transition of the zebrafish blastoderm can be explained by analyzing the cell–cell contact topology using the percolation theory, while a few parameters, including cell–cell connectivity, cell–cell adhesion, and meta-synchronous cell cleavages, were characterized [14]. Even though the percolation model can appropriately explain the rigidity phase transition under multiple experimental disturbances such as cell fate and cell adhesion, it cannot predict the absolute value of viscosity or the changes in the force field that regulate the topological changes mechanically. To elucidate the non-linear and time-dependent mechanics in morphogenesis, it is essential to find out the basic mechanisms of the fluid mechanics involved and the functions of viscous, inertia, and capillary forces.

Capillary forces mediate the inter- and intra-cellular processes through interfacial phase separation and phase transitions, which may act as a general driving force for a wide range of interfacial phenomena, e.g., wetting, adhesion, coalescence, and Pickering emulsion. However, current measurement techniques may not support the precise in vitro or in vivo quantification of the inter- and intra-cellular capillary forces due to the difficulties in determining the strength, location, and time scale of the force. For example, the strength of the surface tension of the biomolecular condensates is estimated to be in the range of 10^{-4} N m^{-1} to 10^{-7} N m^{-1} [15,16], which is significantly weaker than the surface tension of water (γ = 0.072 N m^{-1}) [17]. Despite the tiny scale, the capillarity in cells and tissues shares the same mechanism as the fluidics that naturally occur at the liquid–liquid interface. For example, the reorganization of genomes can be achieved by the coalescence of droplets, which are induced to form at the targeted loci of the genome [18]; the cylindrical thread formed by condensation and crosslinking of actin may break up into discontinuous pieces of tactoids driven by Rayleigh–Plateau instability [19]. Moreover, nucleolar surfactants, such as proteins NO145 and Ki-67, can adsorb and enrich at the nucleolar periphery, mediating the size of the nucleolus by impeding coalescence or Ostwald ripening [20,21]. Therefore, the capillaries at the inter- and intra-cellular interfaces can be definitely studied using the fluidic models at the visible droplet scale. Moreover, the capillary strength, location, and time scale can be measured more precisely using simple apparatuses at this scale, which may realize accurate interpretation and resolution of the unsolved mechanisms of fluid mechanics in general, such as coalescence cascade, Marangoni flow, and Stokes flow.

Interfacial liquid droplets and marbles [22], which can be formed by adding droplets at the interface between two immiscible liquid phases, can be an ideal fluidic model to study the biological capillaries. A schematic illustration of the interfacial liquid droplet is shown in Scheme 1. Specifically speaking, interfacial liquid droplets and marbles have obtained the following features, which make them outstanding in modelling the biological capillaries: (i) The phase-separated liquids mimick the phase-separated liquid-in-liquid environments existed in the cytoplasm and interstitial fluid; (ii) The viscosities of the liquids can be facilely tuned; (iii) The size of the droplet can be effectively controlled to elucidate the effect of viscous, inertia, and gravitational forces on coalescence separately; (iv) Amphiphilic surfactants with different molecular structure, size, and thermodynamics can be used to modify the droplet surface, in order to investigate the Marangoni flow at the interface and mediate the coalescence process; (v) The formation of a transparent and geometrically simple interface is advantageous for observation and modelling; the interface is concealed by the upper liquid phase as the external disturbances, such as wind, dirt, and acoustics, can be minimized.

Herein we use interfacial liquid droplets and marbles as fluidic models to elucidate the mechanisms underlying the general fluidic phenomena existing in the scale of biological capillaries, i.e., coalescence cascade, the role of surface-active particles, and mass transport at the liquid–liquid interface, which may provide useful insights for asymptotic flows and interfacial thermodynamics in biological capillaries.

The general mechanism of the coalescence cascade (CC) of a droplet has been long debated in the science community as a result of the intrinsic complexity in space and time of the process. Numerous internal and external parameters, such as droplet size, surface tension, viscosity, density, coalescence time, and the size of the daughter droplet, may mediate the CC process, affecting its stages and duration. Moreover, discrepancies in the CC phenomena at the air–liquid and liquid–liquid interfaces have been widely observed and investigated, e.g., damped CC [23], second-stage coalescence [24], migrating partial coalescence [25], and multi-stage coalescence [26], making it extremely difficult to elucidate a general mechanism. However, the central idea of the problem is how the droplet generates smaller daughter droplets during the coalescence process. Charles et al. have investigated the CC of a liquid droplet (phase 1) with a viscosity of η_1 and surrounded by another liquid (phase 2) with a viscosity of η_2, at the interface formed between phase 1 and phase 2 [27]. They attributed the formation of daughter droplets to the Rayleigh instability of the liquid column formed during the drainage of the droplet by the underlying liquid reservoir, and the diameter ratio of secondary to primary droplets changed with the viscosity ratio $p = \eta_1/\eta_2$, reaching a maximum near $p = 1$. According to their observation, no daughter droplets could form in the cases of $p < 0.02$ or $p > 11$, which was in line with the prediction of Tomotika [28], who extended the Rayleigh theory on the instability of jets [29] to the case of a cylindrical thread of a viscous liquid in the infinite mass of another viscous liquid. On the other hand, contradictory results are obtained for liquid droplet coalescence at the air–liquid interface with $p \gg 11$. Thoroddsen et al. have observed the CC of a water droplet at the air–water interface, which can proceed for up to six steps [30]. They attributed the termination of CC to the viscous effect, which may become dominant as the Reynolds number (Re) of the droplet is sufficiently small, e.g., Re < 20. Blanchette et al. have investigated the CC of ethanol droplets at the air–ethanol interface, finding that there exists a critical Ohnesorge number (*Oh*) for droplets with small bond numbers (*Bo* < 0.2), above which total coalescence may happen [31]. Chen et al. have investigated the CC of water droplets at the interface formed by decane and water-glycerol, and the results indicate that there exists a maximum *Bo* number of droplets for CC to occur [32].

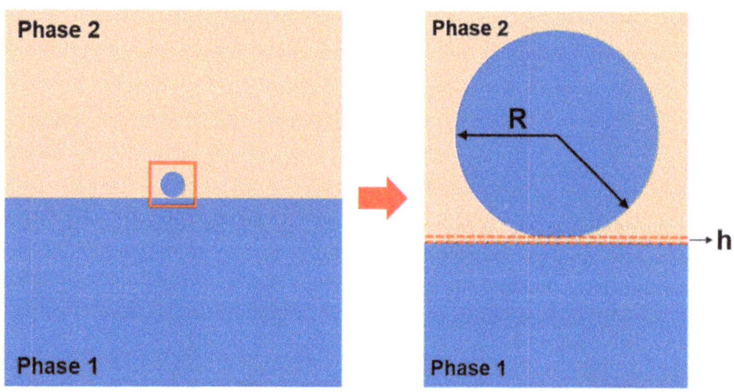

Scheme 1. Illustration of a liquid droplet composed of the same liquid as phase 1, sitting at the liquid–liquid interface formed by two immiscible liquids. The droplet is assumed to have a spherical shape as its size is substantially small and the effect of gravity is negligible. The radius of the liquid droplet is denoted as *R*, and there may exist a thin layer of phase 2 liquid with a thickness of h between the droplet and the reservoir before the initiation of coalescence.

2. Results and Discussion

2.1. Mechanism of Coalescence Cascade in Interfacial Liquid Droplets

From the previous research, it can be inferred that the CC of a liquid droplet on the reservoir of the same liquid can be determined by multidimensional parameters, including time (t), viscosity (η), velocity (v), radius (R), density (ρ), and surface tension (γ). The driving force of the CC process is the mechanical potential of the liquid droplet with radius R, which possesses an internal pressure $\Delta P = 2\gamma/R$. The Gibbs free energy of the system can be written as $\Delta G = -2\gamma V_{mol}/R$, where V_{mol} is the molar volume. The value of Gibbs free energy may imply the maximum work that can be conducted during the

CC process; however, the actual CC process may reflect a balance between the viscous, inertial, and capillary forces, which can be essentially described by t, γ, η, ρ, and R. During each stage of CC, the flow from the droplet to the reservoir is driven by the formation of a highly curved liquid bridge with a length of $2\pi r_B$ and a width of ω, which can be solved two-dimensionally [33], as shown in Scheme 2. In the 2D circumstance, the dimensionless velocity can be written as $v = \gamma/\eta$, and the corresponding coalescence time t can be written as $t = R\eta/\gamma$. Assuming the coalescence time of a liquid droplet at the air–liquid interface to be t and the coalescence time of the same liquid droplet at the liquid–liquid interface to be t^*, there is always $t^* > t$ since the viscous force of the surrounding fluid exerted on the droplet surface may remarkably reduce the flow velocity, indicating liquid droplets at the liquid–liquid interface are intrinsically more stable than the liquid droplets at the air–liquid interface. However, according to our observation, the interfacial liquid droplets, which rest at the interface that composed of an upper viscous liquid and a lower liquid the same with the droplet, are much more stable compared to the existing theoretical prediction.

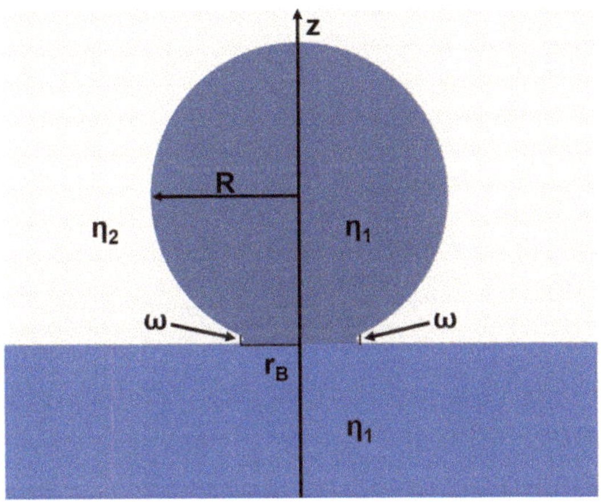

Scheme 2. Illustration of the two-dimensional geometry of a liquid droplet with a viscosity η_1 and a radius of R in the early stages of the coalescence with a liquid reservoir of the same liquid. The liquid droplet is sitting at the interface formed by the liquid reservoir and an external liquid with a viscosity of η_2. The liquid bridge that joins the droplet and the reservoir has a radius of r_B and a width of ω. The line of symmetry is represented by the z-axis.

Specifically speaking, we investigate the CC of a water droplet at the interface formed by tetradecane (phase 2) and water (phase 1). The water droplet is generated by using a micropipette, which is subsequently released and deposited onto the tetradecane/water interface gently from a height of ~3 mm. The CC process of the droplet is recorded by a high-speed camera. To minimize the viscous and gravity forces, the size of the water droplet is set at 15 μL, corresponding to a Bo number of ~0.076 and an Oh number of ~0.003, and a precision of ca. 0.1 μL can be achieved. The radius, Bo number, and Oh number of the droplet are calculated by using Equation (1), Equation (2) and Equation (3), respectively.

$$R = \sqrt[3]{\frac{3V}{4\pi}} \quad (1)$$

$$Bo = \frac{(\rho_1 - \rho_2)gR^2}{\gamma_1} \quad (2)$$

$$Oh = \frac{\eta_1}{\sqrt{\rho_1 R \gamma_1}} \quad (3)$$

Both of the *Bo* and *Oh* numbers are dimensional-less and are used as indicators of the importance of surface tension force against gravitational force and viscous force in general fluidic mechanics, respectively. The *Bo* number is defined as the ratio of the gravitational force to the surface tension force, and the *Oh* number is defined as the ratio of the viscous force to the surface tension force. Both of the *Bo* and *Oh* numbers are important to the CC process. For a liquid droplet of finite size, the values of its *Bo* and *Oh* numbers may determine whether partial coalescence or total coalescence will occur. For large drops where the gravitational force is dominant, the *Bo* number is the major determinant; for small drops where the viscous force is dominant, the *Oh* number is the major determinant.

Typical coalescence processes of a 15 μL water droplet at the tetradecane/water interface are shown in Figure 1. When being cast at the tetradecane/water interface, the water droplet would rest at the interface as a result of the presence of a thin tetradecane layer between the bottom of the droplet and the water reservoir (Figure 1a), which is illustrated in Scheme 1 and can be a few microns thick [34]. The first step of the coalescence cascade (CC) event initiates with the formation of highly curved liquid meniscuses between the bottom of the droplet and the water surface, which are also called "liquid bridges" (Figure 1b). The liquid bridges then drive the drainage of the water droplet into the water reservoir, resulting in the widening of the meniscuses, which in turn generates capillary waves converged at the summit of the droplet and can be seen as the small protrusion at the top of the droplet (Figure 1b). Subsequently, the droplet is stretched upward, forming a liquid cylinder, while the liquid bridges quickly narrow and vanish (Figure 1c), leaving a daughter droplet with a smaller size resting at the interface (Figure 1d). In stage 1, it takes ~60 ms for the original water droplet to coalesce and form the daughter droplet. However, the daughter droplet may repeat the same coalescence process until fully coalescence occurs; due to the reduced size of the daughter droplet (hence enhanced ΔP), the coalescence may proceed in a faster and faster manner (Figure 1: Stage 2 and Stage 3). For example, it takes ~40 ms for stage 2 to complete and ~20 ms for stage 3 to complete. After all, a total of six stages can be observed; however, details of the rest of the stages are not recorded due to limited frame rates. Compared to the CC of water droplets at the air–water interface, the CC of water droplets at the oil–water interface shows three distinct aspects: (i) elliptical shape; (ii) significantly longer coalescence time; and (iii) no pinch-off. The shape of the 15 μL water droplet is remarkably deformed from sphericity at the tetradecane/water interface, forming an elliptical shape at equilibrium (Figure 1a). However, the sphericity of the water droplet is restored for the daughter droplets (Figure 1e,h). The phenomenon indicates that the deformation is caused by the under-liquid pressure, as water droplets with smaller sizes possess higher internal pressure to counterbalance the hydrostatic pressure exerted by the surrounding fluid (tetradecane). On the other hand, gravity is not the cause of the deformation, as the small *Bo* number of the 15 μL water droplet makes it negligible, and the surrounding tetradecane may assist in counter-balancing gravity [35]. Moreover, it can be predicted that water droplets with larger sizes may become more elliptical compared to the smaller ones.

The propagating stages of CC can be readily observed by naked eyes as the 15 μL water droplet is cast at the tetradecane/water interface (Video S1). However, when a water droplet with the same volume is cast at the air/water interface, naked eyes can only observe apparently full coalescence (Video S2). Indeed, the water droplet still goes through a CC process at the air/water interface, but at a speed much faster than that at the tetradecane/water interface. It has been reported that the coalescence time of the first CC stage for water droplets with volumes around 15 μL at the air/water interface is in the magnitude of a few milliseconds [26]. On the other hand, it may take a few tens of milliseconds for the first CC stage of a 15 μL water droplet at the tetradecane/water interface to complete, which is an order of magnitude longer than that at the air/water interface. The coalescence between a liquid droplet and a liquid reservoir can be described asymptotically by applying numerical methods to study the length scales and structures of the liquid bridge formed at the early stage of the coalescence, which was proposed by Eggers and coworkers [33]. They concluded that there existed $\omega \propto r_B^{\alpha}$, and r_B can be described as a function of the coalescence time t as:

$$r_B(t) \sim -\frac{(\alpha - 1)}{2\pi} t ln t \quad (4)$$

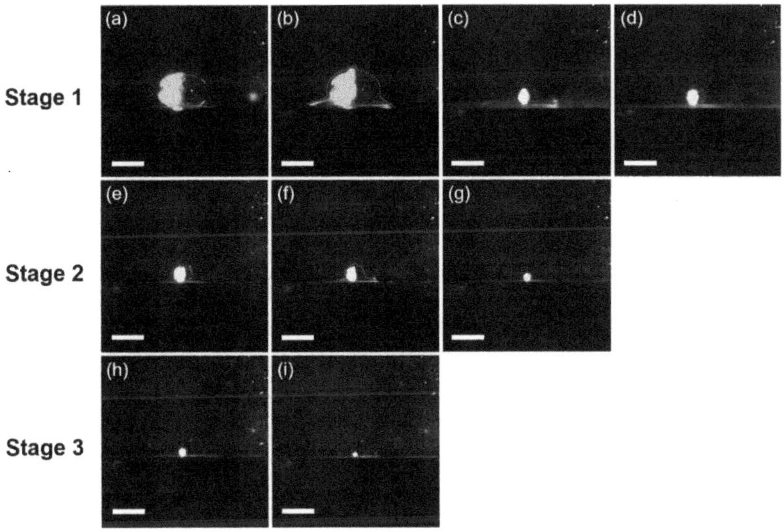

Figure 1. The first three stages of the coalescence cascade (CC) process of a 15 μL water droplet at the tetradecane/water interface. The first stage of CC is shown from (**a**–**d**), while the second and third stages are shown from (**e**–**g**) and (**h**,**i**), respectively. The contours of the coalescing water droplets are shown in (**b**,**f**), where the formation of liquid bridges between the droplet and the reservoir at the tetradecane/water interface can be readily observed. The time gap between each frame is 20 ms. Scale bar: 1 mm.

And there is $\alpha = \frac{3}{2}$ when the external fluid obtained a finite viscosity. Based on their analysis, it can be estimated that the coalescence time of a liquid droplet may increase by a factor of 4 in the presence of an external fluid with a finite viscosity. However, the coalescence time of the water droplet at the tetradecane/water interface would be remarkably underestimated by using this model, which can be attributed to the underestimation of α and r_B. The fact that $\omega \sim r_B^{\frac{3}{2}}$ does not hold as $r_B \to 0$ for any external liquid of finite viscosity indicates that the Stokes equations cannot describe the asymptotic dynamics when $r_B \to 0$, where the surface tension force must be large enough to counterbalance the inertial force and initiate the coalescence. To better describe the flow dynamics as $r_B \to 0$, Paulsen et al. proposed a third inertially limited viscous regime between the inertial and viscous regimes and further formulated the boundary condition of the transition from the inertially limited viscous regime to the Stokes regime as [36]:

$$Oh \propto \left| ln\left(\frac{r_B}{8R}\right) \right| / \sqrt{4\pi} \quad (5)$$

They verified the inertially limited viscous regime both numerically and experimentally, showing that r_B/R should be sufficiently large to induce the inertially limited viscous-to-Stokes transition, which can be satisfied when $Oh \gg 1$. This theory is coincident with the experimental results that the coalescence time of a 15 μL water droplet at the tetradecane/water interface is much longer than the coalescence time predicted using Stokes equations. Moreover, it is observed that the water droplets coalescence at the tetradecane/water interface do not pinch off, which can be attributed to the damping of the up-propagating capillary waves from liquid bridges to the top of the droplet by the viscous and inertial forces of the external tetradecane [23]. The existence of the inertially limited viscous regime is further evidenced by replacing the water droplet with a viscous droplet. The viscous droplet is pipetted from a viscous solution containing 0.4% polyethylene glycol 400 and 0.3% propylene glycol, and a viscosity of ~12.4 mPa·s and a surface tension of ~40 mN m^{-1} can be obtained at 25 °C. Compared to the viscosity of water (~0.89 mPa·s at 25 °C), the viscosity of the viscous droplet is one order of magnitude higher. In the meantime, it is observed that the average resident time of the 15 μL viscous droplets at the tetradecane/water interface is one order of magnitude higher than the resident time of the water droplets with the same volume (~2.8 s), which can last for ~25 s before coalescence (Figure 2a,b). However, as the length of the liquid bridge (r_B) grows large enough, the coalescence of the viscous droplet happens instantaneously, and the total coalescence process finishes in 3.7 ms without leaving any trace of daughter droplets. This phenomenon indicates the transition from the inertially limited viscous regime to the Stokes (viscous) regime. The abrupt increase in the flow velocity during the transition can be explained by the cross-over of the phase boundary between the inertially limited viscous regime to the Stokes regime as r_B is sufficiently large, and the creeping flow in the inertially limited viscous regime would suddenly change to Stokes flow. The velocity scaling in the two regimes can be properly described by the equations proposed by Paulsen et al. [37]:

$$v_{Inertially-limited} \approx \frac{3\eta}{4R^3\rho} r_B^2 \tag{6}$$

$$v_{Stokes} \approx \frac{\gamma}{2\pi\eta}\left(\frac{r_B}{R}\right)\left|\ln\left(\frac{r_B}{8R}\right)\right| \tag{7}$$

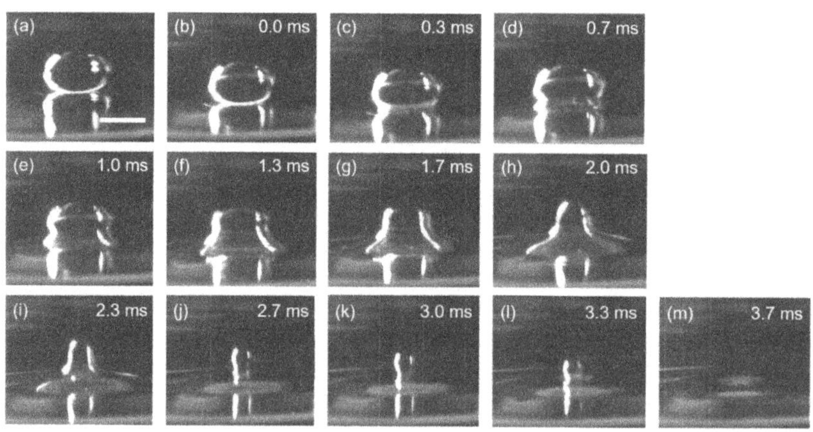

Figure 2. Coalescence of a viscous droplet at the tetradecane/water interface. Two distinct stages can be identified for the coalescence process: (i) The creeping stage, where the liquid bridge is growing slowly and the droplet is residing at the interface for a long duration without shape deformation, as shown in (**a**,**b**). The time gap between (**a**) and (**b**) is ~25 s. (ii) The coalescence stage, where the droplet suddenly deforms and merges into the water reservoir, leaving no trace of daughter droplets, as shown from (**b**–**m**). Scale bar: 1 mm.

As a result of the relatively high viscosity, the horizontal pull at the tetradecane/water interface is restricted, and the capillary wave traveling upwards is damped, while the vertical collapse of the droplet is merely affected, as shown in Figure 2b–m. In this regard, no daughter droplets are formed, and the whole coalescence process is completed in a few milliseconds due to the high velocity of the Stokes flow.

2.2. Mass Transport in Interfacial Liquid Marbles at the Oil–Water Interface

To investigate the mass transport mechanism of interfacial liquid marbles, a 1 M ferrous chloride ($FeCl_2$) solution is used as the liquid phase of the marbles. Interfacial liquid marbles are constructed according to reference [22]. Briefly speaking, 40 mL of deionized (D.I.) water and 20 mL of tetradecane are subsequently added into a clean 100 mL glass beaker. Bubbles and curved interfaces may form during the mixing process of the two immiscible liquid phases, which are removed and flattened using a clean glass rod, and eventually a smooth and flat interface is formed between water and tetradecane. A 30 or 60 μL droplet of 1 M $FeCl_2$ aqueous solution is transferred to the surface of a hydrophobic polyvinylidene fluoride (PVDF) layer, rolling gently on the layer surface until the droplet surface is fully covered by the PVDF powders. The PVDF-coated $FeCl_2$ liquid marbles show a white color, which is subsequently transferred to the tetradecane/water interface. Upon entering tetradecane, the PVDF coating quickly changes to transparent, making the original yellowish-green color of the $FeCl_2$ droplet readily observable (Figure 3a). Six $FeCl_2$ interfacial marbles are laid at the tetradecane–water interface, forming a triangular array. $FeCl_2$ marbles are chosen as the experimental models for investigating the mass transport mechanisms at the tetradecane–water interface, as they can react with oxygen molecules and generate reddish brown $Fe(OH)_3$, acting as an indicator of oxygen. While exposed to the ambient conditions, the $FeCl_2$ droplets quickly change their color from yellowish-green to yellowish-brown within 24 h, and reddish-brown precipitates of $Fe(OH)_3$ are readily observed at the bottom of the droplets (Figure S1a,b). Eventually the $FeCl_2$ droplets become dried, leaving reddish brown precipitates (Figure S1c–f).

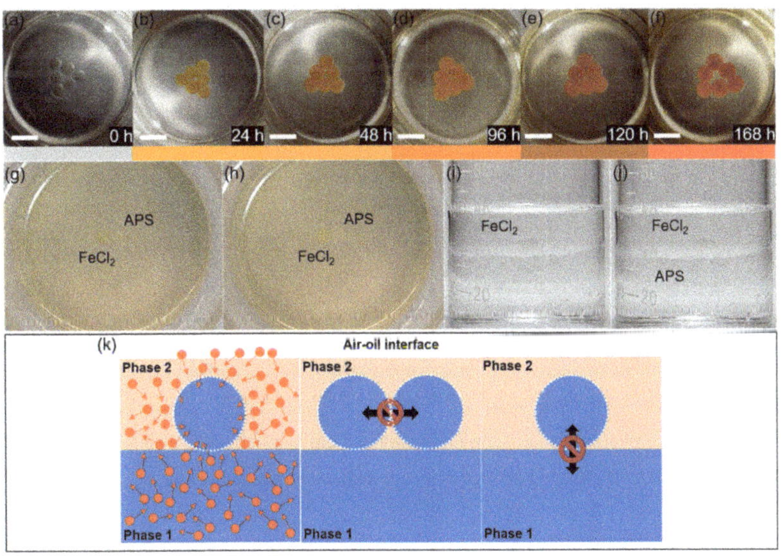

Figure 3. Mass transport phenomena of interfacial liquid marbles at the tetradecane/water interface. (**a–f**) Oxidation of an array of $FeCl_2$ interfacial liquid marbles at the deoxygenated water/tetradecane interface in 168 h, images are taken at an interval of 24 h; the color bars below indicate the apparent colors of the marbles in the above images. Scale bar: 1 cm. (**g**,**h**) No liquid transport takes place between a $FeCl_2$ interfacial marble and an APS interfacial marble laid adjacent to each other from 0 h (**g**) to 1.5 (**h**). (**i**,**j**) No liquid transports are observed between a $FeCl_2$ marble and the underneath water phase before (**i**) and after (**j**) the addition of 1 M APS aqueous solution into the water phase. (**k**) A schematic illustration of the mass transport mechanisms of an interfacial liquid marble at the oil/water interface. The white layer represents the PVDF coating, red circles represent the gas molecules, and the arrows indicate the mass-transport directions.

To evaluate the air permeability at the tetradecane/water interface, $FeCl_2$ marbles are placed at the interface, where the water phase is deoxygenated prior to bubbling argon for 30 min. However, it does not prevent the oxidation of the $FeCl_2$ into $Fe(OH)_3$ inside the liquid marbles, as the marbles resting at the interface gradually change their colors from yellowish-green to brown and eventually reddish brown in 7 days, indicating the supply of oxygen from tetradecane and the atmosphere (Figure 3a–f). On the other hand, a slower oxidation process is observed for the $FeCl_2$ marbles placed at the interface, where both tetradecane and water are deoxygenated by bubbling argon (Figure S2), compared to the control group, where both tetradecane and water are used as received (Figure S3). The mass transport between interfacial liquid marbles is evaluated by placing a 30 μL $FeCl_2$ marble and a 30 μL ammonium persulfate (APS) marble at the tetradecane/water interface side-by-side without distorting the original curvature of the marbles (Figure 3g). Solid contacts are established between the surfaces of the $FeCl_2$ and APS marbles, and the contacting marbles are maintained static at the interface during the experimental observation due to the presence of the protective coating layer composed of PVDF powders and the adsorbed tetradecane molecules. The APS marbles are chosen because $FeCl_2$ can be readily oxidized by APS. If there is any mass transport between the APS and $FeCl_2$ marbles, the color of the $FeCl_2$ marble would change quickly. No mass transports are observed between the $FeCl_2$ and APS marbles during the experiment (~3.5 h), and the two marbles show no observable changes during the first 1.5 h (Figure 3h). However, the color of the $FeCl_2$ marble turns quickly yellow upon the injection of 30 μL of 1 M aqueous APS solution, but the color does not diffuse into the neighboring APS marble (Video S3). To evaluate the mass transfer between the water reservoir and interfacial liquid marble, a $FeCl_2$ marble is placed at the tetradecane/water interface (Figure 3i), and subsequently 1 mL of 1 M APS is added into the lower water phase. No transports of APS from the water phase to the $FeCl_2$ marble are observed during the experiment (Figure 3j), except the $FeCl_2$ marble gradually becomes yellow after 3 h (Figure S4), as a result of the diffusion of oxygen molecules from tetradecane and the atmosphere into the marble.

From the observations of the mass transport experiments, the mechanisms of mass transport at the tetradecane/water interface can be clearly revealed: gas molecules, e.g., O_2, can diffuse into the interfacial liquid marbles through Brownian motion; at static conditions, no liquid transport could take place between two adjacent interfacial liquid marbles, and no liquid transport could take place between the interfacial liquid marble and the underneath liquid phase. Moreover, the curvature of liquid marbles may have a significant impact on the interfacial mass transport process. Providing the *Bo* number of the marble is small and the effect of gravity is negligible, the degree of curvature of the marble is inversely proportional to its radius and size. Therefore, an interfacial liquid marble with a high curvature may have a small size, thus a high specific surface area. In other words, the high-curvature marble would have more interfacial area, where water molecules are oriented in a low-entropy tetrahedral-like structure [38], and the diffusion of molecules and ions through this water structure is impeded compared to the bulk water [39]. A schematic illustration of the mass transport mechanisms at the tetradecane/water interface is shown in Figure 3k.

2.3. Programmable and Versatile Microreactors Using Interfacial Liquid Marbles

Liquid marbles, which are constructed at the air–solid or air–liquid interface, have found substantial applications in miniaturized microreactors and microbioreactors [40–43], demonstrating unique features such as high interfacial area [44], oxygen permeability [45], tunable shell function [46], and stimuli-responsiveness upon applied potential and heat [47]. Compared to liquid marbles, interfacial liquid marbles may obtain extraordinary morphology and composition stability, durability, self-assembly, and transparency; the presence of the upper liquid phase may seal the interfacial liquid marbles with the ambience, prohibiting the evaporation of the encapsulated liquid and lowering the oxygen concentration at the surface of the marbles, making the marbles ideal microreactors for investigating reduction reactions.

Various reduction reactions can be accomplished by the interfacial liquid marbles, while state-of-the-art nanomaterials with structural and compositional uniformity can be obtained. For example, graphene and palladium nanoparticles are obtained by reducing graphene oxide (GO) and palladium chloride ($PdCl_2$) in interfacial liquid marbles. For the reduction in GO, a hexagonal marble array consisting of 42 interfacial liquid marbles is fabricated on a hexane–water interface with an area of 19.5 cm^2 (Figure 4a). Each marble contains 10 μL of 3 mg mL^{-1} aqueous GO dispersion. When being cast at the hexane–water interface, the marbles seek their positions spontaneously, driven by gravity and surface tension, eventually assembling into an ordered hexagonal array. The reduction reaction is triggered by adding 0.1 mL of hydrazine monohydrate into the lower water phase, which generates reducing hydrogen molecules through the following reaction [48]:

$$H_2NNH_2 \rightarrow N_2(g) + 2H_2(g)$$

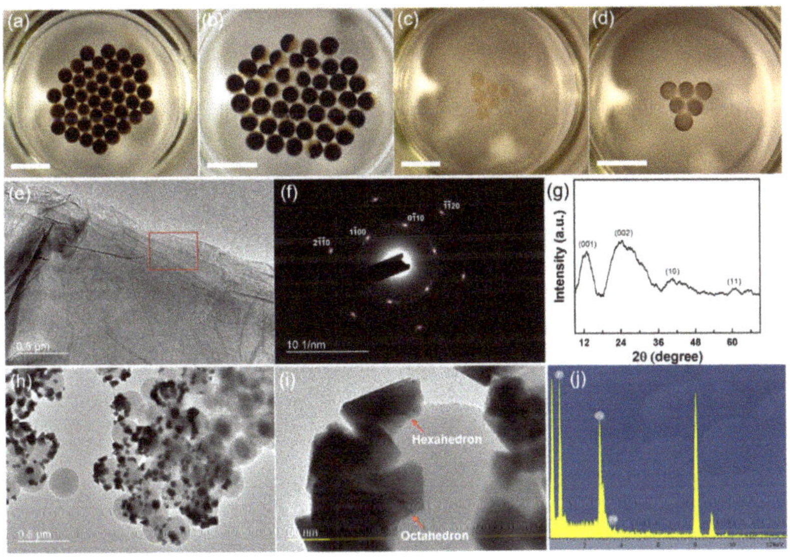

Figure 4. Interfacial liquid marbles as microreactors to synthesize nanomaterials with unique characteristics. (**a**) Formation of a hexagonal array of interfacial liquid marbles at the hexane–water interface by self-assembly. Each marble contains 10 μL of 3 mg mL^{-1} aqueous GO dispersion. (**b**) A hexagonal array of Janus marbles containing rGO is formed after a 24 h reduction. (**c**) Formation of a triangular array of interfacial liquid marbles at the hexane–water interface through self-assembly. Each marble contains 10 μL of aqueous 0.1 mg mL^{-1} $PdCl_2$ dispersion. (**d**) Dispersed Pd nanoparticles are formed inside the marbles after an 18 h static reduction in the ambient conditions. Scale bar: 1 cm. (**e**–**g**) The TEM image, SAED pattern, and XRD spectrum of the as-obtained rGO using interfacial liquid marbles as the microreactors. The SAED pattern is taken from the red rectangular area in (**e**). (**h**) A TEM image of the Pd nanoparticles obtained from the interfacial liquid marbles showed well-dispersed Pd nanoparticles on spherical PVDF particles. (**i**) Magnified TEM image of Pd nanoparticles on PVDF, showing the coexistence of hexagonal and octahedral nanoparticles. (**j**) EDX spectrum of the Pd nanoparticles on PVDF.

The hydrogen gas can permeate into the interfacial liquid marbles and reduce the GO into rGO (reduced graphene oxide). As shown in Figure 4b, phase separation and formation of Janus marbles are observed in the interfacial liquid marbles after a 24 h reduction in the ambient and static conditions. As the hydrophilic functional groups in the GO basal planes are replaced by hydrogen during the reduction, the as-obtained rGO becomes more hydrophobic, which tends to nucleate and aggregate at the surface of the interfacial liquid marbles, causing the formation of Janus marbles. Moreover, the unique reducing environment of interfacial liquid marbles can be used to synthesize metallic nanoparticles. For example, a triangular array containing six interfacial liquid marbles is assembled at the hexane–water interface (Figure 4c). Each marble contains 10 μL of aqueous 0.1 mg mL^{-1} $PdCl_2$ dispersion, which is subsequently reduced by adding hydrazine monohydrate into the lower water phase. Interestingly, no phase separations occur in the interfacial liquid marbles after an 18 h reduction, despite the color of the marbles changing from light yellow to black, as shown in Figure 4d. However, the absence of phase separation also indicates a good dispersibility of the as-synthesized metal nanoparticles in the marbles. rGO nanosheets with the coexistence of oxidized and unoxidized domains and highly dispersed Pd nanoparticles can be readily obtained using the interfacial liquid marbles as the microreactors. As shown in Figure 4e, few-layered rGO nanosheets with a planar and uniform morphology can be obtained using the interfacial liquid marbles. From the selected area electron diffraction (SAED) pattern in Figure 4f, the sharp spots corresponding to the $1\bar{1}00$-type and $2\bar{1}\bar{1}0$-type reflections of single-layer GO sheets can be readily observed with high contrast, indicating the existence of short-range order in the as-obtained rGO sheets over the length scale of a few nanometers [49]. On the other hand, the shallower ring patterns connecting the sharp spots indicate that the rGO sheets are stacking randomly with respect to each other. The XRD spectrum of the as-obtained rGO sheets shows four characteristic peaks at 2θ = 12.9°, 24.7°, 40.5°, and 60.2°, respectively (Figure 4g). The peak located at 2θ = 12.9° corresponds to the (001) reflection of the hexagonal lattice of GO, while the peak located at 2θ = 24.7° corresponds to the (002) reflection of the hexagonal lattice of rGO. Compared with ref. [50], the peak position of the (001) reflection of the as-obtained rGO shifts to higher degrees, while the peak position of the (002) reflection shifts to lower degrees, indicating a portion of GO is reduced to form rGO, accompanied by the shrinkage of the lattice structure. On the other hand, the broadening of the (002) reflection can be attributed to the large interlayer d-spacing of the rGO sheets and the intralayer micro-strains [51]. Moreover, the splitting of the (10) and (11) reflections of the graphene-like honeycomb lattice reveals the existence of a heterostructure, where oxidized and reduced domains with sizes of a few nanometers co-exist in the rGO sheets [52]. Well-dispersed Pd nanoparticles can also be obtained from the interfacial liquid marbles in ambient conditions without using any surfactants or external energies. As shown in Figure 4h, Pd nanoparticles dispersed on the surface of the spherical PVDF particles can be separated from the interfacial liquid marbles after an 18 h reduction process. The TEM image shown in Figure 4i reveals that both hexahedral and octahedral nanoparticles co-exist in the products of the reduction reaction, indicating inhomogeneous nucleation and growth processes of the nanoparticles as compared to the uniform Pd polyhedrons synthesized using capping agents [53]. The inhomogeneous growth of the Pd nanocrystals in the interfacial liquid marbles can possibly be attributed to the local hydrophobic interaction between the PVDF shell layer and Pd^0 nuclei, resulting in spontaneous, unregulated nucleation and inhomogeneous growth by Ostwald ripening. The energy dispersive X-ray spectrum (EDX) of the nanoparticles (Figure 4j) shows that they are solely composed of Pd atoms, despite the signals from PVDF and the copper grid (peaks at 8 and 9 keV). The as-obtained Pd nanoparticles are polycrystalline, as indicated by the ringed SAED pattern (Figure S5).

3. Materials and Methods

3.1. Materials

Polyvinylidene fluoride (Kynar® 761A, Arkema, Shanghai, China), tetradecane (99%, Rhawn™, Shanghai, China), n-hexane (99%, Acros Organics, Shanghai, China), iron (II) chloride tetrahydrate (99%, Sinopharm Chemical Reagent, Shanghai, China), palladium chloride (99.9%, Alfa Aesar, Shanghai, China), and ammonium persulfate (98%, Sigma Chemical, Shanghai, China). All chemicals are used as received.

3.2. Coalescence Cascade of Water Droplets at the Tetradecane/Water Interface

Before each measurement, a micropipette (Eppendorf, Shanghai, China, 2–20 μL) is calibrated to obtain a precision of ca. 0.1 μL. The water droplet is generated by using the calibrated micropipette, which is subsequently released and deposited onto the tetradecane/water interface gently from a height of ~3 mm. The tetradecane/water interface is constructed by placing 20 mL of tetradecane on top of 50 mL of D.I. water in a precleaned glass beaker. The CC process of the droplet is recorded by a high-speed camera with a frame rate of 4000 frames per second. The temperature of the surrounding environment is ~20 °C.

3.3. Reduction in Graphene Oxide and Palladium Chloride in Interfacial Liquid Marbles

Graphene oxide is prepared by using a modified Hummers' method [54]. The as-obtained graphene oxide powder is weighted and dispersed in D.I. water to make a 3 mg mL^{-1} dispersion. A total of 10 μL of the graphene oxide dispersion is fetched by using a calibrated micropipette, which is subsequently transferred to the surface of a PVDF powder layer, rolling gently on the layer until the droplet surface is fully covered by PVDF and no liquid residues are formed. The liquid marble is then loaded onto a stainless-steel lab spoon and transferred to the hexane/water interface by gently releasing it from near the surface of the liquid phase. The liquid marble would head towards the hexane/water interface under gravity while its surface became transparent due to the formation of a thin PVDF coating. Multiple liquid marbles can be cast simultaneously, which may spontaneously self-assemble into ordered patterns at the hexane/water interface, driven by the interfacial tension. Afterwards, 0.1 mL of hydrazine monohydrate is injected into the water reservoir to initiate the reduction reaction, which is allowed to proceed for 24 h at room temperature under static conditions. The reaction is terminated after 24 h, and the products inside the interfacial liquid marbles are extracted by using medical syringes for further characterization. The reduction process of $PdCl_2$ in the interfacial liquid marbles is identical to that of graphene oxide, except that (i) the dispersion is obtained by 1 h sonication and (ii) the reaction time is 18 h.

4. Conclusions

In conclusion, interfacial liquid marbles possess unique internal and external features, making them promising models and versatile tools for the investigation of different interfacial phenomena. The extremely long life expectancy of the interfacial liquid marbles originated from the unexpectedly long coalescence time of the liquid droplet at the liquid–liquid interface, revealing the existence of a third inertially limited viscous regime when $r_B \to 0$. The mass transport from the droplet to the underlying reservoir can be prohibited by coating the droplet with hydrophobic or amphiphilic particles. While the transport of liquid mass is prohibited, the interfacial liquid marbles are permeable to gases. Moreover, the oxygen density at the surface of the interfacial liquid marbles is much lower than that at the air–liquid interface, making the marbles a green and cost-effective solution for conducting reduction reactions. After all, the miniaturized, transparent, versatile, and long-lasting interfacial liquid marbles may find important roles in a wide range of applications, from fluid physics to microreactors to biomolecular condensates.

Supplementary Materials: The following supporting information can be downloaded at: https://www.mdpi.com/article/10.3390/molecules29132986/s1, Figure S1: Oxidation and evaporation of aqueous $FeCl_2$ droplets in the ambient conditions; Figure S2: Oxidation of the $FeCl_2$ interfacial liquid marbles at the deoxygenated tetradecane/water interface; Figure S3: Oxidation of the $FeCl_2$ interfacial liquid marbles at the pristine tetradecane/water interface; Figure S4: Photograph of the $FeCl_2$ interfacial liquid marble sitting at the tetradecane/water interface with the addition of ammonium persulfate; Figure S5: TEM image and SAED pattern of the as-synthesized Pd nanoparticles. Video S1: Coalescence cascade of a 15 μL water droplet at the tetradecane/water interface; Video S2: Coalescence of a 15 μL water droplet at the air/water interface; Video S3: Color change of a 30 μL $FeCl_2$ marble upon injecting 30 μL 1 M APS solution.

Author Contributions: Conceptualization, Y.L.; methodology, Y.L. and Y.W.; validation, Y.L. and Y.W.; formal analysis, Y.L.; resources, Y.L. and J.H.X.; writing—original draft preparation, Y.L.; writing—review and editing, Y.L. and Y.W.; supervision, J.H.X.; funding acquisition, Y.L. and J.H.X. All authors have read and agreed to the published version of the manuscript.

Funding: This research was funded by Science and Technology Innovation Commission of Shenzhen Municipality, grant number JCYJ20220530145603007.

Institutional Review Board Statement: Not applicable.

Informed Consent Statement: Not applicable.

Data Availability Statement: The original contributions presented in the study are included in the article/Supplementary Material, further inquiries can be directed to the corresponding author.

Acknowledgments: The authors gratefully acknowledge the aim from Zhishu Chen in conducting the mass-transport experiments.

Conflicts of Interest: The authors declare no conflicts of interest.

References

1. Hayashi, T.; Carthew, R.W. Surface mechanics mediate pattern formation in the developing retina. *Nature* **2004**, *431*, 647–652. [CrossRef] [PubMed]
2. Gleghorn, J.P.; Manivannan, S.; Nelson, C.M. Quantitative approaches to uncover physical mechanisms of tissue morphogenesis. *Curr. Opin. Biotechnol.* **2013**, *24*, 954–961. [CrossRef] [PubMed]
3. Valet, M.; Siggia, E.D.; Brivanlou, A.H. Mechanical regulation of early vertebrate embryogenesis. *Nat. Rev. Mol. Cell Biol.* **2022**, *23*, 169–184. [CrossRef]
4. Maître, J.L.; Heisenberg, C.P. Three functions of cadherins in cell adhesion. *Curr. Biol.* **2013**, *23*, R626–R633. [CrossRef] [PubMed]
5. Gouveia, B.; Kim, Y.; Shaevitz, J.W.; Petry, S.; Stone, H.A.; Brangwynne, C. Capillary force generated by biomolecular condensates. *Nature* **2022**, *609*, 255–264. [CrossRef]
6. Kooij, S.; Astefanei, A.; Corthals, G.L.; Bonn, D. Size distributions of droplets produced by ultrasonic nebulizers. *Sci. Rep.* **2019**, *9*, 6128. [CrossRef]
7. Castrejón-Pita, J.R.; Castrejón-Pita, A.A.; Thete, S.S.; Basaran, O.A. Plethora of transitions during breakup of liquid filaments. *Proc. Natl. Acad. Sci. USA* **2015**, *112*, 4582–4587. [CrossRef]
8. Ersoy, N.E.; Eslamian, M. Capillary surface wave formation and mixing of miscible liquids during droplet impact onto a liquid film. *Phys. Fluids* **2019**, *31*, 012107. [CrossRef]
9. Ozawa, K.; Nakamura, H.; Shimamura, K.; Dietze, G.F.; Yoshikawa, H.N.; Zoueshtiagh, F.; Kurose, K.; Mu, L.; Ueno, I. Capillary-driven horseshoe vortex forming around a micro-pillar. *J. Colloid Interface Sci.* **2023**, *642*, 227–234. [CrossRef]
10. Quintin, S.; Gally, C.; Labouesse, M. Epithelial morphogenesis in embryos: Asymmetries, motors, and brakes. *Trends Genet.* **2008**, *24*, 221–230. [CrossRef]
11. Porazinski, S.; Wang, H.; Asaoka, Y.; Behrndt, M.; Miyamoto, T.; Morita, H.; Hata, S.; Sasaki, T.; Krens, S.F.G.; Osada, Y.; et al. YAP is essential for tissue tension to ensure vertebrate 3D body shape. *Nature* **2015**, *521*, 217–221. [CrossRef] [PubMed]
12. Dumortier, J.G.; Verge-Serandour, M.L.; Tortorelli, A.F.; Mielke, A.; Plater, L.D.; Turlier, H.; Maître, J. Hydraulic fracturing and active coarsening position the lumen of the mouse blastocyst. *Science* **2019**, *365*, 465–468. [CrossRef] [PubMed]
13. Mongera, A.; Rowghanian, P.; Gustafson, H.J.; Shelton, E.; Kealhofer, D.A.; Carn, E.K.; Serwane, F.; Lucio, A.A.; Giammona, J.; Campás, O. A fluid-to-solid jamming transition underlies vertebrate body axis elongation. *Nature* **2018**, *561*, 401–405. [CrossRef] [PubMed]
14. Petridou, N.I.; Corominas-Murtra, B.; Heisenberg, C.; Hannezo, E. Rigidity percolation uncovers a structural basis for embryonic tissue phase transitions. *Cell* **2021**, *184*, 1914–1928. [CrossRef] [PubMed]
15. Feric, M.; Vaidya, N.; Harmon, T.S.; Mitrea, D.M.; Zhu, L.; Richardson, T.M.; Kriwacki, R.W.; Pappu, R.V.; Brangwynne, C.P. Coexisting liquid phases underlie nucleolar subcompartments. *Cell* **2016**, *165*, 1686–1697. [CrossRef]
16. Caragine, C.M.; Haley, S.C.; Zidovska, A. Surface fluctuations and coalescence of nucleolar droplets in the human cell nucleus. *Phys. Rev. Lett.* **2018**, *121*, 148101. [CrossRef]
17. Pallas, N.R.; Pethica, B.A. The surface tension of water. *Colloids Surf.* **1983**, *6*, 221–227. [CrossRef]
18. Shin, Y.; Chang, Y.; Lee, D.S.W.; Berry, J.; Sanders, D.W.; Ronceray, P.; Wingreen, N.S.; Haataja, M.; Brangwynne, C.P. Liquid nuclear condensates mechanically sense and restructure the genome. *Cell* **2018**, *175*, 1481–1491. [CrossRef]
19. Weirich, K.L.; Banerjee, S.; Dasbiswas, K.; Witten, T.A.; Vaikuntanathan, S.; Gardel, M.L. Liquid behavior of cross-linked actin bundles. *Proc. Natl. Acad. Sci. USA* **2017**, *114*, 2131–2136. [CrossRef] [PubMed]

20. Brangwynne, C.P.; Mitchison, T.J.; Hyman, A.A. Active liquid-like behavior of nucleoli determines their size and shape in Xenopus laevis oocytes. *Proc. Natl. Acad. USA* **2011**, *108*, 4334–4339. [CrossRef]
21. Cuylen, S.; Blaukopf, C.; Politi, A.Z.; Müller-Reichert, T.; Neumann, B.; Poser, I.; Ellenberg, J.; Hyman, A.A.; Gerlich, D.W. Ki-67 act as a biological surfactant to disperse mitotic chromosomes. *Nature* **2016**, *535*, 308–312. [CrossRef] [PubMed]
22. Liu, Y.; Zhang, X.; Poyraz, S.; Zhang, C.; Xin, J.H. One-step synthesis of multifunctional zinc-iron-oxide hybrid carbon nanowires by chemical fusion for supercapacitors and interfacial water marbles. *ChemNanoMat* **2018**, *4*, 546–556. [CrossRef]
23. Shim, S.; Stone, H.A. Damped coalescence cascade of liquid drops. *Phys. Rev. Fluid.* **2017**, *2*, 044001. [CrossRef]
24. Alhareth, A.A.; Thoroddsen, S.T. Partial coalescence of a drop on a large-viscosity pool. *Phys. Fluids* **2020**, *32*, 122115. [CrossRef]
25. Kirar, P.K.; Kolhe, P.S.; Sahu, K.C. Coalescence and migration of a droplet on a liquid pool with an inclined bottom wall. *Phys. Rev. Fluid.* **2022**, *7*, 094001. [CrossRef]
26. Blanchette, F.; Messio, L.; Bush, J.W.M. The influence of surface tension gradients on drop coalescence. *Phys. Fluids* **2009**, *21*, 072107. [CrossRef]
27. Charles, G.E.; Mason, S.G. The mechanism of partial coalescence of liquid drops at liquid/liquid interfaces. *J. Colloid Sci.* **1960**, *15*, 105–122. [CrossRef]
28. Tomotika, S. On the instability of a cylindrical thread of a viscous liquid surrounded by another viscous fluid. *Proc. Roy. Soc. A* **1935**, *150*, 322–337.
29. Rayleigh, F.R.S. On the instability of jets. *Proc. London Math. Soc.* **1878**, *s1-10*, 4–13. [CrossRef]
30. Thoroddsen, S.T.; Takehara, K. The coalescence cascade of a drop. *Phys. Fluids* **2000**, *12*, 1265–1267. [CrossRef]
31. Blanchette, F.; Bigioni, T.P. Partial coalescence of drops at liquid interfaces. *Nat. Phys.* **2006**, *2*, 254–257. [CrossRef]
32. Chen, X.; Mandre, S.; Feng, J.J. Partial coalescence between a drop and a liquid-liquid interface. *Phys. Fluids* **2006**, *18*, 051705. [CrossRef]
33. Eggers, J.; Lister, J.R.; Stone, H.A. Coalescence of liquid drops. *J. Fluid Mech.* **1999**, *401*, 293–310. [CrossRef]
34. Ruiter, J.; Mugele, F.; Ende, D. Air cushioning in droplet impact. I. Dynamics of thin films studied by dual wavelength reflection interference microscopy. *Phys. Fluids* **2015**, *27*, 012104. [CrossRef]
35. Ren, H.; Xu, S.; Wu, S. Effects of gravity on the shape of liquid droplets. *Opt. Commun.* **2010**, *283*, 3255–3258. [CrossRef]
36. Paulsen, J.D.; Burton, J.C.; Nagel, S.R.; Appathurai, S.; Harris, M.T.; Basaran, O. The inexorable resistance of inertia determines the initial regime of drop coalescence. *Proc. Natl. Acad. Sci. USA* **2012**, *109*, 6857–6861. [CrossRef] [PubMed]
37. Paulsen, J.D. Approach and coalescence of liquid drops in air. *Phys. Rev. E* **2013**, *88*, 063010. [CrossRef] [PubMed]
38. Huang, C.; Wikfeldt, K.T.; Tokushima, T.; Nordlund, D.; Harada, Y.; Bergmann, U.; Niebuhr, M.; Weiss, T.M.; Horikawa, Y.; Leetmaa, M.; et al. The inhomogeneous structure of water at ambient conditions. *Proc. Natl. Acad. Sci. USA* **2009**, *106*, 15214–15218. [CrossRef]
39. Cuddeback, R.B.; Koeller, R.C.; Drickamer, H.G. The effect of pressure on diffusion in water and in sulfate solutions. *J. Chem. Phys.* **1953**, *21*, 589–597. [CrossRef]
40. Lin, K.; Chen, R.; Zhang, L.; Zang, D.; Geng, X.; Shen, W. Transparent bioreactors based on nanoparticle-coated liquid marbles for in situ observation of suspending embryonic body formation and differentiation. *ACS Appl. Mater. Interfaces* **2019**, *11*, 8789–8796. [CrossRef]
41. Sheng, Y.; Sun, G.; Wu, J.; Ma, G.; Ngai, T. Silica-based liquid marbles as microreactors for the silver mirror reaction. *Angew. Chem.* **2015**, *127*, 7118–7123. [CrossRef]
42. Han, X.; Lee, H.K.; Lee, Y.H.; Hao, W.; Liu, Y.; Phang, I.Y.; Li, S.; Ling, X.Y. Identifying enclosed chemical reaction and dynamics at the molecular level using shell-isolated miniaturized plasmonic liquid marble. *J. Phys. Chem. Lett.* **2016**, *7*, 1501–1506. [CrossRef] [PubMed]
43. Tian, J.; Fu, N.; Chen, X.D.; Shen, W. Respirable liquid marble for the cultivation of microorganisms. *Colloids Surf. B* **2013**, *106*, 187–190. [CrossRef]
44. Ng, L.S.; Chong, C.; Lok, X.Y.; Pereira, V.; Ang, Z.Z.; Han, X.; Li, H.; Lee, H.K. Dynamic liquid-liquid interface: Applying a spinning interfacial microreactor to actively converge biphasic reactants for the interfacial reaction. *ACS Appl. Mater. Interfaces* **2022**, *14*, 45005–45012. [CrossRef] [PubMed]
45. Sato, E.; Yuri, M.; Fujii, S.; Nishiyama, T.; Nakamura, Y.; Horibe, H. Liquid marbles as a micro-reactor for efficient radical alternating copolymerization of diene monomer and oxygen. *Chem. Commun.* **2015**, *51*, 17241. [CrossRef] [PubMed]
46. Tsumura, Y.; Oyama, K.; Fameau, A.; Seike, M.; Ohtaka, A.; Hirai, T.; Nakamura, Y.; Fujii, S. Photo/thermo dual stimulus-responsive liquid marbles stabilized with polypyrrole-coated stearic acid particles. *ACS Appl. Mater. Interfaces* **2022**, *14*, 41618–41628. [CrossRef]
47. Liu, Z.; Yang, T.; Huang, Y.; Liu, Y.; Chen, L.; Deng, L.; Shum, H.C.; Kong, T. Electrocontrolled liquid marbles for rapid miniaturized organic reactions. *Adv. Funct. Mater.* **2019**, *29*, 1901101. [CrossRef]
48. Singh, S.K.; Xu, Q. Complete conversion of hydrous hydrazine to hydrogen at room temperature for chemical hydrogen storage. *J. Am. Chem. Soc.* **2009**, *131*, 18032–18033. [CrossRef]
49. Wilson, N.R.; Pandey, P.A.; Beanland, R.; Young, R.J.; Kinloch, I.A.; Gong, L.; Liu, Z.; Suenaga, K.; Rourke, J.P.; York, S.J.; et al. Graphene oxide: Structural analysis and application as a highly transparent support for electron microscopy. *ACS Nano* **2009**, *3*, 2547–2556. [CrossRef] [PubMed]

50. Al-Gaashani, R.; Najjar, A.; Zakaria, Y.; Mansour, S.; Atieh, M.A. XPS and structural studies of high quality graphene oxide and reduced graphene oxide prepared by different chemical oxidation methods. *Ceram. Int.* **2019**, *45*, 14439–14448. [CrossRef]
51. Taniguchi, T.; Nurdiwijayanto, L.; Sakai, N.; Tsukagoshi, K.; Sasaki, T.; Tsugawa, T.; Koinuma, M.; Hatakeyama, K.; Ida, S. Revisiting the two-dimensional structure and reduction process of graphene oxide with in-plane X-ray diffraction. *Carbon* **2023**, *202*, 26–35. [CrossRef]
52. Mouhat, F.; Coudert, F.; Bocquet, M. Structure and chemistry of graphene oxide in liquid water from first principles. *Nat. Commun.* **2020**, *11*, 1566. [CrossRef] [PubMed]
53. Jin, M.; Zhang, H.; Xie, Z.; Xia, Y. Palladium nanocrystals enclosed by {100} and {111} facets in controlled proportions and their catalytic activities for formic acid oxidation. *Energy Environ. Sci.* **2012**, *5*, 6352. [CrossRef]
54. Liu, Y.; Wang, X.; Fei, B.; Hu, H.; Lai, C.; Xin, J.H. Bioinspired, stimuli-responsive, multifunctional superhydrophobic surface with directional wetting, adhesion, and transport of water. *Adv. Funct. Mater.* **2015**, *25*, 5047–5056. [CrossRef]

molecules

MDPI

Article

Aggregation Behavior and Application Properties of Novel Glycosylamide Quaternary Ammonium Salts in Aqueous Solution

Yunkai Wang [1], Zeyu Chen [1], Erzhuang Zhang [1], Lifei Zhi [1,*], Martino Di Serio [2], Guoyong Wang [1], Yan Wang [1], Xiaoming Li [1], Xudong Liu [3] and Ying Huang [4,5]

1 College of Chemical Engineering and Technology, Taiyuan University of Science and Technology, Taiyuan 030024, China; wyk990306@163.com (Y.W.); czy15535371231@163.com (Z.C.); aa1162791419@163.com (E.Z.); wanggy@tyust.edu.cn (G.W.); kdwyan@126.com (Y.W.); lxmlily2014@126.com (X.L.)

2 Department of Chemical Sciences, University of Naples Federico II, 80138 Napoli, Italy; martion.diserion@gmail.com

3 Research Institute of Livestock and Aquatic Product Inspection, Shanxi Inspection and Testing Center, Taiyuan 030006, China; 13835179079@163.com

4 Taiyuan Hengdeyuan Animal Protection Technology Development Co., Ltd., Taiyuan 030003, China; huangying0351@163.com

5 Shanxi Livestock and Poultry Breeding Co., Ltd., Jinzhong 031800, China

* Correspondence: lifeizhi@yeah.net or lifeizhi@tyust.edu.cn

Abstract: Amidation of lactobionic acid with N,N-dimethylaminopropyltriamine was conducted to obtain N-(3′-dimethylaminopropyl)-lactamido-3-aminopropane (DDLPD), which was quaternized with bromoalkanes of different carbon chain lengths to synthesize double-stranded lactosylamide quaternary ammonium salt N-[N′[3-(lactosylamide)]propyl-N′-alkyl] propyl-N,N-dimethyl-N-alkylammonium bromide (C_nDDLPB, n = 8, 10, 12, 14, 16). The surface activity and the adsorption and aggregation behaviors of the surfactants were investigated via equilibrium surface tension, dynamic light scattering, and cryo-electron microscopy measurements in an aqueous solution. The application properties of the products in terms of wettability, emulsification, foam properties, antistatic, salt resistance, and bacteriostatic properties were tested. C_nDDLPB exhibited a low equilibrium surface tension of 27.82 mN/m. With an increase in the carbon chain length, the critical micellar concentration of C_nDDLPBD decreased. Cryo-electron microscopy revealed that all products except C_8DDLPB formed stable monolayer, multi-layer, and multi-compartmental vesicle structures in an aqueous solution. C_{14}DDLPB has the best emulsification performance on soybean oil, with a time of 16.6 min; C_{14}DDLPB has good wetting and spreading properties on polytetrafluoroethylene (PTFE) when the length of carbon chain is from 8 to 14, and the contact angle can be lowered to 33°~40°, C_nDDLPB has low foam, which is typical of low-foaming products; C_8DDLPB and C_{10}DDLPB both show good antistatic properties. C_8DDLPB and C_{14}DDLPB have good salt resistance, and C_{12}DDLPB has the best antimicrobial property, with the inhibition rate of 99.29% and 95.28% for E. coli and Gluconococcus aureus, respectively, at a concentration of 350 ppm.

Keywords: surfactant; lactose amide; quaternary ammonium salts; surface activity; vesicle

1. Introduction

Among global environmental issues, carbon emissions have become a key issue that has received a lot of attention [1,2]. Of particular note is China's high level of carbon emissions between 2009 and 2022 [3]. To address this challenge, China made a solemn commitment at the 75th session of the United Nations General Assembly to adopt dual carbon targets, i.e., carbon peaking and carbon neutrality, as a part of its core carbon reduction strategy. Owing to their green, environmentally friendly, and resource-saving characteristics, biobased chemicals are in alignment with the "dual-carbon" policy development; these chemicals are gradually becoming a new leading industry in the contemporary world of scientific and technological innovation and economic development. China's "13th Five-Year Plan" and "14th Five-Year Plan" also clearly reveal that focus will be placed on the development of biobased industries [4]. Biobased chemicals, which are derived from renewable resources, are bulk chemicals and fine chemicals prepared from biomass as raw materials, exhibiting several advantages such as carbon reduction and sustainability [5–7]. Biobased chemicals produce very little waste during their production process, and most of this waste can be recycled due to their green and healthy characteristics. Their easily degradable nature also provides an effective way to solve the pollution problem of petroleum-based plastics [8,9]. Sugars are typical representatives of biomass raw materials and prepared from natural renewable resources such as starch, exhibiting not only abundant sources but also cost-effectiveness [10]. The use of sugar groups as hydrophilic groups in surfactants endows them with several advantages: (1) Sugar groups endow surfactants with considerably good biodegradation performance and toxicological properties. Moreover, they are mild and non-irritating to the skin and eyes as well as easily biodegradable under anaerobic and aerobic conditions [11–13]. (2) Sugar groups contain multiple hydroxyl groups, the oleophobicity and hydrophilicity of which is greater than those of conventional polyoxyethylene ether surfactants; the oleophobicity of a sec-hydroxyl group is 4.5 times greater than that of polyethylene oxide. In oil–water systems, sugar-based surfactants exhibit better interfacial chemistry [14]. (3) Sugar groups exhibit higher resistance to hard water because they can form complexes with metal ions such as Ca^{2+} to form water-soluble complex ions. (4) Sugar groups also endow surfactants with insecticidal and herbicidal or antimicrobial activity [15–17]. Introducing sugar groups into the molecular structure of cationic surfactants may improve their irritation, toxicity, biodegradability, water solubility, compatibility, etc. [18]. Various cationic surfactants derived from sugar groups exhibit advantages, including green characteristics, natural and renewable raw materials, facile biodegradability, and multifunctionality; these surfactants are also safe and mild for human body. Therefore, these surfactants have globally attracted considerable research attention [19–21].

Herein, a double-chain lactide amide quaternary surfactant (C_nDDLPB), which combines the advantages of a sugar group, an amide bond, and a cationic surfactant, was synthesized via the amine ester reaction of lactobionic acid with an alkyl amine and then quaternized with bromoalkanes (the reaction and synthesis routes are shown in Figure 1), and the structures of the intermediates and products were characterized via Fourier transform infrared (IR) spectroscopy, nuclear magnetic resonance (NMR) spectroscopy, and mass spectrometry(MS). In addition, aggregation behavior and application properties of C_nDDLPB were also comprehensively investigated.

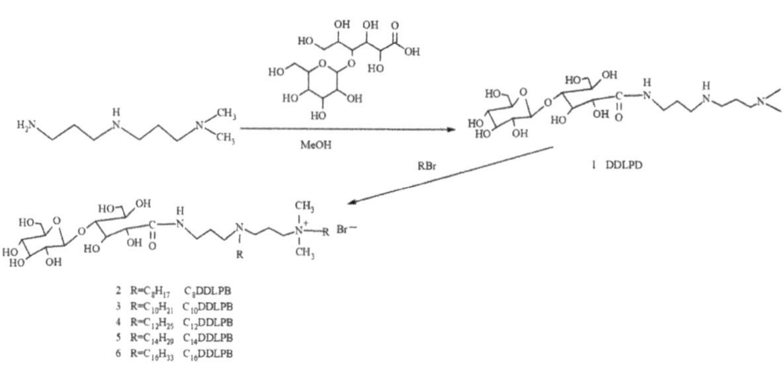

Figure 1. Synthesis roadmap of double-chain lactosamine quaternary ammonium salts (C_nDDLPB).

2. Results and Discussion

2.1. Structure Identification

The chemical structures of the raw materials, synthesized intermediates, and target compounds were characterized using FTIR, ^{1}H NMR, and ^{13}C NMR spectroscopies as well as other characterization methods. The spectra of all products are in the Supporting Information.

Figure S1a shows the FTIR spectrum of N,N-dimethyldipropylenetriamine. The peak observed at 2931 cm^{-1} corresponds to the symmetrical telescopic vibration of C–H of the methyl group, while the peak observed at 2762 cm^{-1} corresponds to the symmetrical telescopic vibration of C–H of the methylene group [22]. Figure S1b shows the FTIR spectrum of lactose. A broad and strong absorption peak observed at 3400–3200 cm^{-1} corresponds to the stretching vibration of the O–H group in lactobionic acid, the absorption peak observed at 1730 cm^{-1} corresponds to the stretching vibration of the C=O bond in lactobionic acid, and the strong peak observed at 1030 cm^{-1} corresponds to the C–O bond stretching vibration.

In the spectrum of DDLPD in Figure S1c, absorption peaks corresponding to the O–H group in lactobionic acid and C–H of N, N-dimethyldipropenyltriamine are observed; the two strong characteristic absorption peaks observed at 1650 and 1540 cm^{-1} correspond to the telescopic vibration of the C=O bond in amide groups and the bending vibration of N–H, respectively, indicating that amide bonds are generated.

2.2. Electrospray Mass Spectrometry (EMS)

The synthesized glycosylamide quaternary ammonium salt was analyzed via ESI-MS, and its composition was analyzed and identified according to its molecular weight: the relative molecular mass of the expected product C_{12}DDLPB is 916. Figure S16 shows the ESI-MS spectrum of the target product C_{12}DDLPB. The quaternary ammonium surfactant C_{12}DDLPB exhibits an $(M–H)^+$ quasimolecular ion peak at 915.35 (m/z). The quasimolecular ion is relatively stable after losing one electron; Br^- is bombarded using the positive-ion mode, and a strong $[M–Br]^+$ fragmentation peak is observed at m/z 836.86. The fragmentation peak of $[M\text{-}Br\text{-} OH\text{-}OH]^+$ is observed at m/z 746.88, the fragmentation peak of $[M\text{-}Br\text{-}CH_3O\text{-}C_2H_5O_2\text{-}O\text{-}O]^+$ is observed at m/z 612.55, and the fragmentation peak of $[M\text{-}Br\text{-}CH_3O\text{-}C_2H_5O_2\text{-}OH\text{-}OH\ \text{-}H\text{-}H]^+$ is observed at m/z 608.85. The m/z values of the molecular ion peaks of C_8DDLPB, C_{10}DDLPB, C_{14}DDLPB, and C_{16}DDLPB are 803.49, 862.51, 974.38, and 1029.45, respectively, which are consistent with the expected molecular masses of the products (Supplementary Material Figures S14–S18) [23].

The combined FTIR spectroscopy, ^{1}H NMR spectroscopy, ^{13}CNMR spectroscopy, and ESI-MS results reveal that the double-stranded lactose amide quaternary surfactants are successfully synthesized.

2.3. Surface Tension

The surface tension of C_nDDLPB was determined using the hanging drop method at 25 °C ± 0.1 °C. The surfactant solution was prepared from RO water; the surface tension of RO water was 72 ± 0.3 mN/m. The surface tension versus concentration curve reveals an inflection point in the concentration, corresponding to the critical micellar concentration (CMC). Before the turning point is reached (Figure 2), a linear relationship between the increase in the surfactant concentration and the decrease in surface tension is observed, indicating that the surfactant molecules begin to arrange closely at the interface. After reaching the CMC, the surface tension remains constant.

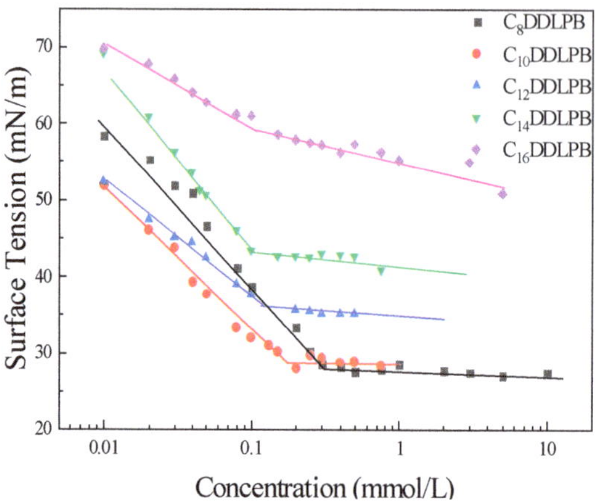

Figure 2. Surface tension of C_nDDLPB in an aqueous solution as a function of concentration.

According to Equations (1)–(3) [24,25], the amount of adsorption at saturation at the gas/liquid interface (Γ_{max}), cross-sectional area per molecule A_{min}, and surface activity efficiency pC_{20} can be estimated:

$$\Gamma_{max} = -\frac{1}{2.303nRT}\left(\frac{\partial\gamma}{\partial lgc}\right)_T \quad (1)$$

$$A_{min} = \frac{10^{16}}{N_A\Gamma_{max}} \quad (2)$$

$$pC_{20} = -logC_{20} \quad (3)$$

where γ represents the surface tension, mN/m; T represents the absolute temperature, K; and R represents the gas constant, 8.314 J/(mol·K). Moreover, the value of n for an ionic surfactant type 1-1 is considered to be 2, N_A represents the Avogadro constant, and C_{20} represents the concentration of the surfactant required to reduce the surface tension of water by 20 mN/m.

CMC/C_{20} indicates the difficulty and simplicity of the surfactant adsorption and micellization process. The higher CMC/C_{20}, the easier the tendency for the surfactant to adsorb on the interface than that to form a micelle.

The CMC values of the five double-chain lactose amide quaternary ammonium salts (C_nDDLPB) decrease in the order of C_8DDLPB > $C_{10}DDLPB$ > $C_{12}DDLPB$ > $C_{14}DDLPB$ > $C_{16}DDLPB$ (Figure 3). Their CMC values decrease linearly with the growth of the hydrophobic carbon chain [26]. This result is attributed to the fact that, when two carbon chains of the surfactant molecule become longer, the space occupied by the molecules arranged at the interface is saturated more rapidly. The larger the space occupied by the two carbon chains of the surfactant molecules, the faster the molecules arranged at the interface reach saturation; at the same time, the longer the carbon chains, the hydrophobicity of the hydrocarbon bonds is gradually enhanced, indicating that the separation formed between the hydrophilic head groups under the action of electrostatic repulsive force is hindered, rendering a significant promotion effect on the development of aggregates in the surfactants. The shorter the carbon chain, the lower the surface tension. The surface tension of the C_8DDLPB aqueous solution can be reduced to 27.82 mN/m. The inflection point of CMC on the curve of the change in the surface tension with the $C_{16}DDLPB$ concentration is not extremely clear, which may be attributed to the fact that the two hydrophobic carbon chains are extremely long and lead to poor solubility and a low surface activity.

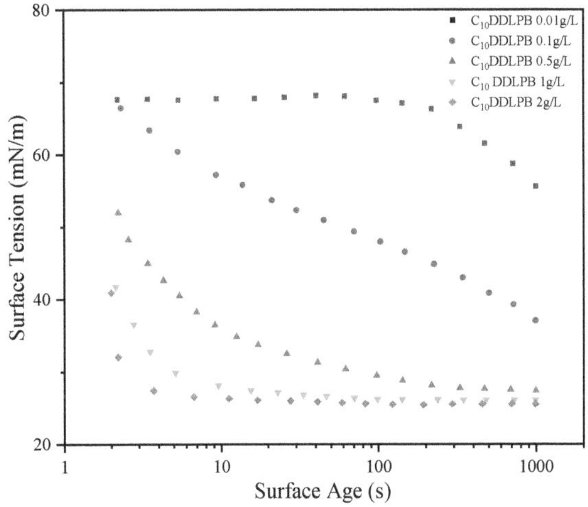

Figure 3. Dynamic surface tensions vs. surface age for 0.01 g/L–2 g/L C_nDDDLPB.

With an increase in the length of the carbon chain, Γ_{max} decreases (Table 1). When the carbon chain becomes longer, the obstruction of the molecules to occupy a position in space increases. The possibility of the dense arrangement of molecules within the unit surface decreases, and the cross-sectional area Amin per molecule increases. The pC_{20} value of C_{10}DDLPB is the highest, indicative of the higher adsorption efficiency at the interface.

Table 1. Surface adsorption and aggregation parameters of C_nDDLPB in aqueous solution.

Surfactant	CMC (mmol/L)	Γ_{cmc} (mN/m)	pC_{20}	CMC/C_{20}	Γ_{max} ($mol \cdot cm^{-2}$)	A_{min} ($Å^2$)	$\Delta G^0{}_{mic}$ ($KJ \cdot mol^{-1}$)	$\Delta G^0{}_{ads}$ ($KJ \cdot mol^{-1}$)
C_8DDLPB	0.333	27.82	4.585	12.808	1.890×10^{-10}	87.857	−59.96	−64.85
C_{10}DDLPB	0.136	30.27	5.046	15.111	1.675×10^{-10}	99.15	−62.75	−67.05
C_{12}DDLPB	0.12	36.11	4.971	11.215	1.275×10^{-10}	130.266	−64.66	−67.74
C_{14}DDLPB	0.106	43.1	4.343	2.335	1.198×10^{-10}	138.603	−65.33	−67.12
C_{16}DDLPB	0.103	59.09	2.356	0.023	8.365×10^{-11}	198.509	−65.28	−66.81

2.4. Dynamic Surface Tension

Dynamic surface tension can be used to study the adsorption–diffusion kinetics of surfactants at the air-water interface. In this study, the dynamic surface tension changes in C_{10}DDLPB were measured in the concentration range of 0.01 g/L–2 g/L by the suspended droplet method, and the dynamic surface tension of C_nDDLPB with five different carbon chain lengths at a concentration of 1 g/L (25 °C) was measured in detail.

Figure 3 demonstrates the dynamic surface tension curves of C_{10}DDLPB at different concentrations, from which it can be seen that the surface activity increases with increasing concentration, and the higher the concentration, the faster the adsorption rate.

Figure 4 demonstrates the surface tension versus surface age plots of C_nDDLPB with different carbon chain lengths at 1 g/L concentration. With the increase in surface age, the surface tension of each surfactant gradually decreased and finally reached a stable value. This indicates that the C_nDDLPB molecules undergo the process of diffusion–adsorption, reaching a steady state at the liquid surface. The ability to reduce the surface tension has a tendency to increase with the decrease in carbon chain length, which may be related to the fact that the shorter carbon chain molecules are easier to be arranged and form a compact structure at the interface. However, it is noteworthy that the adsorption rates of the surfactants increased with decreasing carbon chain length, except for C_8DDLPB, for which this anomaly may be attributed to the shorter hydrophobic chain.

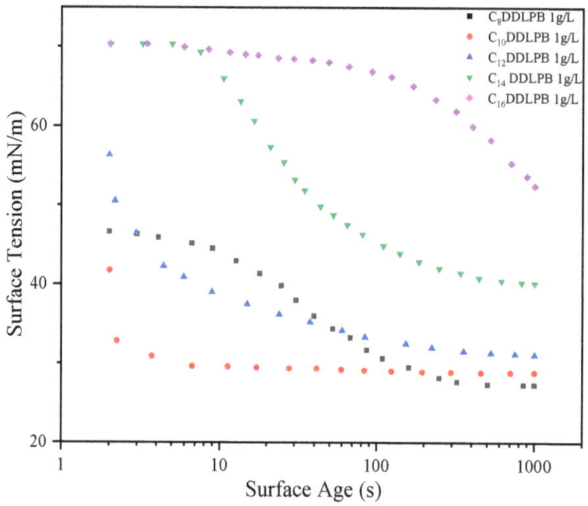

Figure 4. Dynamic surface tensions vs. surface age for 1 g/L C_nDDDLPB.

2.5. Aggregation Behavior in Aqueous Solutions

The aggregation behavior of C_nDDLPB at a concentration of 5 g/L in aqueous solution was observed using cryo-transmission electron microscopy (Cryo-TEM) at a magnification of 92,000 times, as shown in Figure 5. Except for C_8DDLPB, which forms ordinary micelles with a diameter of approximately 20 nm in an aqueous solution, the other samples form single-layered, multi-layered, or multi-compartmental vesicle structures, with vesicle diameters ranging from approximately 100 to 400 nm. Among them, C_{10}DDLPB and C_{12}DDLPB formed multi-compartmental vesicles, and C_{14}DDLPB and C_{16}DDLPB formed single-layer vesicle structures. Figure 5b,c and Supporting Information Figure S19b,c show the cryogenic transmission electron microscopy images of C_{10}DDLPB samples kept for 1 month and 24 h, respectively. The magnification of the cryo-electron microscope in the Supporting Information is 45,000×. With an increase in the time of placing the samples, more vesicles are formed, indicating that the vesicle-forming system is more stable with an increase in the sample storage time and that more vesicles are aggregated to form a multi-compartmental vesicle structure. The results reveal that the vesicle-forming system is more stable with an increase in the sample storage time.

The appearance of the prepared solutions also reveals that C_8DDLPB is a translucent light gray solution and that C_{10}DDLPB-C_{16}DDLPB is a light blue solution. During vesicle formation, surfactant molecules aggregate at the water–oil interface and form a closed structure with hydrophobic chains inside and hydrophilic chains outside. The length of the hydrophobic chains is critical to the stability of the vesicles. Surfactant molecular structures with two hydrocarbon chains and large head groups are prone to spontaneous vesicle formation [27–29]. The molecular structure of the surfactant synthesized herein (C_nDDLPB) is in agreement with such features. However, in case of a short-carbon chain, sufficient hydrophobic forces cannot be provided to maintain the stability of vesicles, resulting in the facile rupture of vesicles. In addition, surfactant molecules with a short-carbon chains exhibit a high degree of expansion in an aqueous solution, indicating that the formation of tight aggregation structures by these short-carbon chains is difficult and that water molecules cannot be wrapped effectively to form complete vesicle structures.

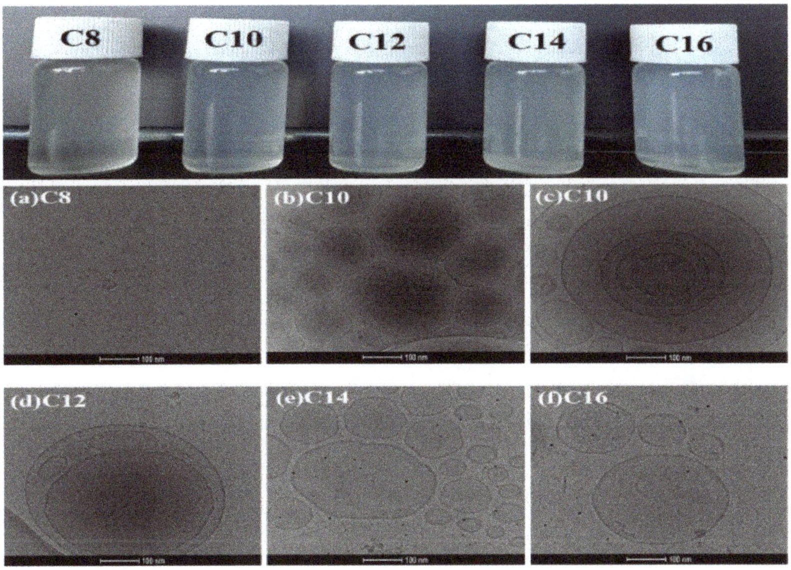

Figure 5. Aggregation behavior of C_nDDLPB under cryo-electron microscopy.

The mechanism of C_nDDLPB solution vesicle formation is shown in Figure 6: (1) First, C_nDDLPB surfactant molecules are dissolved in water, and the hydrophilic head interacts with water molecules and is surrounded by them, which are dispersed in solution as monomers. (2) Formation of micelles: at the water/oil interface or water/air interface, the hydrophobic tails of the surfactant molecules interact with the oil or air phase, leading to the adsorption of the molecules on the interface. Formation of a stable interface: when the solution concentration reaches the CMC, the active molecules form monolayer micelles [30]. (3) Formation of flexible bilayers: With an increase in the concentration, the micelles begin to disperse and reorganize into flexible bilayer structures. With an increase in the size of the molecular films, they begin to inwardly bend spontaneously to form curved structures to reduce edge energy; this closure can be facilitated by hydrophobic forces between the hydrophobic tails [31,32]. (4) Unstable vesicles: when the bilayer is completely closed, small unstable vesicles are formed [33]. (5) Formation of stable large vesicles: Over time, multiple small vesicles merge with each other to form a stable large vesicle [34,35]. The ultimate stability of a vesicle depends on the balance between hydrophobic effects and surface tension. The stronger the interaction force of the hydrophobic tail, the higher the stability of the vesicle. At the same time, the film formed by surfactant molecules at the interface can also reduce the surface energy of the system and make the vesicle structure more stable. C_{10}DDLPB and C_{12}DDLPB exhibit the best balance of the hydrophobic effect and surface tension, which are more likely to aggregate on the membrane surface and induce the neighboring layers to come closer together, leading to the gradual stacking of the membrane layers to form multi-layer and multi-compartmental structures.

Most cationic surfactants cannot spontaneously form vesicles. Therefore, C_nDDLPB with a long-carbon chain demonstrates good application prospects in slow-release drug carriers, template agents, biofilm mimicry, microreactors, and the cosmetic and food industries.

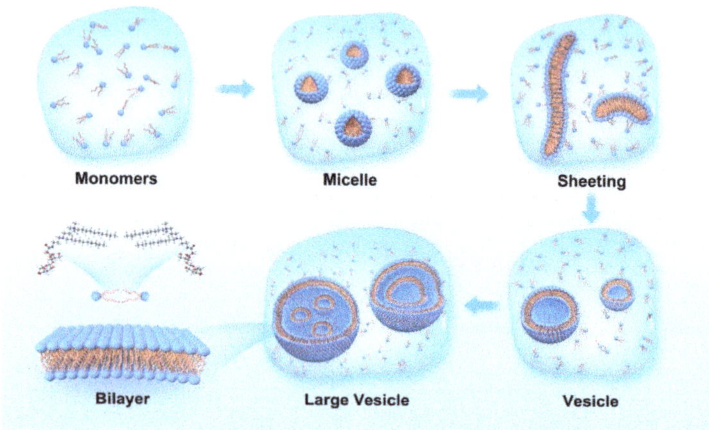

Figure 6. Schematic diagram of the vesicle formation of C_nDDLPB.

Figure 7 shows the particle size distribution of the 5 g/L C_nDDLPB sample. C_nDDLPB exhibits a single-peak particle distribution, and the overall particle sizes range from 20 nm to 1000 nm. The particle sizes of $C_{16}DDLPB$ and $C_{10}DDLPB$ exhibit a narrower distribution between 50 nm and 400 nm (Figure 7a,d), indicating that the diameters of the formed aggregates are mainly concentrated in this range. The particle size distribution of the $C_{14}DDLPB$ solution becomes narrower between 50 nm and 380 nm (Figure 7b), indicating that the distribution is more concentrated. The particle size distribution shown in Figure 7c is between 50 nm and 700 nm, which is a wider particle size distribution than those of other samples, indicating that the size of the formed aggregates is larger. As shown in Figure 7e, the particle size of this solution is mainly distributed between 20 nm and 80 nm, indicating that larger aggregates are not formed, which is consistent with the TEM results, and the system only forms micelles without vesicle formation.

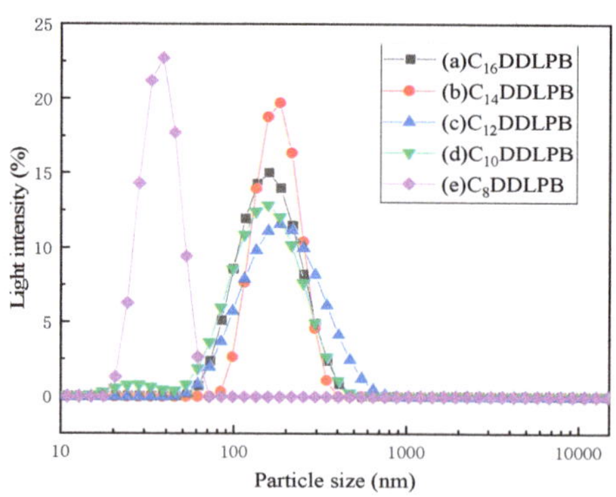

Figure 7. Particle size distributions of (**a**) $C_{16}DDLPB$, (**b**) $C_{14}DDLPB$, (**c**) $C_{12}DDLPB$, (**d**) $C_{10}DDLPB$, and (**e**) C_8DDLPB.

2.6. Wetting Ability

The changes in contact angles between different concentrations of C_nDDLPB and PTFE over time are illustrated in Figure 8. It is evident from the graph that the contact angles of C_nDDLPB on the surface of PTFE decrease with the increasing concentration.

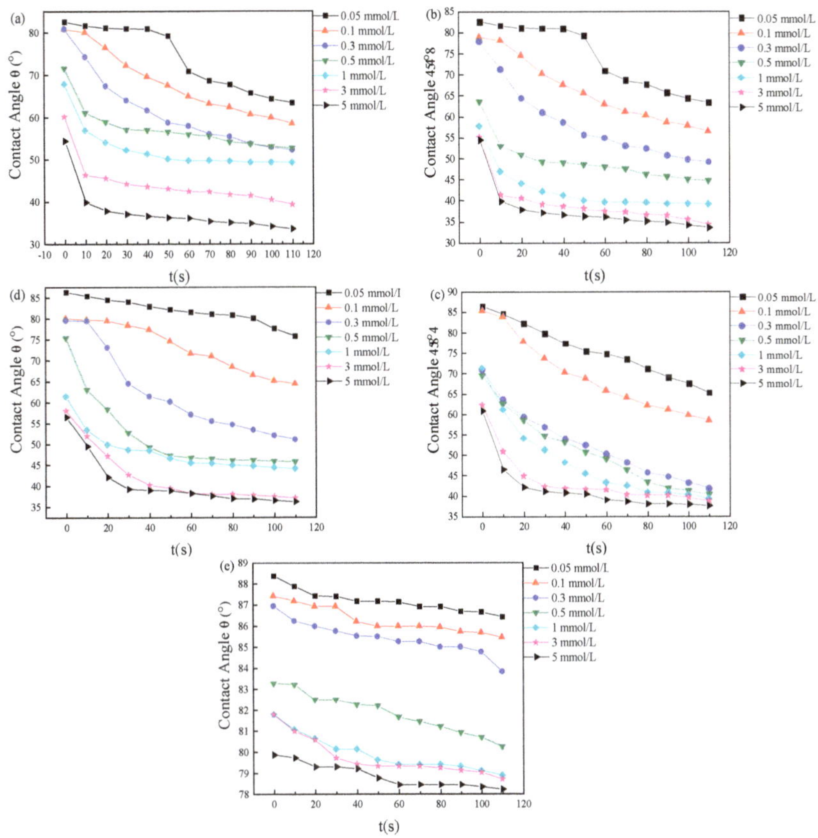

Figure 8. The effect of surfactant concentration on the contact angle of PTFE surface: (a) C_8DDLPB, (b) C10DDLPB, (c) C12DDLPB, (d) C14DDLPB, and (e) C16DDLPB.

In this study, C_8DDLPB, C_{10}DDLPB, C_{12}DDLPB, and C_{14}DDLPB have demonstrated favorable wetting properties on the PTFE surface, with contact angles reducing to 33°~38°. Particularly, C_8DDLPB and C_{10}DDLPB exhibit the ability to reduce the surface tension of water to 27.82 and 30.27 mN/m, respectively. Conversely, C_{16}DDLPB shows the least wetting ability, with a contact angle only decreasing to 78° within 110 s. Its equilibrium surface tension value is measured at 59 mN/m. These findings suggest that the wetting effect of C_nDDLPB on the PTFE surface is closely associated with its capability to lower the surface tension of water. Moreover, the impact of hydrophobic chain length in the hydrophobic group is notable [36,37].

2.7. Emulsifying Ability

The emulsification capacity test results for soybean oil are shown in Figure 9. As the carbon chain length increases, the emulsification capacity initially increases and then decreases. Specifically, C_{14}DDLPB exhibits the best emulsification performance with an emulsification time of 16.6 min, while the emulsification effect is poorest with a carbon chain length of eight. Furthermore, when the hydrophobic chain length becomes too long, as in the case of C_{16}DDLPB, the emulsification ability also weakens. This trend parallels that observed for the single-chain glucose acylamide quaternary ammonium salt C_nDGMAPB synthesized by our research group [38]. The hydrophobic chain length of C_nDDLPB significantly influences the emulsification capacity of soybean oil. Poor emulsification performance is observed with hydrophobic chain lengths of 8 and 16, whereas a moderate hydrophobic chain length demonstrates good emulsification.

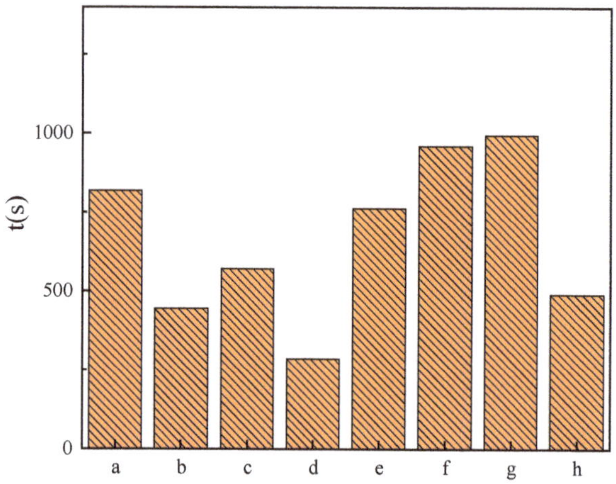

Figure 9. Emulsification time of soybean oil by surfactants: (**a**) 1227, (**b**) 1231, (**c**) HACC, (**d**) C_8DDLPB, (**e**) C_{10}DDLPB, (**f**) C_{12}DDLPB, (**g**) C_{14}DDLPB, and (**h**) C_{16}DDLPB.

2.8. Foam Properties

The foam performance of C_nDDLPB surfactants in deionized water was evaluated using the Ross–Miles method, and the results are shown in Figure 10. From the graph, it is evident that C_{12}DDLPB surfactant exhibits superior foamability, with a foam height reaching 67 mm. After 30 s, the foam height decreases by only 1 mm, and there is no further change in foam height over the subsequent 5 min, indicating excellent foam stability. However, compared to common anionic surfactants (foam height of >100 mm), it belongs to the category of low-foaming surfactants. C_8DDLPB, C_{10}DDLPB, and C_{14}DDLPB initially exhibit similar foam heights of 25 mm, 25 mm, and 22 mm, respectively, indicating relatively poor foamability. On the other hand, the foam generated by C_8DDLPB and C_{16}DDLPB dissipates rapidly, classifying them as low-foaming surfactants overall. These surfactants hold promise for applications in low-foam detergents and related products.

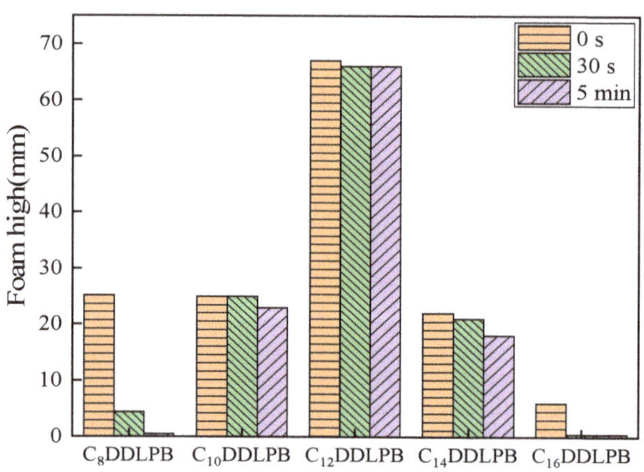

Figure 10. Graph of the foam properties of C_nDDLPB surfactants.

2.9. Antistatic Performance

The surface resistivity of polyester fabric before and after treatment was measured using a high-resistance meter, with the effectiveness of the surfactant's antistatic properties evaluated based on the magnitude of the decrease in surface resistivity logarithm values ($\triangle lg\rho_s$). A higher $\triangle lg\rho_s$ value indicates better antistatic performance. A series of synthesized lactose-based quaternary ammonium salt surfactants and commercially available antistatic agent, dihydroxyethyl stearylamine nitrate (SN), were tested, and the experimental results are presented in Table 2: at a concentration of 0.35 g/L, SN > C_8DDLPB > C_{10}DDLPB > C_{12}DDLPB > C_{14}DDLPB > C_{16}DDLPB, with C_8DDLPB exhibiting the largest $\triangle lg\rho_s$ among the C_nDDLPB surfactants, thereby indicating the best antistatic effect. The antistatic effects of C_8DDLPB and C_{10}DDLPB are approaching those of SN. Additionally, there is a trend of decreasing antistatic performance with increasing carbon chain length in C_nDDLPB.

Table 2. Antistatic properties of C_nDDLPB surfactant.

Sample Name	Concentration (g/L)	R_s/Ω	Lg (R_s/Ω)	$\Delta lg\rho_s$
Blank Control	0.35	2.70×10^{-13}	13.43	0
C_8DDLPB	0.35	2.52×10^{-10}	10.32	3.115
C_{10}DDLPB	0.35	2.29×10^{-10}	10.36	3.075
C_{12}DDLPB	0.35	4.93×10^{-12}	12.68	0.75
C_{14}DDLPB	0.35	1.99×10^{-13}	13.2	0.235
C_{16}DDLPB	0.35	2.63×10^{-13}	13.42	0.01
SN	0.35	3.41×10^{-10}	10.3	3.13

From Table 2, it can be observed that the maximum adsorption (Γ_{max}) of C_nDDLPB decreases with the increase in carbon chain length. Among the synthesized surfactants, C_8DDLPB and C_{10}DDLPB exhibit the highest Γ_{max} values, which are 1.890×10^{-10} mol·cm^{-2} and 1.675×10^{-10} mol·cm^{-2}, respectively. These results suggest that their antistatic effects may be related to their adsorption on the fabric surface.

2.10. Salt-Resistant Performance

This study investigated the salt resistance of DDAC and synthesized C_nDDLPB surfactants at different concentrations in various salt environments, as shown in Tables 3–5. When the concentration of sodium chloride added was 50 g/L, C_8DDLPB and C_{14}DDLPB exhibited transmittance rates of 94.19% and 99.08%, respectively, with excellent transmittance and no precipitation, indicating good salt resistance. In the presence of 35 g/L sodium chloride in the C_{12}DDLPB solution, the transmittance was 93.76%, demonstrating some degree of salt resistance.

Table 3. Permeability of surfactant solutions with NaCl concentration.

Concentration (g/L)	DDAC	C_8DDLPB	C_{10}DDLPB	C_{12}DDLPB	C_{14}DDLPB	C_{16}DDLPB
0	100%	100%	100%	100%	100%	100%
10	precipitates	96.83%	precipitates	100%	100%	90.16%
20	precipitates	96.38%	precipitates	100%	100%	69.98%
25	precipitates	95.72%	precipitates	98.86%	100%	precipitates
30	precipitates	95.72%	precipitates	98.17%	100%	precipitates
35	precipitates	95.06%	precipitates	93.76%	99.08%	precipitates
40	precipitates	95.06%	precipitates	64.57%	99.08%	precipitates
45	precipitates	94.62%	precipitates	precipitates	99.08%	precipitates
50	precipitates	94.19%	precipitates	precipitates	99.08%	precipitates

Table 4. Permeability of surfactant solutions with $CaCl_2$ concentration.

Concentration (g/L)	DDAC	C_8DDLPB	C_{10}DDLPB	C_{12}DDLPB	C_{14}DDLPB	C_{16}DDLPB
0	100%	100%	100%	100%	100%	100%
10	precipitates	96.83%	precipitates	100%	100%	90.16%
20	precipitates	96.38%	precipitates	100%	100%	69.98%
25	precipitates	95.72%	precipitates	98.86%	100%	precipitates
30	precipitates	95.72%	precipitates	98.17%	100%	precipitates
35	precipitates	95.06%	precipitates	93.76%	99.08%	precipitates
40	precipitates	95.06%	precipitates	64.57%	99.08%	precipitates
45	precipitates	94.62%	precipitates	precipitates	99.08%	precipitates
50	precipitates	94.19%	precipitates	precipitates	99.08%	precipitates

Table 5. Permeability of surfactant solutions with $MgSO_4$ concentration.

Concentration (g/L)	DDAC	C_8DDLPB	C_{10}DDLPB	C_{12}DDLPB	C_{14}DDLPB	C_{16}DDLPB
0	100%	100.00%	100.00%	100.00%	100.00%	100.00%
10	99.77%	99.54%	precipitates	precipitates	85.70%	precipitates
20	67.92%	99.54%	precipitates	precipitates	86.30%	precipitates
25	precipitates	99.54%	precipitates	precipitates	86.30%	precipitates
30	precipitates	99.54%	precipitates	precipitates	86.50%	precipitates
35	precipitates	99.54%	precipitates	precipitates	87.30%	precipitates
40	precipitates	99.54%	precipitates	precipitates	84.53%	precipitates
45	precipitates	99.54%	precipitates	precipitates	87.90%	precipitates
50	precipitates	99.54%	precipitates	precipitates	86.90%	precipitates

At a concentration of 50 g/L magnesium sulfate, C_8DDLPB and C_{14}DDLPB maintained transmittance rates of 99.54% and 86.90%, respectively, indicating strong salt resistance. In the presence of NaCl, $CaCl_2$, and $MgSO_4$, C_8DDLPB consistently exhibited excellent salt resistance.

In comparison with DDAC, the series of C_nDDLPB synthesized in this study have lactose-based hydrophilic groups containing numerous hydroxyl groups with strong negative polarity. Compared to micelles, these groups can readily attract more metal cations, resulting in a significant reduction in the concentration of unbound counterions in the solution, thereby hindering the formation of precipitates to some extent.

2.11. Antibacterial Performance

In this study, Gram-positive bacteria (*Staphylococcus aureus*) and Gram-negative bacteria (*Escherichia coli*) were used as samples to assess the antibacterial efficacy of C_{12}DDLPB and C_{14}DDLPB. Each group selected an appropriate dilution gradient for antibacterial rate calculation, utilizing the following formula:

$$\text{Antibacterial Rate}(\%) = \frac{\text{C - E}}{\text{C}} \times 100\% \qquad (4)$$

where C denotes the average bacterial count of the control sample, and E denotes the average bacterial count of the experimental sample.

Quaternary ammonium salt surfactants carry a positive charge in water, allowing them to interact with negatively charged microbial surfaces, leading to adsorption and the formation of small aggregates that adhere to the cell wall. This process inhibits microbial growth. Additionally, the hydrophobic groups of these surfactants interact with the hydrophilic groups of microbial cells, altering the permeability of the cell membrane. This results in membrane damage and leakage of intracellular substances. Furthermore, the abundance of positive charges can coagulate and denature proteins within microbial cells, affecting cell metabolism and achieving disinfection and antibacterial effects [39].

From Tables 6 and 7, it can be observed that at a concentration of 350 ppm, C_{12}DDLPB exhibits excellent antibacterial performance against *Escherichia coli* and *Staphylococcus aureus*, with antibacterial rates reaching 99.29% and 95.28%, respectively. Conversely, C_{14}DDLPB demonstrates poorer antibacterial efficacy, with antibacterial rates against *Escherichia coli* and *Staphylococcus aureus* at 73.20% and 76.91%, respectively. The antibacterial effect is more pronounced when the hydrophobic chain length is 12, which aligns with the previous literature [40]. Therefore, C_{12}DDLPB holds promise as a novel disinfectant product for potential applications.

Table 6. Inhibition of *Escherichia coli* by C_{12}DDLPB and C_{14}DDLPB.

Sample Name	Sample Number	Dilution Factor	CFU Prorata	Average Value	Antibacterial Rate
Control Group	*Escherichia coli*1	1.00×10^8	2.15×10^{10}	2.26×10^{10}	0
	*Escherichia coli*2	1.00×10^8	2.43×10^{10}		
	*Escherichia coli*3	1.00×10^8	2.20×10^{10}		
C_{12}DDLPB	*Escherichia coli* C12-1	1.00×10^6	1.47×10^8	1.60×10^8	99.29%
	Escherichia coli C12-2	1.00×10^6	2.01×10^8		
	Escherichia coli C12-3	1.00×10^6	1.31×10^8		
C_{14}DDLPB	*Escherichia coli* C14-1	1.00×10^7	4.89×10^9	6.06×10^9	73.20%
	Escherichia coli C14-2	1.00×10^7	6.86×10^9		
	Escherichia coli C14-3	1.00×10^7	6.42×10^9		

Table 7. Inhibition of *Gluconococcus aureus* by C_{12}DDLPB and C_{14}DDLPB.

Sample Name	Sample Number	Dilution Factor	CFU Prorata	Average Value	Antibacterial Rate
Control Group	*Staphylococcus aureus*1	1.00×10^7	3.07×10^9	3.06×10^9	0
	*Staphylococcus aureus*2	1.00×10^7	3.01×10^9		
	*Staphylococcus aureus*3	1.00×10^7	3.10×10^9		
C_{12}DDLPB	*Staphylococcus aureus* C12-1	1.00×10^6	1.41×10^8	1.44×10^8	95.28%
	Staphylococcus aureus C12-2	1.00×10^6	1.37×10^8		
	Staphylococcus aureus C12-3	1.00×10^6	1.55×10^8		
C_{14}DDLPB	*Staphylococcus aureus* C14-1	1.00×10^7	5.40×10^8	7.07×10^8	76.91%
	Staphylococcus aureus C14-2	1.00×10^7	7.50×10^8		
	Staphylococcus aureus C14-3	1.00×10^7	8.30×10^8		

3. Experimental

3.1. Materials and Instruments

Lactobionic acid (99%) was purchased from Aldrich, and chemically pure octyl bromide, decyl bromide, dodecyl bromide, tetradecyl bromide, and hexadecyl bromide were purchased from Shanghai Bohua Biochemical Reagent Co. Ltd., Shanghai, China, and N,N-dimethyl dipropylenetriamine (99.15%) was purchased from Guangdong Swell River Chemical Reagent Co. Ltd., Wengjiang, China. Deuterated dimethyl sulfoxide [DMSO] (99.9%) was purchased from Shanghai McLean Biochemical Technology Co., Shanghai, China. Didecyldimethylammonium chloride [DDAC] (95%) was purchased from Shanghai McLean Biochemical Science and Technology Co., Shanghai, China. Dodecyldimethylbenzylammonium chloride [1227] (99%) and dodecyltrimethylammonium chloride [1231] (99%) were purchased from Shanghai McLean Biochemical Science and Technology Co., Shanghai, China.

FT-IR spectrometer, model VERTEX 70 (Bruker, Saarbrücken, Germany); nuclear magnetic resonance (NMR) instrument, model AVANCE III (Bruker, Germany); electrospray ionization mass spectrometer, model Q Exactive (Thermofisher, Waltham, MA, USA); vacuum drying oven, model DZF-0B (China Yuejin Medical Equipment Co., Ltd., Shanghai, China); automatic surface tension meter, model KRÜSS-Tensiío (KRÜSS, Germany); drop shape analyzer, model DSA25B (KRÜSS, Hamburg, Germany); cryogenic transmission electron microscopy, model Talos F200C (Thermo Fisher, Waltham, MA, USA); and dynamic light scattering instrument, model JEM-1011EX (China Baxter Instruments Co., Ltd., Dandong, China), were utilized.

3.2. Synthesis of the Intermediate N-(3′-dimethylaminopropyl)-lactamido-3-aminop-ropane (DDLPD)

Lactobionic acid (0.1 mol) and N,N-dimethyldipropylenetriamine (0.12 mol) were added to 200 mL of methanol, and the reaction was conducted under reflux conditions for 2 h. After completion of the reaction, the heating was stopped, and the reaction was allowed to cool. The solvent was evaporated using a rotary evaporator, and the resulting product was washed thrice with ether to remove the residual N,N-dimethyldipropylenetriamine, followed by drying under vacuum until a constant weight was obtained, affording the intermediate DDLPD.

3.3. Synthesis of N-[N′[3-(lactosyl amide)]propyl-N′-alkyl]propyl-N,N-dimethyl-N-alkylammonium Bromide (C_nDDLPB)

First, DDLPD (0.05 mol), bromoalkane (0.15 mol), and 200 mL of anhydrous methanol were added in a 250 mL three-neck round-bottom flask equipped with a thermometer and a spherical condenser tube. Second, the reaction was performed under reflux conditions for 10 h. After completion of the reaction, the heating was stopped, and the reaction was allowed to cool. The solvent was evaporated using a rotary evaporator, and the product was washed thrice with ether and dried under vacuum until a constant weight was obtained, affording the product C_nDDLPB.

3.4. Characterization of C_nDDLPB

The samples to be tested were tested using a VERTEX 70 Fourier transform infrared (FTIR) spectral analyzer, and the samples to be tested were pressed with KBr, which was mixed homogeneously with an appropriate amount of the samples to be tested. The raw materials N,N-dimethyldipropylenetriamine, lactobionic acid, DDLPD, and C_nDDLPB were scanned in the wavelength range of 500–4000 cm^{-1}, and the structure of the products was characterized by observing the characteristic peaks of the IR spectra generated after scanning.

The synthesized products were subjected to ^1H-NMR and ^{13}C-NMR measurements (internal standard was tetramethylsilane (TMS), and solvent was deuterated dimethylsulfoxide (DMSO)) using an AVANCE III NMR spectrometer, and the molecular structure of the synthesized compounds was analyzed by NMR spectroscopy.

The synthesized compounds were detected using Q Exactive electrospray ionization mass spectrometer and identified and analyzed by ESI-MS spectra.

3.5. Surface Tension

The surface tension of C_nDDLPB was determined using a completely automated surface tension meter (KRÜSS-Tensiío type) according to the hanging drop method. The prepared solutions to be tested with different concentration gradients were allowed to stand for at least 24 h before the test and then measured after a constant temperature of 20 min was achieved at 25 °C [41].

3.6. Dynamic Surface Tension (DST)

A DSA25B tensiometer (Krüss Company, Hamburg, Germany) was used at 25.0 ± 0.1 °C to record dynamic surface tension (DST) data. The effective surface ages were within the range of 10 ms to 200 s, with an accuracy of ±0.01 mN/m.

3.7. Cryogenic Transmission Electron Microscopy (Cryo-TEM)

First, an aliquot of 3.5 μL solution (3.0 mg/mL or 1.0 mg/mL) was applied to glow-discharged Quantifiol R 2/1 holey carbon grids and blotted for 3 s under a humidity of 100% at 4 °C before being plunged into liquid ethane using a Vitrobot Mark IV (Thermo

Fisher Scientific, Waltham, MA, USA). The sample structure was immobilized in ice in a glassy state, and then, the sample attached to the copper mesh was transferred to a cryogenic transfer sample rod, followed by TEM imaging to observe the morphology of the 5 g/L solution of C_nDDLPB during a stationary 2-week period [42].

3.8. Dynamic Light Scattering (DLS)

Reverse osmosis (RO) water was used to prepare 5 g/L of C_nDDLPB (n = 8, 10, 12, 14, and 16) samples for measurement. The configured solution needed to be stable and homogeneous, and the temperature was set to 25 °C. An appropriate cuvette was selected, the angle of incident light and other parameters of the instrument were set, the solution (height of 1–1.5 cm) to be measured was added into the cuvette, and it was placed into the sample tank to start the measurement [43,44].

3.9. Wettability Study

According to the seated-drop method, the C_nDDLPB solutions of various concentrations were aspirated using a micro-syringe at 25 °C and dropped onto a polytetrafluoroethylene (PTFE) film. The contact angle was measured using a contact angle meter, and photographs were taken, and the contact angle was recorded at 10 s intervals [45].

3.10. Emulsifying Performance

Emulsification performance analysis testing is commonly conducted using the cylinder method. In a 500 mL iodine flask, 40 mL of the test solution (1.0 g/L) and an equal volume of edible soybean oil are added. The mixture is vigorously shaken up and down five times, followed by a 1 min static period. This process is repeated five times. After completing the aforementioned steps, the mixture is quickly transferred to a 100 mL graduated cylinder, and timing begins immediately. The timing stops when 10 mL of water appears in the lower layer. The time taken for this process is recorded. This experiment is repeated five times to obtain the average value [46].

3.11. Foam Morphology Characterization

According to the Ross–Miles method, the foam performance of C_nDDLPB was evaluated. The prepared solution (1 g/L) was placed in a constant-temperature water bath at (30 ± 0.5) °C for preheating for 30 min. Subsequently, the prepared solution was measured using a foam analyzer. The foam stability was determined based on the initial foam height and the subsequent change in foam height over a specified duration [47].

3.12. Antistatic Performance

The antistatic performance can be tested and analyzed according to GB/T 16801-2013. First, it is necessary to take the fabric finishing agent test solution to pre-treat the polyester fabric samples. Before and after the fabric samples are pre-treated, the surface resistivity is measured using a high resistance meter. The antistatic performance of the fabric finishing agent is evaluated based on the change in surface resistivity or its logarithmic value.

3.13. Salt-Resistant Performance

Prepare a 0.5 g/L solution of the surfactant. Transfer 10 mL of the solution into a test tube, followed by the addition of varying amounts of salt to create solutions of different concentrations. Allow the solutions to stand for 24 h. Subsequently, measure the transmittance of the solutions at 700 nm using a UV-Vis spectrophotometer [48].

3.14. Antibacterial Performance

Prepare a 35% drug stock solution using sterile water and sterilize it by filtration for subsequent use. Streak the tested *Staphylococcus aureus* and *Escherichia coli* on TSA plates and incubate them overnight at 37 °C until visible colonies appear. Select an appropriate quantity of colonies and transfer them to 5 mL of TSB liquid medium for incubation at a constant temperature of 37 °C for approximately 7 h, or until reaching an OD600 of approximately 0.6. Once the OD600 reaches 0.6 after adjustment with physiological saline, transfer the culture at a 0.1% ratio to TSB solution and incubate it at 37 °C for 18 h. Following incubation, dilute the treated bacterial suspension with PBS, and apply 100 µL of the diluted suspension onto TSA plates. Incubate the plates at 37 °C for 16 h, then photograph and count the colonies for quantification. Select an appropriate dilution gradient from each group for calculating the antibacterial rate.

4. Conclusions

In this study, double-chain lactobionic amide quaternary ammonium salts were synthesized by the amidation of lactobionic acid with N-N-dimethyldipropyltriamine to obtain glycosylamides, followed by quaternization with bromoalkanes of different chain lengths. The raw materials, intermediates, and target products were analyzed by Fourier transform infrared (FTIR) spectroscopy, proton nuclear magnetic resonance (^{1}H NMR) spectroscopy, (^{13}C NMR) spectroscopy, and electrospray ionization mass spectrometry (ESI-MS). The results indicated the successful synthesis of the target product. Through measurements including equilibrium surface tension, dynamic light scattering, and transmission electron microscopy, the surface activity, adsorption, and aggregation behavior of these compounds in aqueous solutions were investigated. Additionally, their application properties such as wetting ability, emulsification capability, foamability, antistatic performance, salt tolerance, and antibacterial activity were analyzed.

The CMC values of the five double-chain lactose amide quaternary ammonium salts (C_nDDLPB) decreased in the order of C_8DDLPB > C_{10}DDLPB > C_{12}DDLPB > C_{14}DDLPB > C_{16}DDLPB. With the growth of the carbon chain, the CMC of the lactose amide quaternary ammonium salts (C_nDDLPB) decreased, and the products exhibited a good surface activity, which can reduce the surface tension of water to 27.82 mN/m. The solutions of compounds with carbon chain lengths ranging from 8 to 14 demonstrate favorable wetting and spreading properties on PTFE, with contact angles decreasing to 33°~40°.TEM images revealed that, except for C_8DDLPB, the other products could form stable vesicle systems in an aqueous solution.

In terms of applications, C_{14}DDLPB exhibits the best emulsification performance on soybean oil, with a time of 16.6 min. The foaming properties of C_nDDLPB are generally low, characteristic of typical low-foaming products. Both C_8DDLPB and C_{10}DDLPB demonstrate excellent antistatic properties, comparable to the commonly used antistatic agent SN. C_8DDLPB and C_{14}DDLPB show good salt tolerance to NaCl, $CaCl_2$, and $MgSO_4$, with light transmittance exceeding 85% at a salt concentration of 50 g/L. Particularly, C_{12}DDLPB displays excellent antibacterial activity against *Escherichia coli* and *Staphylococcus aureus*, with inhibition rates reaching 99.29% and 95.28%, respectively, at a concentration of 350 ppm.

Therefore, this product is a novel glucosamine-based cationic surfactant characterized by low foaming, antibacterial properties, antistatic properties, salt resistance, and the ability to form stable vesicular systems. It holds promise pertaining to applications in various fields such as drug delivery carriers, biomimetic membranes, microreactors, daily chemical industry, and food industry in the future.

Supplementary Materials: The following Supporting Information can be downloaded at: https://www.mdpi.com/article/10.3390/molecules29122749/s1, Figure S1: FT-IR spectra of (a) N-N-Dimethyldipropyltriamine, (b) Lactobionic Acid, (c) DDLPD; Figure S2: ^{1}H-NMR spectra of DDLPD; Figure S3: ^{1}H-NMR spectra of C_8DDLPB ; Figure S4: ^{1}H-NMR spectra of C_{10}DDLPB ; Figure S5: ^{1}H-NMR spectra of C_{12}DDLPB ; Figure S6: ^{1}H-NMR spectra of C_{14}DDLPB ; Figure S7: ^{1}H-NMR spectra of C_{16}DDLPB; Figure S8: ^{13}C-NMR spectra of DDLPD ; Figure S9: ^{13}C-NMR spectra of C_8DDLPB ; Figure S10: ^{13}C-NMR spectra of C_{10}DDLPB ; Figure S11: ^{13}C-NMR spectra of C_{12}DDLPB ; Figure S12: ^{13}C-NMR spectra of C_{14}DDLPB ; Figure S13: ^{13}C-NMR spectra of C_{16}DDLPB ; Figure S14: ESI-MS spectra of C_8DDLPB ; Figure S15: ESI-MS spectra of C_{10}DDLPB ; Figure S16: ESI-MS spectra of C_{12}DDLPB ; Figure S17: ESI-MS spectra of C_{14}DDLPB; Figure S18: ESI-MS spectra of C_{16}DDLPB; Figure S19: cryo-EM spectra of C_nDDLPB.

Author Contributions: Conceptualization, G.W. and Y.W. (Yan Wang); methodology, L.Z.; software, Z.C.; validation, X.L. (Xiaoming Li); formal analysis, X.L. (Xudong Liu); investigation, Y.W. (Yunkai Wang); resources, E.Z.; data curation, Y.H.; writing—original draft preparation, Y.W. (Yunkai Wang); writing—review and editing, Y.W. (Yunkai Wang); visualization, L.Z.; supervision, M.D.S.; project administration, Z.C.; funding acquisition, L.Z. All authors have read and agreed to the published version of the manuscript.

Funding: This project was funded by the the Taiyuan University of Science and Technology Scientific Initial Funding (No. 20222010), Taiyuan University of Science and Technology Graduate Education Innovation Programme (No. SY2023023), The Nation Natural Science Foundation of China (No. 22178240) for the financial support.

Institutional Review Board Statement: Not applicable.

Informed Consent Statement: Not applicable.

Data Availability Statement: Data are contained within the article and Supplementary Materials.

Conflicts of Interest: Author Ying Huang was employed by the company Taiyuan Hengdeyuan Animal Protection Technology Development and Shanxi Livestock and Poultry Breeding. The remaining authors declare that the research was conducted in the absence of any commercial or financial relationships that could be construed as a potential conflict of interest.

References

1. Du, Q.; Li, Z.; Du, M.; Yang, T. Tianle. Government venture capital and innovation performance in alternative energy production: The moderating role of environmental regulation and capital market activity. *Energy Econ.* **2024**, *129*, 107196. [CrossRef]
2. Yang, T.; Fang, S.; Du, A.; Du, Q. Navigating the nexus: Geopolitical risk, fossil energy prices, and European utility stock returns—Implications for environmental management and energy security in a conflict-ridden global landscape. *J. Environ. Manag.* **2024**, *352*, 120086.
3. Yang, T.; Dong, Q.; Du, Q.; Du, Q.; Du, M. Carbon dioxide emissions and Chinese OFDI: From the perspective of carbon neutrality targets and environmental management of home country. *J. Environ. Manag.* **2021**, *295*, 113120. [CrossRef] [PubMed]
4. Tong, S.; Hu, Y.; Yu, J. Research Progress on Bio-based Vitrimer Materials. *Biomass Chem. Eng.* **2023**, *57*, 37–46.
5. Xue, X.; Wang, S.; Yang, T.; He, M.; Tang, X. Preparation and photocatalytic performance of Cu/cellulose-based TiO_2 composite. *New Chem. Mater.* **2022**, *50*, 195–200.
6. Gericke, M.; Amaral, A.J.R.; Budtova, T.; Tatiana, B.; Pieter, D.; Thomas, G.; Thomas, H.; Herman, H.; Anton, H.; Olli, I.; et al. The European Polysaccharide Network of Excellence (EPNOE) research roadmap 2040: Advanced strategies for exploiting the vast potential of polysaccharides as renewable bioresources. *Carbohydr. Polym.* **2024**, *326*, 121633. [CrossRef] [PubMed]
7. Martínez, A.; Magallanes, L.; Tarditto, L.; María C, P.; María F, G. María. Fatty acids methyl esters from soybean oil for biobased surfactants industry: Obtention C16 C18 concentrate for use as feedstock. *Ind. Crops Prod.* **2022**, *190*, 115892. [CrossRef]
8. Reid, A.G.; Eswara Rao, C.A.; Abhay, A.; Taylor, U.; Ravikumar R, G.; Avantika, S.; Jason S, D.; Gregg T, B.; Eugene, C. Bio-based lactone acrylic plastics with performance and recyclability advantages. *Cell Rep. Phys. Sci.* **2024**, *5*, 101938.
9. Thuy, T.H.N.; Wahyu, S.P.; Jun-Chul, C.; Norihisa, F.; Satoshi, T.; Takehiro, Y.; Nobuo, H.; Sho, K. Design and Evaluation of Bio-Based Industrial Symbiosis System Producing Energy and Chemicals Using Regionally Available Crop Residue. *Resour. Conserv. Recycl.* **2024**, *204*, 107509.
10. Lu, H.; Pezron, I.; Gaudin, T.; Audrey, D. Non-equilibrium micelles formed by sugar-based surfactants under their Krafft temperature. *Colloids Surf. A Physicochem. Eng. Asp.* **2018**, *540*, 167–176. [CrossRef]
11. Syguda, A.; Ławniczak, Ł.; Wróbel, P.; Filip, W.; Grzegorz, F.; Anna, P.; Marta, W.; Michał, N.; Aleksandra, G.; Łukasz, C. Biodegradable amidequats, derivatives of caprylic and pelargonic acids as cationic surfactants for agricultural applications. *J. Mol. Liq.* **2023**, *391*, 123221. [CrossRef]
12. Siddiqui, U.; Aslam, J.; Ansari, W.; Din, K. Micellization and aggregation behavior of a series of cationic gemini surfactants (m-s-m type) on their interaction with a biodegradable sugar-based surfactant (octyl-β-D-glucopyranoside). *Colloids Surf. A Physicochem. Eng. Asp.* **2013**, *421*, 164–172. [CrossRef]
13. Lee, S.; Lee, J.; Yu, H.; JongChoo, L. Synthesis of environment friendly nonionic surfactants from sugar base and characterization of interfacial properties for detergent application. *J. Ind. Eng. Chem.* **2016**, *38*, 157–166. [CrossRef]
14. Guido, V.; Pierluigi, Q.; Claudia, B.; Savarino, P.; Barni, E.; Fisicaro, E. Synthesis and Surface and Antimicrobial Properties of Novel Cationic Surfactants. *J. Org. Chem.* **2000**, *65*, 8197–8203.
15. Wojciech, S.; Natalia, B.; Michal, H. Evaluation of surface active and antimicrobial properties of alkyl D-lyxosides and alkyl L-rhamnosides as green surfactants. *Chemosphere* **2021**, *271*, 129818.
16. Zhi, L.; Zhang, E.; Shi, X. Research progress of glycosyl cationic surfactants. *Huaxue Tongbao* **2022**, *85*, 790–801.

17. Lu, B.; Vayssade, M.; Miao, Y.; Vincen, C.; Eric, G.; Anne, W.; Denis, P.; Audrey, D.; Christophe, E.; Isabelle, P. Physico-chemical properties and cytotoxic effects of sugar-based surfactants: Impact of structural variations. *Colloids Surf. B Biointerfaces* **2016**, *145*, 79–86. [CrossRef] [PubMed]
18. Cheng, W.; Yu, L.; Cheng, G. Synthesis characterization and application of Bola silicone quaternary ammonium salt. *CIESC J.* **2021**, *72*, 2837–2848.
19. Vargas-Ruiz, S.; Soltwedel, O.; Micciulla, S. Sugar Surfactant Based Microemulsions at Solid Surfaces: Influence of the Oil Type and Surface Polarity. *Langmuir ACS J. Surf. Colloids* **2016**, *32*, 11928–11938. [CrossRef]
20. Salomé, V.; Jana, L.; Regine, V.; Thomas, H.; Stefan, W. Wetting of planar solid surfaces by bicontinuous sugar surfactant-based microemulsions. *Colloid Polym. Sci.* **2017**, *295*, 2183–2190.
21. Wang, X.; Hao, J. Ionogels of Sugar Surfactant in Ethylammonium Nitrate: Phase Transition from Closely Packed Bilayers to Right-Handed Twisted Ribbons. *J. Phys. Chem. B* **2015**, *119*, 13321–13329. [CrossRef] [PubMed]
22. Wen, P.; Sun, Y.; Sun, J. Properties of ethoxylated tris-ester-based quaternary ammonium salts. *Deterg. Cosmet.* **2014**, *37*, 69–71.
23. Dufour, A.; Thiébaut, D.; Ligiero, L.; Matthieu, L.; Vial, J. Chromatographic behavior and characterization of polydisperse surfactants using Ultra-High-Performance Liquid Chromatography hyphenated to High-Resolution Mass Spectrometry. *J. Chromatogr. A* **2020**, *1614*, 460731. [CrossRef] [PubMed]
24. Rosen, M.J.; Kunjappu, J.T. *Surfactants and Interfacial Phenomena*, 4th ed.; Wiley: Hoboken, NJ, USA, 2012; pp. 336–367.
25. Fainerman, V.B.; Miller, R.; Aksenenko, E.V.; Makievski, A.V. 3. Equilibrium adsorption properties of single and mixed surfactant solutions. *Stud. Interface Sci.* **2001**, *13*, 189–285.
26. Wang, T.; Xu, P.; Ma, A. Effect of hydrophobic chain length on surface activity of gemini betaine. *Chem. Res. Appl.* **2021**, *33*, 2210–2216.
27. Aratono, M.; Onimaru, N.; Yoshikai, Y.; Makiko, S.; Ikuyo, K.; Kanda, W.; Akio, O.; Takanori, T.; Belkoura, L.; Reinhard, S.; et al. Spontaneous vesicle formation of single chain and double chain cationic surfactant mixtures. *J. Phys. Chem. B* **2007**, *111*, 107–115. [CrossRef]
28. Johnsson, M.; Wagenaar, A.; Engberts, J.B. Sugar-based gemini surfactant with a vesicle-to-micelle transition at acidic pH and a reversible vesicle flocculation near neutral pH. *J. Am. Chem. Soc.* **2003**, *125*, 757–760. [CrossRef]
29. Yun, Y.; Xiong, W.; Li, X.; Lu, T.; Huang, G.; Li, Z.; Fu, H. Molecular Packing Parameter in Bolaamphiphile Solutions: Adjustment of Aggregate Morphology by Modifying the Solution Conditions. *J. Phys. Chem. B* **2007**, *111*, 2225–2230. [CrossRef]
30. Gong, J.; Song, Y.; Sun, Y.; Sun, Q.; Liu, C.; Tan, J.; Zhao, L.; Xu, B. Vesicle-to-micelle transition in a double chain quaternary ammonium surfactant system: Interfacial behavior and molecular insights. *J. Mol. Liq.* **2024**, *394*, 123714. [CrossRef]
31. Xia, Y.; Goldmints, I.; Johnson, P.; Alan, T.; Arijit, B. Temporal Evolution of Microstructures in Aqueous CTAB/SOS and CTAB/HDBS Solutions. *Langmuir ACS J. Surf. Colloids* **2002**, *18*, 3822–3828. [CrossRef]
32. Hsieh, A.; Franses, E.; Corti, D. Formation of gem-like dispersions of soft crystallites in water by vesicles of a cationic surfactant. *Colloids Surf. A Physicochem. Eng. Asp.* **2022**, *652*, 129822. [CrossRef]
33. Surajit, G.; Chiranjib, G.; Chiranjib, B.; Sarthak, M.; Jagannath, K.; Nilmoni, S. Spontaneous Transition of Micelle–Vesicle–Micelle in a Mixture of Cationic Surfactant and Anionic Surfactant-like Ionic Liquid: A Pure Nonlipid Small Unilamellar Vesicular Template Used for Solvent and Rotational Relaxation Study. *Langmuir ACS J. Surf. Colloids* **2013**, *29*, 10066–10076.
34. Akihisa, S.; Alan, H. Model for Formation and Growth of Vesicles in Mixed Anionic/Cationic (SOS/CTAB) Surfactant Systems. *Langmuir ACS J. Surf. Colloids* **2002**, *18*, 7341–7348.
35. Zhang, B.; Yao, X.; Li, P.; Guo, C.; Ren, X.; Li, J. Preparation of chitosan sulfate and vesicle formation with a conventional cationic surfactant. *Carbohydr. Polym.* **2018**, *183*, 240–245. [CrossRef] [PubMed]
36. Li, Z.; Zhao, K.; Wang, Y.; Zheng, Z.; Zhang, C.; Gao, Y.; Du, F. Droplet splash and spread on superhydrophobic lotus leaves: Direct regulation by tuning the chain length of surfactant. *Colloids Surf. A Physicochem. Eng. Asp.* **2022**, *648*, 129178. [CrossRef]
37. Katarzyna, M.; Katarzyna, S.; Daniela, G.; Daria, W. Synthesis, Surface and Antimicrobial Activity of New Lactose-Based Surfactants. *Molecules* **2019**, *24*, 4010.
38. Zhi, L. *Synthesis and Properties of Novel Gluconamide-Type Surfactants*; ShanXi University: Taiyuan, China, 2014.
39. Jiang, J.; Zou, Y.; Sun, Q. Copolymers functionalized with quaternary ammonium compounds under template chain exhibit simultaneously efficient bactericidal and flocculation properties: Characterization, performance and mechanism. *J. Hazard. Mater.* **2024**, *465*, 133476. [CrossRef] [PubMed]
40. Wang, J.; Yang, X.; Chen, Y.; Zou, W. Bactericidal activity of sugar-based cationic gemini surfactants. *China Surfactant Deterg. Cosmet.* **2019**, *49*, 83–86.
41. Dou, J.; Liu, J.; Wang, G. Surface Activity, Wetting, and Aggregation of a Perfluoropolyether Quaternary Ammonium Salt Surfactant with a Hydroxyethyl Group. *Molecules* **2023**, *28*, 7151. [CrossRef]
42. Ram-On, M.; Cohen, Y.; Talmon, Y. Effect of Polyelectrolyte Stiffness and Solution pH on the Nanostructure of Complexes Formed by Cationic Amphiphiles and Negatively Charged Polyelectrolytes. *J. Phys. Chem. B* **2016**, *120*, 5907–5915. [CrossRef]
43. Matthew, L.L.; Tom, K.; Michael, R.W. Anticipating colloidal instabilities in cationic vesicle dispersions by measuring collective motions with dynamic light scattering. *J. Colloid Interface Sci.* **2006**, *296*, 599–607.
44. Milad, E.; Karin, S.; Stoyan, I.; Nikolay, G.; Kerstin, E. Oppositely charged surfactants and nanoparticles at the air-water interface: Influence of surfactant to nanoparticle ratio. *J. Colloid Interface Sci.* **2024**, *653*, 1388–1401.

45. Shi, X.; Zeng, M.; Xu, X.; Liu, Y.; Kou, J.; Bian, Q.; Song, H.; Zhang, J.; Wang, Q. Promotion of droplet deposition, diffusion-wetting and retention on hydrophobic surfaces by nonionic star-shaped oligomeric surfactants. *J. Mol. Liq.* **2023**, *386*, 122521. [CrossRef]
46. Cao, Y.; Yang, W.; Jiang, Y. Synthesis and properties of a zwitterionic Gemini surfactant. *Fine Chem.* **2021**, *38*, 335–340.
47. Ning, B.; Zhang, M.; Bai, Y.; Wang, W.; Wang, G. Comparison of the properties of perfluoroalkyl polyoxyethylene ether and alkyl polyoxyethylene ether. *Colloid Polym. Sci.* **2020**, *298*, 1389–1399. [CrossRef]
48. Hu, Y.; Xu, Z.; Li, Y. Synthesis and application performance of branched chain alcohol ether sodium sulfate based on SO3 sulfonation. *Text. Aux.* **2022**, *39*, 37–40.

molecules

MDPI

Article

Dispersive Solid Phase Extraction of Melatonin with Graphene/Clay Mixtures and Fluorescence Analysis in Surfactant Aqueous Solutions

Lucía Gutiérrez-Fernández [1], Ana M. Díez-Pascual [1,2] and María Paz San Andrés [1,2,*]

[1] Universidad de Alcalá, Facultad de Ciencias, Departamento de Química Analítica, Química Física e Ingeniería Química, Ctra. Madrid-Barcelona Km. 33.6, 28805 Alcalá de Henares, Madrid, Spain; lucia.gutierrezf@uah.es (L.G.-F.); am.diez@uah.es (A.M.D.-P.)

[2] Instituto de Investigación Química Andrés M. del Río (IQAR), Universidad de Alcalá, Ctra. Madrid-Barcelona Km. 33.6, 28805 Alcalá de Henares, Madrid, Spain

* Correspondence: mpaz.sanandres@uah.es

Abstract: In this work, the dispersive solid phase extraction (dSPE) of melatonin using graphene (G) mixtures with sepiolite (SEP) and bentonite (BEN) clays as sorbents combined with fluorescence detection has been investigated. The retention was found to be quantitative for both G/SEP and G/BEN 4/96 and 10/90 *w*/*w* mixtures. G/clay 4/96 *w*/*w* mixtures were selected to study the desorption process since the retention was weaker, thus leading to easier desorption. MeOH and aqueous solutions of the nonionic surfactant Brij L23 were tested as desorbents. For both clays and an initial sample volume of 25 mL, a percentage of melatonin recovery close to 100% was obtained using 10 or 25 mL of MeOH as desorbent. Further, using a G/SEP mixture, 25 mL as the initial sample volume and 5 mL of MeOH or 60 mM Brij L23 solution as the desorbent, recoveries of 98.3% and 90% were attained, respectively. The whole method was applied to herbal tea samples containing melatonin, and the percentage of agreement with the labeled value was 86.5%. It was also applied to herbal samples without melatonin by spiking them with two concentrations of this compound, leading to recoveries of 100 and 102%.

Keywords: melatonin; graphene; clays; dispersive solid phase extraction; fluorescence; surfactant

1. Introduction

Melatonin (*N*-acetyl-5-methoxytryptamine), a derivative of the essential amino acid tryptophan, is a hormone secreted by the enigmatic pineal gland in response to darkness. It has numerous roles related to circadian rhythm control, and it directly influences metabolism, the immune system, and the somnus [1]. It has also been proved that melatonin alleviates neurodegenerative illnesses such as Alzheimer's and Parkinson's [2] and has anti-inflammatory effects and anticancer properties [3]. In addition, it is a powerful antioxidant agent; it can scavenge free radical species (both reactive oxygen species (ROS) and reactive nitrogen species (RNS)) and stimulate the activity of antioxidant enzymes [4]. Outside the Animal Kingdom, melatonin was first discovered in an alga in 1991 [5]. Since then, many studies have reported the presence of melatonin in plants in concentrations ranging from picograms to micrograms per gram of plant [6]. It has also been found in fruits, including grapes, apples, pineapples, tomatoes, bananas, and cherries [7], in olive oil, and in beverages such as wine. Given the importance of melatonin, accurate analytical methods together with suitable extraction protocols to determine it in fruits and plants are needed. Nonetheless, the analysis of melatonin in these samples is challenging due to the broad range of concentrations among the different fruit (or plant) types, the difficulty in choosing the appropriate extraction solvent, and the instability of melatonin owed to its strong antioxidant capability since it reacts speedily with other food components [4].

Melatonin can be detected by several methods, including radioimmunoassay (RIA) [8], enzyme-linked immunosorbent assay (ELISA), gas chromatography–mass spectroscopy (GS–MS) [9], and high-performance liquid chromatography (HPLC) with an electrochemical detector (HPLC–ECD) [10], fluorescence detector (HPLC–FD) [11], or coupled to mass spectrometry (HPLC–MS) [12]. These methods differ in their sensitivity and specificity. Chromatographic methods have been the most widely used for melatonin determination in recent years since they offer high sensitivity and excellent detection selectivity. Some results have been reported when analyzing the same fruit varieties using different detection methods [10,13]. The concentrations of melatonin found in Montmorency are 12.3 and 13–19 ng g^{-1}, and in Balaton Tart Cherries, are 2.03 and 2.9 ng g^{-1}.

Also, melatonin shows native fluorescence due to the presence of the indole group in its chemical structure, which makes it suitable for direct analysis by fluorescence. This technique has been widely used owing to its high sensitivity and selectivity, easiness, speediness, and low solvent consumption, and has been applied to different types of samples, such as pharmaceutical preparations and urine [14], biological [15], and food/beverage samples [16]. In the presence of cyclodextrins, the fluorescence of melatonin is enhanced [17], while reagents such as 2,3-naphthalenedialdehyde provide a fluorescent product of the reaction with melatonin, which is measured by absorbance and fluorescence [15]. Other systems, such as metal–organic frameworks (MOFs) encapsulated into molecularly imprinted polymers (MIPs) (MOF@MIP), have been utilized for developing a sensor with luminescence detection [16]. Thus, Zr-based MOF as nanoparticles were incorporated into the MIP to obtain a composite with the target analyte in its structure (the template). Upon extraction of the melatonin, a selective material for the molecular recognition of this compound was obtained.

Due to its low concentration (usually from picograms to micrograms) and serious matrix interferences in plants, foods, and biological samples, solid–liquid extraction (SLE) with Na_2CO_3 and diethyl ether has been typically used for the analysis of melatonin by HPLC in complex matrices. Further, SLE with $HClO_4$ or acetone/methanol followed by solid phase extraction (SPE) in a C18 cartridge for purification and subsequent measurement via HPLC and ELISA immunoassay has also been applied [18]. SPE has been applied using different cartridges and solvents for elution [19]. On the other hand, melatonin in mulberry leaves has been determined via both SPE and liquid–liquid extraction (LLE) prior to its quantification by HPLC with fluorescence detection (HPLC-FD) [20]. Other solid phase extraction modalities have been developed, such as dispersive solid phase extraction (dSPE), which is based on the addition of a sorbent directly into the analytical solution followed by shaking, thus favoring the contact between the sorbent and the analytes [21]. Once the dispersion process is completed, the sorbent, with the analytes retained on its surface, is separated by a mechanical process, such as centrifugation or filtration. The most important advantage of dSPE is that the contact between the solid sorbent and the analyte is significantly more efficient than that of the SPE method. After the adsorption step, an appropriate solvent is needed for the analyte desorption. In this regard, the proper selection of the sorbent in SPE and dSPE is decisive to attain high selectivity. Many materials have been tested as sorbents, including MIPs, MOFs, mesoporous silica, and magnetic nanoparticles [22]. Carbon nanomaterials such as fullerenes, quantum dots (QDs), carbon nanotubes (CNTs), and graphene (G) have been used as sorbents in sample pretreatment [23] owed to their high specific surface area for analyte adsorption.

Graphene (G), a 2D carbon nanomaterial, has attracted significant attention due to its outstanding physical and chemical properties since it was first introduced in 2004 [24]. Its huge specific surface area and high chemical and thermal stability, combined with its electron-rich and hydrophobic properties, have made graphene an excellent sorbent for cleaning and preconcentration of target analytes [25]. Furthermore, G can interact strongly via π-π stacking interactions with compounds that contain aromatic rings owed of its large delocalized π-electron system. However, it is typically challenging to use G in this technique since it is an ultra-light material. In addition, it easily experiences irreversible aggregation in the packed cartridges during the SPE process, which would lead to reduced sorption capacity and extraction efficiency [26]. To address the abovementioned issues, novel approaches have been designed, such as the preparation of G-based magnetic materials [25] and the immobilization of G on the surface of silica [27].

Nanoclays such as sepiolite (SEP) and bentonite (BEN) are highly porous materials that have also been used as nanometric sorbents in dSPE. SEP ($Si1_2Mg_8O_{30}(OH)_4(H_2O)_4{\cdot}8H_2O$) is a hydrated magnesium silicate with a fibrous morphology, very high aspect ratio, and good mechanical properties, making it ideal for the development of hybrid materials [28]. Conversely, BEN ($Al_2H_2Na_2O_{13}Si_4$) is composed mainly of the clay mineral smectite. Its structure comprises octahedral sheets of aluminum and tetrahedral sheets of silica, and it shows a layer-like or tissue-like texture. In previous studies, our research group has developed methods for the determination of riboflavin, polycyclic aromatic hydrocarbons, and tryptophan based on dSPE using sepiolite and mixtures of sepiolite and bentonite clays with graphene as solid sorbents [29–31]. In the present work, G/SEP and G/BEN mixtures at different weight ratios have been prepared in order to obtain nanomaterials with different polarities, and they have been used as sorbents for melatonin extraction prior to their determination by fluorimetry. Methanol, aqueous solutions of hexadecyl trimethyl ammonium bromide (CTAB) (a cationic surfactant), and polyoxiethylen-23-lauryl ether (Brij L23) (a nonionic surfactant) were used for melatonin desorption. The developed method has been successfully applied to the determination of this analyte in herbal tea samples in a fast and sustainable way.

2. Results and Discussion

The solid phase extraction sorbents used in this work (G/SEP and G/BEN mixtures) were prepared as described in the experimental section and were characterized by transmission (TEM) and scanning electron microscopy (SEM).

2.1. Electron Microscopy Analysis of Graphene/Clay Mixtures

To obtain information about the surface morphology of sepiolite, bentonite, and their mixtures with graphene, the samples were observed by SEM and TEM, and representative micrographs are displayed in Figure 1 and Figure S1, respectively.

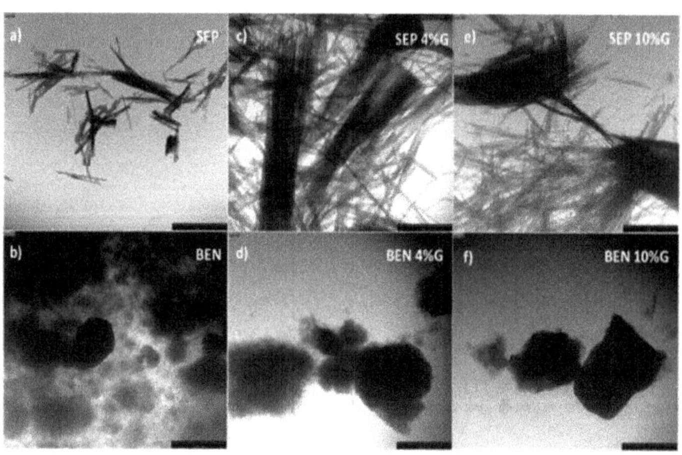

Figure 1. TEM images of neat SEP and BEN as well as G/clay mixtures with G percentages of 4 and 10 wt% at a magnification of 10,000×.

Regarding G/BEN mixtures, randomly alternated sheets of both compounds can be clearly observed. The most rigid platelets correspond to BEN, while the most flexible and thinnest sheets correspond to G. Also, good interaction between both compounds is suggested. In the mixture with 4 wt% G (Figure 1d), most of the flakes show diameters smaller than 30 nm since graphene intercalates within the bentonite lamellar structure, inducing the exfoliation of the flakes. The mixture with 10 wt% G (Figure 1f) shows less dense and more flexible structures, with many graphene sheets wrapping around the bentonite flakes, which are even thinner.

In the TEM image of pristine SEP, smooth, thin, rigid fibers can be observed (Figure 1a). They are aggregated in needles typically 50–100 nm in thickness. On the other hand, the TEM micrograph of neat BEN shows numerous rigid and quasi-spherical-shaped dark flakes (Figure 1b). Upon mixing with graphene, the sepiolite fiber bundles become thinner (Figure 1c,e), which is in agreement with the observations from the SEM analysis. Similarly, the bentonite flakes are more separated in the presence of graphene (Figure 1d,f), though in this case, it is more difficult to observe the graphene flakes since they are much lighter. Overall, microscopic observations reveal a uniform dispersion of both graphene and clay in the mixtures, thereby increasing their specific surface area, which is advantageous for application as sorbents in dSPE.

In the SEM images, pristine SEP fibers show a needle-like morphology, with sizes of 40–70 nm in diameter and >1 μm in length, Figure S1a. The fibers formed bundles-like aggregates by surface interaction between the hydroxyls (silanol groups, Si–OH) located at their external surface. Conversely, bentonite shows a lamellar structure composed of rigid individual nanoplatelets with thicknesses ranging from 30 to 50 nm (Figure S1b).

The images of the G/SEP mixtures show the graphene nanosheets randomly intercalated within the sepiolite nanofibers, suggesting a good interaction between both compounds. In the mixture with 4 wt% G (Figure S1c), the SEP nanofibers show diameters in the range of 30–50 nm, hence appear more separated than in pristine sepiolite due to the presence of graphene that inserts within the fiber aggregates and causes a debundling. This effect is even more pronounced in the sample with 10 wt% G (Figure S1e), in which more graphene flakes and fewer sepiolite fibers can be observed, and the fibers are much more separated.

2.2. Analysis of Melatonin in Different Media by Fluorescence

2.2.1. Fluorescence Spectra of Melatonin in Different Media

The fluorescence spectra of melatonin were recorded in different media, such as ultrapure water, for measuring the retention percentage and in methanol, sodium dodecyl sulfate (SDS), Brij L23, and CTAB as possible desorbents. Fluorescence contour graphs were obtained for a melatonin concentration of 80 μg L^{-1}. The maximum excitation wavelength of melatonin is 280 nm for all the media studied, but the maximum emission wavelength depends on the medium. Figure 2 shows the fluorescence contour graphs of melatonin in the different media, namely water, methanol, and aqueous solutions of surfactants (SDS, CTAB, and Brij L23) at a concentration of 20 mM, which is above their critical micelle concentration (CMC).

As can be observed in Figure 2, the emission wavelength of melatonin in water is 358 nm, about 20 nm higher than that found in methanol (337 nm). This shift in the wavelength is attributed to the different polarity of the solvents. The higher polarity of water results in a higher emission wavelength due to the more polar microenvironment of melatonin in this solvent. The dipole moment of a fluorophore in the excited state is larger than in the ground state. Following excitation, the solvent dipoles can reorient or relax around the molecule dipole, which lowers the energy of the excited state. As the solvent polarity is increased, this effect becomes larger, resulting in emission at lower energies or longer wavelengths [32]. In the presence of surfactants, the fluorescence intensity changes depending on the interaction with micellar aggregates [33]. When the polarity of the molecule is low, it is highly soluble in the micelle aggregates, which results in a higher fluorescence intensity as occurs for pyridoxine and riboflavin and fat-soluble vitamins [34,35]. In the surfactant solutions, the emission wavelength of melatonin is between those of water and methanol, attributed to the interaction of melatonin with the aqueous micelles that leads to a more hydrophobic microenvironment than in water, but less than in methanol. Regarding the fluorescence intensity, the maximum value is observed in Brij L23 aqueous solution, followed by CTAB and methanol, while the minimum is found in water.

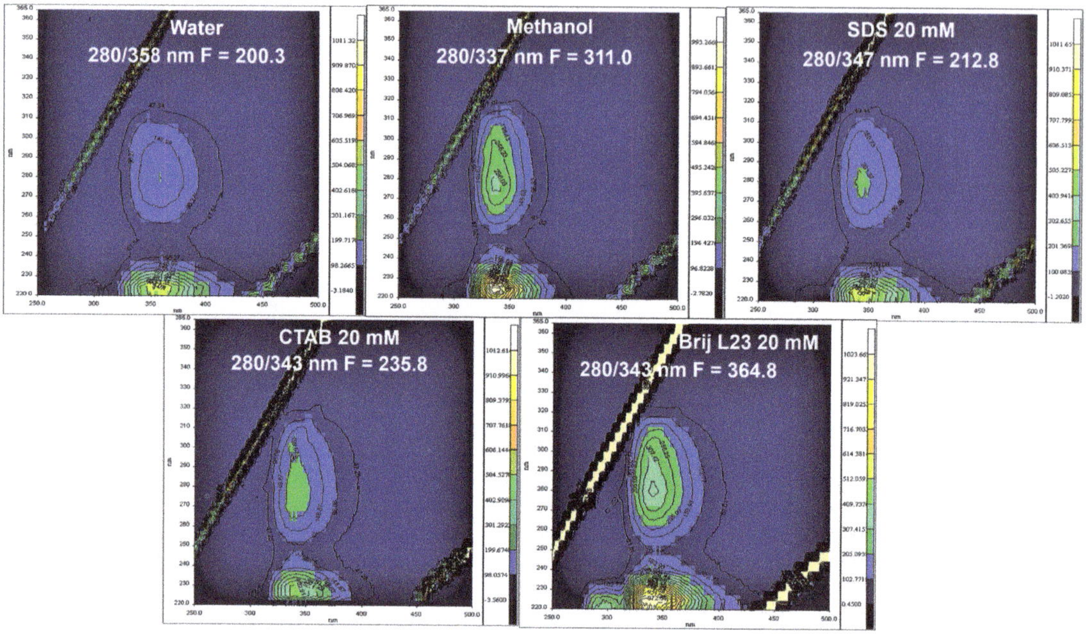

Figure 2. Fluorescence contour graphs of melatonin in water, methanol, and 20 mM aqueous solutions of SDS, CTAB, and Brij L23.

Given that the strongest fluorescence intensity of melatonin is achieved in the nonionic surfactant Brij L23, this medium was chosen to study the influence of the solvent concentration since the maximum fluorescence intensity (F) increases with increasing concentration. The fluorescence contour graphs of melatonin in 40, 60, and 80 mM Brij L23 are shown in Figure S2. As depicted in the figure, all the spectra are very similar, with a slight increase in F upon increasing surfactant concentration.

The fluorescence intensity of melatonin 80 μg L^{-1} was recorded in 1 and 20 mM SDS, in 0.5, 1, 5, 10, and 20 mM CTAB, as well as in 0.05, 1, 5, 10, 20, 30, 40, 60, and 80 mM Brij L23. The excitation and emission wavelengths remain constant for all the concentrations of each surfactant. Figure 3 shows the change in melatonin fluorescence as a function of the surfactant concentration for excitation/emission wavelengths of 280/343 nm.

Regarding Brij L23, F rises by a factor of 1.98 from the value in the water solution ([Brij L23] = 0) up to a concentration of 20 mM and increases slightly at higher surfactant concentrations. In the case of CTAB, it continuously grows with increasing concentration, showing a 1.55-fold increase at a concentration of 20 mM. It is important to note that the CMC of Brij L23 is 0.09 mM while that of CTAB is about 10 times higher; therefore, the F value increases for both surfactants above the CMC. The fluorescence intensity of melatonin is significantly higher in Brij L23 solutions than in CTAB and SDS. The anionic surfactant leads to lower F values. Hence, it was discarded as desorbent in the following stages of this study.

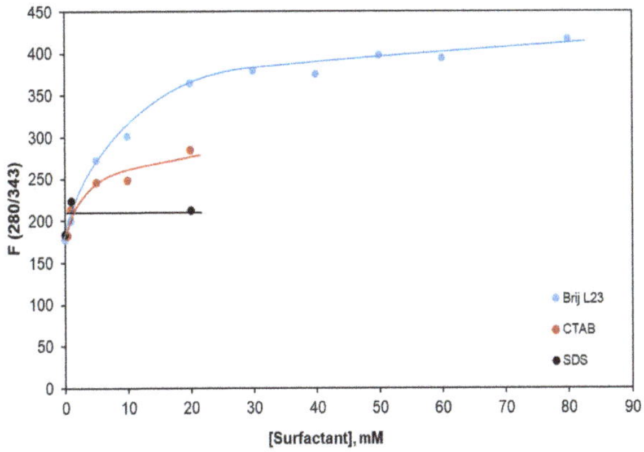

Figure 3. Change in the fluorescence intensity of melatonin as a function of the SDS, CTAB, and Brij L23 concentration. $\lambda_{exc}/\lambda_{em}$ = 280/343 nm.

2.2.2. Analytical Characteristics of Melatonin Analysis in Water and Desorbents

The analytical characteristics of the fluorescence analysis of melatonin were investigated in water to determine the retention of melatonin in the G/C solids as the difference between the initial concentration and that in the supernatant after the retention process. For this purpose, four calibration curves were carried out using the external standard method. Table 1 shows the sensitivity, limit of detection, limit of quantification, robustness, and reproducibility.

Table 1. Analytical characteristics of the fluorescence determination of melatonin in methanol and Brij L23.

	Water	MeOH	Brij L23 60 mM	Brij L23 80 mM
Linear range, μg L^{-1}	12–215	11–215	12–220	13–220
r	0.9996	0.9997	0.9996	0.9995
Sensitivity, L μg^{-1}	2.35	3.61	3.78	3.89
Limit of detection (LOD), μg L^{-1}	4 ± 2	3.2 ± 0.8	3.5 ± 0.4	3.9 ± 0.5
Limit of quantification (LOQ), μg L^{-1}	12 ± 3	11 ± 3	12 ± 1	13 ± 2
Robustness, %RSD (n = 4)	5.27	8.02	1.45	2.76
Reproducibility, %RSD (n = 4) 80 μg L^{-1}	4.15	3.24	3.10	1.58
Reproducibility, %RSD (n = 4) 120 μg L^{-1}	3.05	3.88	3.45	1.81

Prior to determining the accurate recovery percentage of melatonin with the chosen desorbents (MeOH and Brij L23 solutions), the analytical characteristics of the fluorescence analysis in these media (MeOH and 60 and 80 mM Brij L23) were also determined and are shown in Table 1.

The sensitivity was found to be higher in 80 mM Brij L23, which is in agreement with the fluorescence contour graphs. For the three media, the limit of quantification is very low, ranging between 11 and 13 μg L^{-1}. The robustness is systematically lower than 10% and lower than 5% for the two concentrations of Brij L23. Overall, the good analytical characteristics of the method corroborate its suitability for the analysis of melatonin.

A very good sensitivity and a low quantification limit have been obtained, allowing to determine melatonin concentrations higher than 12 μg L^{-1}. Further, the robustness is acceptable (around 5% RSD).

2.3. Retention of Melatonin in Graphene/Clay Solid Mixtures

Melatonin retention was studied in two G/clay mixtures (4/96 and 10/90 *w*/*w*). Its initial concentration in aqueous medium was set as 80 µg L^{-1}. The retention was found to be quantitative for both clays (SEP and BEN) combined with G (4/96 and 10/90 *w*/*w*).

The desorption of melatonin could be carried out using the two mixtures prepared herein since both show a retention percentage of 100%. However, the desorption is easier with the 4/96 *w*/*w* mixture since the retention is weaker due to its higher polarity. Therefore, this solid mixture was chosen to investigate the desorption process.

2.4. Desorption of Melatonin from Graphene/Clay 4/96 w/w Mixture with Different Desorbents

Melatonin desorption after retention in the 4/96 *w*/*w* mixture was carried out using three different media as desorbents: methanol (an organic solvent) as well as aqueous solutions of the cationic surfactant CTAB and the nonionic surfactant Brij L23. Both sepiolite and bentonite clays were used to prepare the solid mixture. The percentage of recovery was calculated by comparing the fluorescence intensity of melatonin in each desorbent with that measured initially (prior to the extraction process).

2.4.1. Desorption with CTAB

Two CTAB solutions (10 and 20 mM) were used for the desorption of melatonin (80 µg L^{-1}), and a recovery of 73% was attained for both concentrations in duplicate. Thus, the amount of melatonin desorbed from the solids hardly changes with increasing surfactant concentration. In addition, CTAB solutions are difficult to prepare due to solid precipitation below 20 °C.

2.4.2. Desorption with Brij L23

In order to study the potential of Brij L23 as a desorbent for melatonin, three different concentrations of this surfactant were tested (40, 60, and 80 mM), and the recoveries obtained are shown in Figure 4. The highest recovery (98.5%) was attained from the mixture comprising sepiolite clay using 60 mM Brij L23 as a desorbent. Thus, this surfactant solution has been chosen for the following experiments. Nonetheless, it should be noted that the recovery from mixtures comprising either SEP or BEN was also high (about 84.3%) for a surfactant concentration of 80 mM.

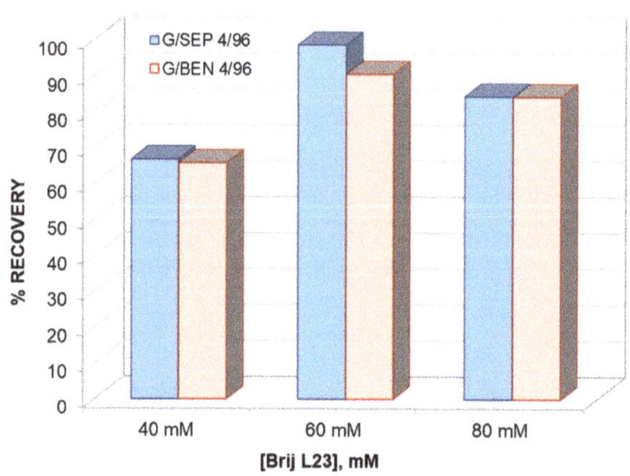

Figure 4. Percentage of melatonin recovery from G/clay (4/96 *w*/*w*) mixtures using different concentrations of Brij L23 as desorbent.

2.4.3. Desorption with Methanol

The fluorescence intensity of melatonin in MeOH is slightly lower than that in Brij L23, though it is also high. The ability of this organic solvent to recover compounds in SPE is very useful due to its intermediate polarity; hence, the desorption with this solvent was carried out and compared with the surfactant solutions. The recovery was quantitative for both G/SEP and G/BEN (4/96 *w*/*w*) mixtures using an initial and final solvent volume of 25 mL.

2.4.4. Comparison of the Extraction Recovery with MeOH, CTAB and Brij L23

The different media used for melatonin desorption were compared. The highest recovery attained for CTAB solutions (10 and 20 mM) was 73%, irrespective of the surfactant concentration. Further, as mentioned above, the recoveries were lower than those obtained with MeOH and Brij L23; hence, CTAB solutions were discarded as desorbents.

Both Brij L23 aqueous solutions and MeOH provide excellent recoveries, up to 98.3%, using 25 mL of desorbent. With the aim to preconcentrate the melatonin in the desorbent, lower volumes of MeOH were used. Figure 5a shows the recoveries from graphene mixtures with both clays, sepiolite, and bentonite, obtained with 25, 10, 5, and 2 mL of MeOH as desorbent. As can be observed, melatonin can be preconcentrated with the extraction process up to 5 and 2.5 times using sepiolite and bentonite mixtures, respectively.

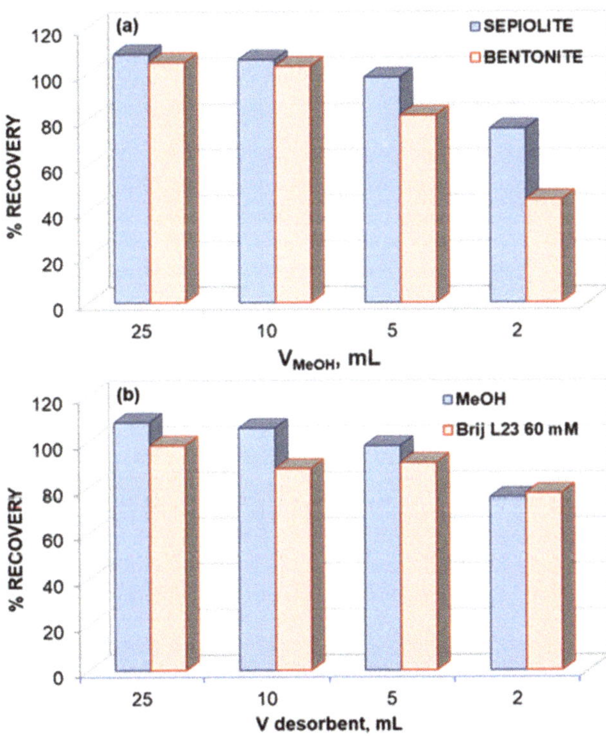

Figure 5. Percentage of melatonin recovery from G/clay 4/96 *w*/*w* mixtures using different volumes of MeOH as desorbent in sepiolite and bentonite (**a**) and using different volumes of MeOH and Brij L23 60 mM (**b**). Initial sample volume of 25 mL.

In the desorption process, for MeOH volumes lower than 5 mL, the percentage of melatonin recovery decreases to 76% and 45% using 2 mL of SEP and BEN mixtures, respectively. A very high recovery (98.3%) from G/SEP (4/96 *w*/*w*) mixture has been attained using 5 mL of MeOH as desorbent, which implies a preconcentration factor of 5.

The percentage of recovery for different volumes (25, 10, 5, and 2 mL) of 60 mM Brij L23 solution as desorbent was comparatively analyzed (Figure 5b). For this purpose,

G/SEP 4/96 *w/w* was also selected since this clay systematically provides better results than BEN. The recovery decreased slightly upon decreasing the desorbent volume to 78% for 2 mL. The comparison of the recoveries obtained for different volumes of MeOH or 60 mM Brij L23 as desorbents, using G/SEP 4/96 *w/w* mixture, is shown in Figure 5b.

The recoveries were high and quite similar for both desorbents. Since methanol is more dangerous than Brij L23, and the surfactant solutions are prepared in water, the most sustainable solvent, these were chosen to carry out the desorption process.

When the desorption volume decreased, the capacity of the desorbent remained unchanged up to a volume at which it could not desorb because of the impossibility of interacting with the molecules retained in the solid. Hence, it cannot solubilize the retained molecules. The process is a sum of different equilibria. On the one hand the interactions between melatonin and G/clay via π-π stacking between the aromatic rings and H-bond formation must be broken because of the stronger interaction between the desorbent and melatonin. In this regard, a higher desorption volume can solubilize the melatonin better (it gives a lower concentration), but in an analytical process, it is interesting to decrease the desorption volume in order to increase the preconcentration factor. Therefore, the influence of the desorption volume was studied, and the one that led to the lowest volume combined with the highest recovery was chosen. The best results were obtained for 25 mL of a sample volume in a G/SEP 4/96 *w/w* mixture as sorbent and 5 mL of 60 mM Brij L23 solution as desorbent. Hence, these were chosen as the optimal conditions for carrying out the experiments with real samples.

2.5. Analysis of Melatonin in Herbal Tea Samples

Melatonin content in herbal tea samples cannot be directly measured by fluorescence in aqueous solutions due to the interferences of the matrix. Therefore, the dSPE developed a method using a G/SEP (4/96 *w/w*) mixture as sorbent needs to be applied to efficiently separate the melatonin from the matrix prior to the fluorescence analysis. The results from the dSPE extraction combined with fluorescence measurement have been compared with those obtained with the dSPE method combined with a reference method (HPLC). This method was chosen as a reference since it can separate melatonin from other extracted compounds and determine its concentration, though it consumes high amounts of organic solvents such as methanol, which is not environmentally friendly.

Melatonin levels in the herbal samples were too high when an entire tea bag was used. Therefore, 8 mg of herbal tea bags were weighed and subsequently prepared as described in the experimental section. The limit of quantification calculated in the solid sample prepared in this way was found to be 0.4 mg g^{-1}. The results obtained for both methods were compared. The standard deviations compared by the F test are statistically equal for a 95% confidence level (F = 0.63). For equal standard deviations, the comparison of the means using the Student's *t*-test demonstrates the absence of statistical differences between them for the same confidence level and a *p*-value of 0.35. The amount of melatonin found in the herbal tea samples via fluorescence in the liquid extract is 50 ± 9 µg L^{-1}, which corresponds to 1.6 ± 0.3 µg g^{-1} in the solid sample and is 44 ± 7 µg L^{-1} through HPLC, which corresponds to 1.4 ± 0.2 µg g^{-1}. The amount labeled is 1.85 µg g^{-1}. Therefore, the percentage of recovery using the fluorescence method considering the labeled value is 86.5%, which can be regarded as very high, taking into account the inhomogeneity of the sample.

Fluorescence measurements are faster, cheaper, and more environmentally friendly than HPLC. Hence, the developed method is a good alternative to determinate melatonin in real samples.

Another herbal tea sample without melatonin was chosen in order to demonstrate the accuracy of the developed method. This sample was spiked with two melatonin concentrations (8 and 10 µg L^{-1}) below the LOQ of the fluorescence method (13 µg L^{-1}). In this solid sample, from 5 mg of the sample extracted with 250 mL of water, the LOQ is 0.63 mg g^{-1}.

The dSPE process combined with fluorescence allows the quantification of low levels of melatonin thanks to the desorption and preconcentration step using 10 mL of 60 mM Brij L23. To check for reproducibility, the experiments were carried out in quadruplicate on different days. For both spiked concentrations, the reproducibility is 5.0 and 5.9 as RSD for 8 and 10 μg L^{-1}, respectively, which demonstrates the good precision of the developed method. The melatonin concentration found for the two concentrations added and the percentage of recovery using the whole method of analysis (dSPE extraction combined with fluorescence) is shown in Table 2. Results demonstrate the high accuracy of the method since the recovery is very high (100 and 102% for spiked concentrations of 8 and 10 μg L^{-1}, respectively).

Table 2. Melatonin concentration in herbal tea samples spiked with two concentrations below the LOQ of the fluorescence method, obtained using 60 mM Brij L23 as desorbent (n = 4).

[MEL] Added, μg L^{-1}	[MEL] Found, μg L^{-1}	% Recovery
8.0	8.0 ± 0.4	100
10.0	10.2 ± 0.6	102

In these conditions (sample mass and water volume for SLE), the whole method can accurately determine melatonin as it has been explained above, but higher sample mass or lower water volumes may make it difficult to determine.

3. Materials and Methods

3.1. Reagents and Chemicals

Sepiolite ($Mg_4(Si_6O_{15})(OH)_2{\cdot}6H_2O$, purity > 95%), a hydrated magnesium silicate with a very large surface area (close to 300 m^2 g^{-1}) comprising 60.2 wt% SiO_2, 26.1% MgO, 1.7% Al_2O_3, and minor amounts of Fe_2O_3, CaO, Na_2O, and K_2O, was provided by Sepiol SA (Azuqueca de Henares, Spain). Few high-purity graphene layers in powder form, with a layer thickness smaller than 2 nm, a surface area close to 500 m^2 g^{-1}, and a low oxygen content (≤5 wt%), were supplied by Avanzare Innovación Tecnológica, SL (Logroño, Spain). Melatonin ($C_{13}H_{16}N_2O_2$, purity ≥ 98%, M_w = 232.3 g mol^{-1}) and hexadecyltrimethylammonium bromide (CTAB), with a formula of $C_{19}H_{42}BrN$, critical micelle concentration of 0.9 mM and M_w of 364.5 g mol^{-1} were purchased from Merck (Darmstadt, Germany). Polyoxyethylene-23-lauryl ether (Brij L23) with a formula of $C_{12}H_{25}(OCH_2CH_2)_{23}OH$, critical micelle concentration of 91 μM and M_w of 1198.6 g mol^{-1}) and sodium dodecylsulfate (SDS), with a formula of $C_{12}H_{25}O_4SNa$, M_w of 288.4 g mol^{-1}, and critical micelle concentration of 8.3 mM) were provided by Sigma (Madrid, Spain). Methanol (CH_3OH) with a density of 0.79 g cm^{-3} and M_w of 32.04 g mol^{-1} was provided by Scharlau (Barcelona, Spain). All the reagents were of high purity grade and used as received. Aqueous solutions were prepared in ultrapure water obtained from a Milli-Q system (Millipore, Milford, MA, USA). A stock solution of melatonin with a concentration of 20 mg L^{-1} was initially prepared, and the subsequent solutions were obtained by dilution and then stored in the dark at 4 °C. Herbal tea samples were obtained from a local store (Alcalá de Henares, Spain).

3.2. Instrumentation

Fluorescence spectra were obtained with an LS-50B fluorescence spectrophotometer (Perkin-Elmer, Hopkinton, MA, USA) incorporating a Xe lamp and thermostatised at 25 ± 1 °C with a Braun Thermomix BU bath (Analytical Instruments LLC, Minneapolis, MN, USA). For the measurements, quartz cuvettes with a conventional 1 cm path length were used at a speed of 500 nm min^{-1}. The thickness of the excitation and emission slits was 5 nm, and FLWin Lab software was used for data treatment. A chromatographic system incorporating a Flexar binary LC Pump (PerkinElmer, Hopkinton, MA, USA) with vacuum degasification, a manual Rheodyne injection valve, 6-port, with a loop of

20 μL, and a Jet-Stream Plus thermostatic column (Knauer, Berlin, Germany), combined with a programmable fluorescence detector (Perkin-Elmer, Hopkinton, MA, USA) was used to perform the reverse-phase high-performance liquid chromatography (RP-HPLC) measurements. Chromatographic data acquisition and treatment were performed with TotalChrom WS software version 6.3.2 (Perkin-Elmer, Hopkinton, MA, USA). The analytical column, with dimensions of 5 μm, 150 × 4.6 mm, was a Chromaphase RP-18 provided by Scharlab (Barcelona, Spain). Measurements were performed using MeOH/H_2O 50/50 *v*/*v* as the mobile phase at 25 °C and a flow rate of 1 mL min^{-1}. The peak of melatonin was observed without any interference at a retention time of 3.5 min. pH measurements were performed using an InoLab pH meter (Mexico DF, Mexico). The mixtures were shaken with a mechanical stirrer (Selecta, Barcelona, Spain) and subsequently centrifuged with a refrigerated centrifuge (Digicen, Ortoalresa, Madrid, Spain). A UP400S ultrasonic probe (Hielscher Ultrasonics GmbH, Teltow, Germany), integrating a titanium sonotrode (ø = 3 mm; l = 100 mm) was used for sample ultrasonication. Transmission electron microscopy (TEM) micrographs at an amplification of 50,000× were obtained with a Zeiss EM-10 C microscope (Oberkochen, Germany) working at an acceleration voltage of 60 kV. Scanning electron microscopy (SEM) images at an amplification of 20,000× were acquired with a Zeiss DSM-950 microscope (Oberkochen, Germany), working at 15.0 kV. Statgraphics Centurion program, version XVII 17.0.16, was used for statistical analysis of the images.

3.3. *Experimental Procedure*

3.3.1. Preparation of Graphene/Clay Mixtures

In order to attain materials with different polarities, several graphene/clay (G/C) mixtures (4/96 and 10/90 *w*/*w*, 100 mg) were prepared by weighing the necessary amounts of both solids and mixing them in 50 mL of distilled water. To attain homogenous dispersions, the mixtures were then subjected to sonication for 10 min with an ultrasonic probe at a power of 160 W, centrifuged for 5 min at a speed of 2598 g, and filtered with a cellulose filter (ø = 0.45 μm). The resulting solids were finally dried under ambient conditions and stored for the subsequent extractions. A diagram showing the entire procedure for the synthesis of the graphene/clay mixtures is shown in Figure S3.

3.3.2. Scanning and Transmission Electron Microscopies

Both the raw clays and the different graphene/clay mixtures mentioned above were analyzed by transmission electron microscopy (TEM) and scanning electron microscopy (SEM). Regarding TEM analysis, a small amount of the solid sample was pulverized and suspended in water. Then, a drop of the dispersion was placed onto a cooper grid with carbon formvar and dried under ambient conditions afterward. With regard to SEM analysis, the pulverized samples were fixed onto a metallic sample holder with double-sided tape and then covered with a thin film of gold to prevent charging upon irradiation.

3.3.3. Fluorescence Analysis of Melatonin Aqueous Solutions

Fluorescence contour graphs on melatonin (MEL) aqueous solutions with a concentration of 80 μg L^{-1} were registered to assess the fluorescence intensity (F) as well as the maximum excitation and emission wavelengths. MEL spectra were also registered in methanol and in aqueous solutions of CTAB and Brij L23 surfactants, which were explored as desorbing agents. The calibration curves in all the indicated media were recorded at the maximum excitation and emission wavelengths. The external standard method was applied to validate the developed methodology. The sensitivity, limit of detection (LOD), limit of quantification (LOQ), robustness, linear range, and precision of the method using Brij L23 and MeOH as desorbing agents were assessed using just the fluorescence method as well as together with the dSPE process. The sensitivity was obtained from the calibration curve as the ratio of the change in the concentration to the change in the fluorescence intensity. The LOQ and LOD were calculated as the concentration matching the intercept plus ten or three times the standard deviation of the intercept, respectively. The robustness was calculated as the relative standard deviation of the slopes of the calibration curve obtained on four different days. The intra-day precision (repeatability) and inter-day precision (reproducibility) were estimated, respectively, as the relative standard deviation of four measurements carried out within the same day or on different days.

3.3.4. Extraction of Melatonin Using Graphene/Clay Mixtures as Sorbents

The extraction of MEL was first carried out by mixing 10 mg of the graphene/clay solid mixture with 25 mL of a melatonin solution at a concentration of 80 $\mu g\ L^{-1}$. The mixture was agitated for 15 min and then centrifuged at a speed of 2598 g for the same period. Upon elimination of the supernatant, fluorescence measurements were performed at excitation and emission wavelengths of 280 and 340 nm, respectively. The retained MEL was calculated as the difference between the final and the first concentration measured. After solid isolation, the tube containing the solid was mixed with 25 mL of the solution containing the desorbing agent (MeOH, CTAB, or Brij L23), agitated for 15 min, and finally centrifuged under the identical conditions to those mentioned for the extraction stage. The desorption volume varied from 25 mL to 5 mL with MeOH and Brij L23, respectively. Finally, the supernatant was measured by fluorescence at the indicated excitation and emission wavelengths.

3.3.5. Melatonin Extraction and Determination in Real Samples

Melatonin concentration in herbal tea samples was determined by fluorescence measurements after the dSPE process, and the results were compared with a reference method (HPLC). 8 mg of the herbal tea bags were treated with 100 mL of ultrapure water and heated for 10 min at the boiling point. The solution was filtered, and the dSPE method was applied using 25 mL as sample volume with 5 mL of desorption volume (preconcentration factor of 5). The supernatant obtained was separated and measured by fluorescence and HPLC. The results from both methods were compared by a statistical t-test comparison.

Another herbal tea sample without melatonin was spiked with low concentrations of the analyte. In total, 5 mg of the herbal tea were weighed, the appropriate volumes of a 20 mg L^{-1} melatonin stock solution were added to the solid, and they were treated with 250 mL of ultrapure water followed by heating for 10 min at the boiling point. The melatonin concentrations in the extract were 8 and 10 $\mu g\ L^{-1}$. In total, 25 mL of the supernatant was extracted and desorbed with 5 mL of Brij L23 60 mM by dSPE and measured by fluorescence. The melatonin concentration was quantified after a preconcentration in the dSPE extraction step with 60 mM Brij L23.

4. Conclusions

In this work, an easy and cheap method for the direct determination of melatonin that combines a dSPE stage with fluorescence analysis has been developed. G/SEP and G/BEN 4/96 and 10/90 w/w mixtures have been tested as sorbents for the extraction step, and the retention has been found to be quantitative for the mixtures with both types of clays. Hence, both can be used in the dSPE process. The desorption was found to be easier for the 4/96 w/w mixtures. Hence, they were chosen to investigate the desorption process. The percentage of melatonin recovery in these mixtures was very high (>90%) using MeOH and 60 mM or 80 mM Brij L23 as desorbents, though it was slightly lower for G/BEN than for G/SEP. Very high recoveries and a preconcentration factor of up to 5 have been achieved using the G/SEP mixture as sorbent and MeOH or 60 mM Brij L23 as desorbents. The optimal extraction conditions were a G/SEP (4/96 w/w) mixture as sorbent and a 60 mM Brij L23 solution as desorbent. A preconcentration factor of up to 5 with very good recovery (90.9%) can be attained by reducing the final volume of the extract to 5 mL.

In herbal tea samples containing melatonin, the results obtained via dSPE extraction combined with fluorescence measurement and with the HPLC method were statistically equal. The percentage of agreement between the melatonin concentration obtained by the whole developed method and the labeled value is 86.5%. Another herbal tea sample without melatonin was spiked with two concentrations below the limit of quantification of the fluorescence method, leading to very high recoveries (100 and 102%) that demonstrate the accuracy of the method developed herein.

Supplementary Materials: The following supporting information can be downloaded at https://www.mdpi.com/article/10.3390/molecules29112699/s1, Figure S1: SEM images of different G/C mixtures with a magnification of 5000×. Figure S2: Fluorescence contour graphs of MEL in different concentrations of Brij L23. Figure S3: Preparation of Graphene/Clay mixtures for dispersive solid phase extraction.

Author Contributions: L.G.-F.: investigation and writing—original draft. A.M.D.-P.: writing—review and editing—and funding acquisition. M.P.S.A.: conceptualization, methodology, writing—original draft—and Supervision. All authors have read and agreed to the published version of the manuscript.

Funding: This research was funded by the Community of Madrid, grant number EPU-INV/2020/012.

Institutional Review Board Statement: Not applicable.

Informed Consent Statement: Not applicable.

Data Availability Statement: Data is contained within the article and Supplementary Materials.

Acknowledgments: The authors want to thank Belen Marcos and the enterprise SEPIOL SA for providing sepiolite and bentonite clays used in this work.

Conflicts of Interest: The authors declare no conflicts of interest.

References

1. Mercolini, L.; Mandrioli, R.; Raggi, M.A. Content of melatonin and other antioxidants in grape-related foodstuffs: Measurement using a MEPS-HPLC-F method. *J. Pineal Res.* **2012**, *53*, 21–28. [CrossRef] [PubMed]
2. Wang, X. The antiapoptotic activity of melatonin in neurodegenerative diseases. *CNS Neurosci. Ther.* **2009**, *15*, 345–357. [CrossRef] [PubMed]
3. Anisimov, V.N.; Popovich, I.G.; Zabezhinski, M.A.; Anisimov, S.V.; Vesnushkin, G.M.; Vinogradova, I.A. Review-melatonin as antioxidant, geroprotector and anticarcinogen. *Biochim. Biophys. Acta* **2006**, *1757*, 573–589. [CrossRef] [PubMed]
4. Poeggeler, B.; Saarela, S.; Reiter, R.J.; Tan, D.X.; Chen, L.D.; Manchester, L.C.; Barlow-Walden, L.R. Melatonin—A highly potent endogenous radical scavenger and electron donor: New aspects of the oxidation chemistry of this indole accessed in vitro. *Ann. N. Y. Acad. Sci.* **1994**, *738*, 419–420. [CrossRef] [PubMed]
5. Poeggeler, B.; Balzer, I.; Hardeland, A.; Lerchl, A. Pineal hormone melatonin oscillates also in the dinoflagellate Gonyaulax polyedra. *Sci. Nat.* **1991**, *78*, 268–269. [CrossRef]
6. Posmyk, M.M.; Janas, K.M. Melatonin in plants. *Acta Physiol. Plant.* **2009**, *31*, 1–11. [CrossRef]
7. González-Gómez, D.; Lozano, M.; Fernández-León, M.F.; Ayuso, M.C.; Bernalte, M.J.; Rodríguez, A.B. Detection and quantification of melatonin and serotonin in eight sweet cherry cultivars (*Prunus avium* L.). *Eur. Food Res. Technol.* **2009**, *229*, 223–229. [CrossRef]
8. Van Tassel, D.L.; Roberts, N.; Lewy, A.; O'Neill, S.D. Melatonin in plant organs. *J. Pineal Res.* **2001**, *31*, 8–15. [CrossRef]
9. Badria, F.A. Melatonin, serotonin, and tryptamine in some Egyptian food and medicinal plants. *J. Med. Food* **2002**, *5*, 153–157. [CrossRef]
10. Burkhardt, S.; Tan, D.X.; Manchester, L.C.; Hardeland, R.; Reiter, R.J. Detection and quantification of the antioxidant melatonin in Montmorency and Balaton tart cherries (*Prunus cerasus*). *J. Agric. Food Chem.* **2001**, *49*, 4898–4902. [CrossRef]
11. Iriti, M.; Rossoni, M.; Faoro, F. Melatonin content in grape: Myth or panacea? *J. Sci. Food Agric.* **2006**, *86*, 1432–1438. [CrossRef]
12. Stürtz, M.; Cerezo, A.B.; Cantos-Villar, E.; Garcia-Parrilla, M.C. Determination of the melatonin content of different varieties of tomatoes (*Lycopersicon esculentum*) and strawberries (*Fragaria ananassa*). *Food Chem.* **2011**, *127*, 1329–1334. [CrossRef]
13. Kirakosyan, A.; Seymour, E.M.; Urcuyo Llanes, D.E.; Kaufman, P.B.; Bolling, S.F. Chemical profile and antioxidant capacities of tart cherry products. *Food Chem.* **2009**, *115*, 20–25. [CrossRef]
14. Pola, M.L.; Algarra, M.; Becerra, A.; Hernández, M. Cyclodextrin enhanced spectrofluorimetric determination of melatonin in pharmaceuticals and urine. *Anal. Lett.* **2000**, *33*, 891–903. [CrossRef]
15. Liu, B.; You, Y.; Lin, D.; Chen, Z.; Qiu, P. Simple colorimetric and fluorometric assay based on 2,3-naphthalenedialdehyde for melatonin in human saliva. *Chem. Afr.* **2020**, *3*, 181–188. [CrossRef]
16. Ashrafzadeh Afshar, E.; Ali Taher, M.; Karimi, F.; Karaman, C.; Moradi, O. Ultrasensitive and highly selective "turn-on" fluorescent sensor for the detection and measurement of melatonin in juice samples. *Chemosphere* **2022**, *295*, 133869. [CrossRef] [PubMed]
17. Galian, R.E.; Veglia, A.V.; de Rossi, R.H. Hydroxypropyl-β-cyclodextrin enhanced fluorimetric method for the determination of melatonin and 5-methoxytryptamine. *Analyst* **2000**, *125*, 1465–1470. [CrossRef] [PubMed]
18. Pape, C.; Lüning, K. Quantification of melatonin in phototrophic organisms. *J. Pineal Res.* **2006**, *41*, 157–165. [CrossRef]

19. El Moussaoui, N.; Bendriss, A. Analysis of Melatonin by High Performance Liquid Chromatography after Solid-Phase Extraction (SPE/HPLC-FD). *Int. J. Eng. Res. Technol.* **2015**, *4*, 988–993.
20. Pothinuch, P.; Tongchitpakdee, S. Melatonin contents in mulberry (*Morus* spp.) leaves: Effects of sample preparation, cultivar, leaf age and tea processing. *Food Chem.* **2011**, *128*, 415–419. [CrossRef]
21. Anastassiades, M.; Lehotay, S.J.; Stajnbaher, D.; Schenck, F.J. Fast and easy multiresidue method employing acetonitrile extraction/partitioning and dispersive solid-phase extraction for the determination of pesticide residues in produce. *J. AOAC Int.* **2003**, *86*, 412–431. [CrossRef] [PubMed]
22. Scigalski, P.; Kosobucki, P. Recent Materials Developed for Dispersive Solid Phase Extraction. *Molecules* **2020**, *25*, 4869. [CrossRef] [PubMed]
23. González-Sálamo, J.; Socas-Rodríguez, B.; Hernández-Borges, J.; Rodríguez-Delgado, M.A. Nanomaterials as sorbents for food sample analysis. *Trends Anal. Chem.* **2016**, *85*, 203–220. [CrossRef]
24. Geim, A.; Novoselov, K. The rise of graphene. *Nat. Mater.* **2007**, *6*, 183–191. [CrossRef] [PubMed]
25. Sitko, R.; Zawisza, B.; Malicka, E. Graphene as a new sorbent in analytical chemistry. *Trends Anal. Chem.* **2013**, *51*, 33–43. [CrossRef]
26. Liu, Q.; Shi, J.; Jiang, G. Application of graphene in analytical sample preparation. *Trends Anal. Chem.* **2012**, *37*, 1–11. [CrossRef]
27. Liu, Q.; Shi, J.; Sun, J.; Wang, T.; Zeng, L.; Jiang, G. Graphene and graphene oxide sheets supported on silica as versatile and high-performance adsorbents for solid-phase extraction. *Angew. Chem. Int. Ed.* **2011**, *50*, 5913–5917. [CrossRef] [PubMed]
28. Aranda, P.; Darder, M.; Wicklein, B.; Rytwo, G.; Ruiz-Hitzky, E. Clay–Organic Interfaces for Design of Functional Hybrid Materials. In *Hybrid Organic-Inorganic Interfaces*; Delville, M.H., Taubert, A., Eds.; Wiley-VCH Verlag GmbH & Co. KGaA: Weinheim, Germany, 2018; pp. 1–84. [CrossRef]
29. Mateos, R.; Vera-López, S.; Díez-Pascual, A.M.; San Andrés, M.P. Dispersive solid phase extraction/fluorescence analysis of riboflavin using sepiolite as sorbent. *Appl. Clay Sci.* **2018**, *163*, 279–290. [CrossRef]
30. Mateos, R.; Vera-López, S.; Saz, M.; Díez-Pascual, A.M.; San Andrés, M.P. Graphene/sepiolite mixtures as dispersive solid-phase extraction sorbents for the analysis of polycyclic aromatic hydrocarbons in wastewater using surfactant aqueous solutions for desorption. *J. Chromatogr. A* **2019**, *1596*, 30–40. [CrossRef]
31. Gutiérrez-Fernández, L.; Vera-López, S.; Díez-Pascual, A.M.; San Andrés, M.P. Easy, fast, and clean fluorescence analysis of tryptophan with clays and graphene/clay mixtures. *J. Food Compos. Anal.* **2022**, *114*, 104858. [CrossRef]
32. Lakowicz, J.R. *Principles of Fluorescence Spectroscopy*; Springer Science + Business Media: New York, NY, USA, 2006; pp. 205–208.
33. Lerner, D.A.; Martin, M.A. Luminescence in organized media and supramolecular interactions: Physicochemical aspects and applications. *Analusis* **2000**, *28*, 649–663. [CrossRef]
34. Ramos-Lledó, P.; Vera, S.; San Andrés, M.P. Determination of vitamins A and E in milk samples by fluorescence in micellar media. *Fres. J. Anal. Chem.* **2001**, *369*, 91–95. [CrossRef] [PubMed]
35. León-Ruiz, V.; Vera, S.; San Andrés, M.P. Validation of a screening method for the simultaneous identification of fat-soluble and water-soluble vitamins (A, E, B_1, B_2 and B_6) in an aqueous micellar medium of hexadecyltrimethylammonium chloride. *Anal. Bioanal. Chem.* **2005**, *381*, 1568–1575. [CrossRef] [PubMed]

molecules

MDPI

Article

A Fusion–Growth Protocell Model Based on Vesicle Interactions with Pyrite Particles

Dong Guo [1], Ziyue Zhang [1], Jichao Sun [1], Hui Zhao [2], Wanguo Hou [1,2] and Na Du [1,*]

1 Key Laboratory of Colloid and Interface Chemistry (Ministry of Education), School of Chemistry and Chemical Engineering, Shandong University, Jinan 250100, China
2 National Engineering Technology Research Center for Colloidal Materials, Shandong University, Jinan 250100, China
* Correspondence: duna@sdu.edu.cn; Tel.: +86-531-88364242

Abstract: Protocell models play a pivotal role in the exploration of the origin of life. Vesicles are one type of protocell model that have attracted much attention. Simple single-chain amphiphiles (SACs) and organic small molecules (OSMs) possess primitive relevance and were most likely the building blocks of protocells on the early Earth. OSM@SAC vesicles have been considered to be plausible protocell models. Pyrite (FeS_2), a mineral with primitive relevance, is ubiquitous in nature and plays a crucial role in the exploration of the origin of life in the mineral–water interface scenario. "How do protocell models based on OSM@SAC vesicles interact with a mineral–water interface scenario that simulates a primitive Earth environment" remains an unresolved question. Hence, we select primitive relevant sodium monododecyl phosphate (SDP), isopentenol (IPN) and pyrite (FeS_2) mineral particles to build a protocell model. The model investigates the basic physical and chemical properties of FeS_2 particles and reveals the effects of the size, content and duration of interaction of FeS_2 particles on IPN@SDP vesicles. This deepens the understanding of protocell growth mechanisms in scenarios of mineral–water interfaces in primitive Earth environments and provides new information for the exploration of the origin of life.

Keywords: vesicle; single-chain amphiphiles; pyrite; solid–liquid interface; protocell

1. Introduction

"How does life originate? How can non-living inorganic matter transform into organic life forms through complex systems? Do humans have the ability to create life from scratch? Can we control complex systems like cells?" were identified as some of the world's 125 most cutting-edge major scientific questions, as published in the journal *Science* in 2021 [1]. The origin of cellular life remains elusive, with no definitive answer. Primitive life depends on the self-organizing properties of its constituent parts, as well as on the input of energy and matter from the environment, to execute the most fundamental cellular processes [2]. Protocells, regarded as the earliest life-like entities, are hypothesized to consist of three essential components: membrane-forming molecules, information molecules and catalytic molecules [3]. In the origin of life, energy supply was imperative. In 1988, Wächtershäuser postulated the hypothesis of the origin of life within an iron–sulfur environment [4,5], arguing that at the origin of life, energy supply was primitive autotrophic metabolism [6]. The energy sources hypothesized to have driven the origin of life included iron sulfides and other minerals (e.g., pyrite). The energy released from the redox reactions of these metal sulfides could have facilitated the synthesis not only of organic molecules but also of oligomers or polymers. These systems could have developed sets of autocatalytic systems capable of self-replication, as well as living and independent entities capable of metabolizing life forms prior to those known to us today [4]. In contemporary geochemical environments, the population of prokaryotic cells within biofilm communities residing at mineral–water interfaces surpasses by orders of magnitude those inhabiting water environments, owing to the manifold metabolic and protective functions these interfaces offer [7–9]. Hence, it is a plausible conjecture that protocells on the primordial Earth were in proximity to mineral–water interfaces [10–13].

Through the compositional analysis of meteorite extracts [14,15] and experiments simulating primordial Earth environmental conditions [10,11], researchers have identified a range of organic molecules pertinent to early Earth conditions: (1) lipid-like compounds capable of forming protocell structures, consisting of simple single-chained amphiphilic molecules (SCAs) such as fatty acids, alkyl keto acids and monoalkyl phosphates [16–22]; (2) organic small molecules (OSMs) containing L-amino acids, purines, pyrimidines, D-ribose and isoprenoid derivatives, etc. [23–25]. It can be deduced that these SCAs and OSMs, demonstrated to possess primitive relevance, were probably constituents of protocells during the early stages of Earth. SCAs have been extensively employed as modeling units for protocell membranes in studies on the origin of life [26–29] It has been shown that specific minerals can play some key roles as catalysts [30–33]. For instance, they have been shown to catalyze the formation of RNA polymers [34] and peptides [35], while minerals can enhance the initial assembly rate of amphiphiles into vesicles, ultimately contributing to the formation of protocell membranes [36,37]. The iron group elements Fe, Co and Ni serve as the most effective and versatile catalysts of life, and Fe is the most prominent and has the greatest geochemical abundance. The most stable compounds of the iron family elements and the most prominent iron compounds under anaerobic conditions are sulfides [38]. The transition from inanimate to living matter may result from the self-assembling properties of organic molecules and their interaction with the chemical diverse inorganic environment [2]. So, what kind of "sparks" can collide between minerals of primordial relevance and the "OSM@SCA vesicle" protocell model?

In this study, we select primitive relevant sodium monododecyl phosphate (SDP), isopentenol (IPN) and pyrite (FeS_2) mineral particles to establish a protocell model. We investigate the fundamental physical and chemical properties of FeS_2 particles and reveal the effects of the FeS_2 size, content and duration of interaction on IPN@SDP vesicles. Through simulating scenarios of the mineral–water interface in the primitive Earth environment, the changing rules of morphology, size and structure of the protocell model system are explored to provide information for the exploration of the origin of life.

2. Results and Discussion

2.1. Basic Physical and Chemical Properties of FeS_2

From scanning electron microscope (SEM) (Figure 1a) and optical microscope (OM) (Figure 1b) observations, FeS_2-1 is an irregularly shaped massive particle with a size of ~1 μm. The peak of its dynamic light scattering (DLS) size distribution is located at ~830 nm (Figure 1c), which is consistent with the results of the electron microscopy observations. FeS_2-60 is an irregularly shaped massive particle (Figure 1d,e) with a size of 60 ± 15 μm. The isoelectric point of FeS_2 particles in water is ~6.6 (Figure 1f), which is consistent with the values reported in the literature [39]. The pK_a (where K_a is the acid dissociation constant) values of SDP are 2.85 (pK_{a1}) and 7.35 (pK_{a2}) [40]. The pH of the SDP/IPN/H_2O homogeneous solution is measured as 6.30–6.50, close to pK_{a2} (7.35). This indicates that under the studied conditions, the main forms of SDP are $C_{12}H_{25}OP(OH)O_2Na$ and $C_{12}H_{25}OPO_3Na_2$, with trace amounts of $C_{12}H_{25}OP(OH)_2$. Under this circumstance, those FeS_2 particles should be slightly positively charged. The contact angles of water on the surfaces of FeS_2-1 and FeS_2-60 particles are 48 ± 3° and 47 ± 3°, respectively (Figure 1g,h), indicating that surfaces of FeS_2 mineral particles are hydrophilic. Type IV adsorption isotherms and H3-type hysteresis loops are obtained through nitrogen adsorption and desorption measurements (Figure 1i,j). Combining the results of the Barrett–Joyner–Halenda (BJH) pore-size distributions, it is revealed that the FeS_2 particles construct few slit mesopores by particle stacking. The Brunauer–Emmett–Teller (BET) specific surface areas of FeS_2-1 and FeS_2-60 are 1.2 ± 0.1 m^2/g and 0.3 ± 0.1 m^2/g, respectively, with the former being ~4 times that of the latter.

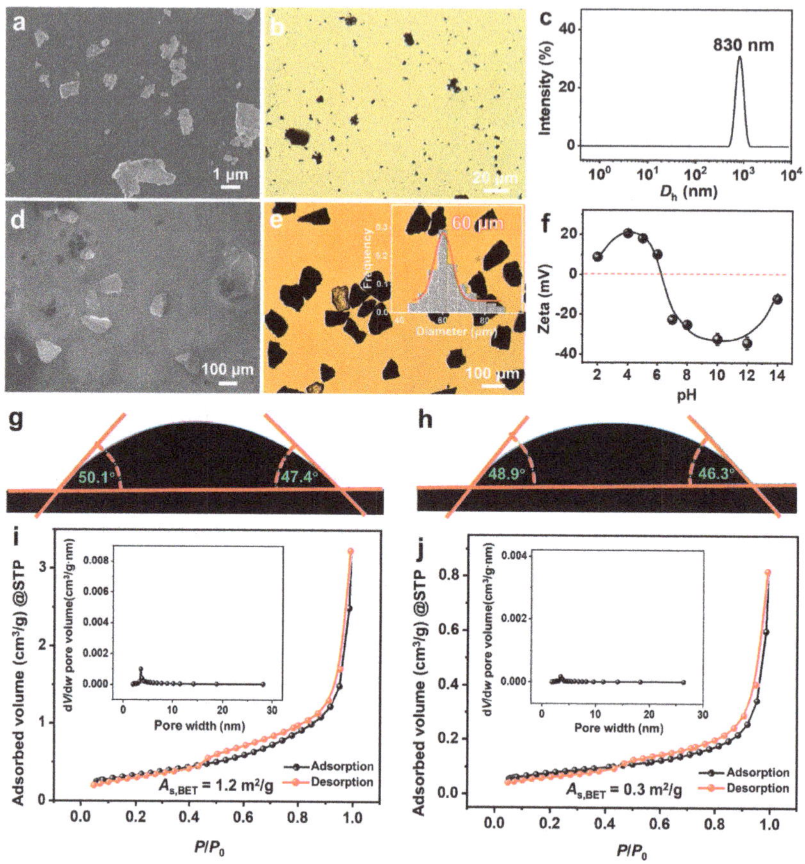

Figure 1. (**a**,**d**) SEM images, (**b**,**e**) OM images, (**c**) DLS size distributions, (**f**) zeta potential versus pH curves (25.0 ± 0.5 °C), (**g**,**h**) contact angles of water on the surface of FeS_2 particles and (**i**,**j**) nitrogen adsorption–desorption isotherms for (**a**–**c**,**f**,**g**,**i**) FeS_2-1 and (**d**,**e**,**h**,**j**) FeS_2-60, with the inset in e being a histogram of the size distribution of FeS_2-60, and the insets in (**i**,**j**) being the pore size distributions of BJH.

2.2. Effect of FeS_2 Particles on Vesicles

The vesicle phase is detected in the SDP/IPN/H_2O ternary system in the water-rich zone (Figure S1). The concentrations of SDP and IPN are quite low, and the appearance of the vesicle solution is clear and transparent (Figure 2a). The appearance of the FeS_2-vesicle solutions turns turbid upon the interaction time increasing (Figure 2a,b). The transmittance of the FeS_2-vesicle solution decreases gradually from ~99% to ~83% and ~94% with the addition of FeS_2-1 and FeS_2-60 (Figure 2c), respectively, and reaches a constant in ~12 h. In contrast, the supernatants of FeS_2-H_2O show no change at all for the transmittance test. FeS_2 particles with smaller size enhance the turbidity of the FeS_2-vesicle solution more significantly than the case of FeS_2-60, which may be attributed to its higher specific surface area (Figure 1i,j).

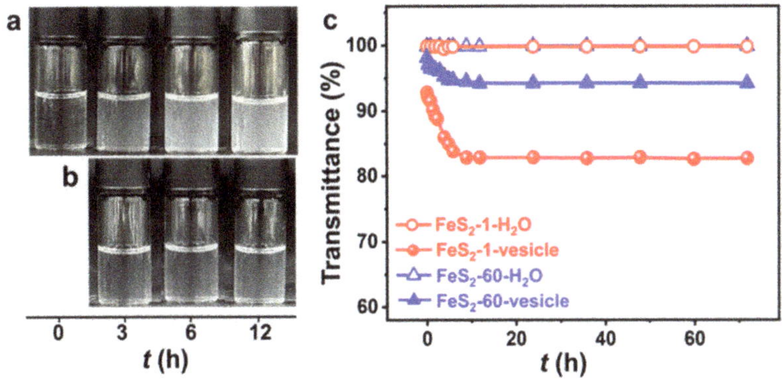

Figure 2. (**a**,**b**) Photographs of the appearance and (**c**) transmittance curves (25.0 ± 0.5 °C) of FeS_2-vesicle solutions at different times. Transmittance is the average of three measurements. (**a**) FeS_2-1-vesicle, (**b**) FeS_2-60-vesicle.

The spherical IPN@SDP vesicles are observed by negative staining transmission electron microscopy (NS-TEM) (Figure 3a). After the addition of FeS_2, the vesicles vary gradually from spherical to dumbbell-shaped, then to ellipsoidal, and finally to spherical vesicles again, but with larger sizes (Figure 3c–e,i–k). According to the results of size distribution, the vesicle size distribution peak is initially located at ~150 nm (Figure 3b). Upon the interaction between FeS_2 particles and vesicles, a new peak appears at 500–600 nm (Figure 3f,l), probably representing the size peak of the dumbbell-shaped vesicles. Then, the peak area of the new peak gradually increases, and the peak of the size distribution is located at ~400 nm. This may result from the fusion of the two spheres in the dumbbell-shaped vesicles, namely ellipsoidal vesicles (Figure 3g,m). Ultimately, the size distribution peak is located at ~300 nm, corresponding to the spherical vesicles with larger sizes than the vesicles without FeS_2 particles (Figure 3h,n). The DLS results generally agree with the NS-TEM observations. In summary, the FeS_2 particles induce the morphology transition of sphere–dumbbell–ellipsoid–larger sphere vesicles, which may be the intrinsic reason for the increase in the turbidity of the vesicular solution under the effect of FeS_2 particles. This transition takes 12 h to establish a dynamic equilibrium state, which is also consistent with the results of turbidity tests (Figure 2). Compared to the original pyrite samples, the water contact angles (θ_w) of the FeS_2 mixed with vesicle solution are decreased from 47–48° to 26–27° (Figure S2a,b), respectively. The decrease in θ_w is probably attributed to the adsorption of the amphiphilic bilayer [41–43] on the particle surface. Moreover, the θ_w values of FeS_2 samples that interact with the vesicle solutions increase to 46–48° when they are subjected to ultrasonic treatment in water (Figure S2c,d), which are almost the same values as the original FeS_2 particle samples (Figure 1g,h). Therefore, it is reasonable to believe that the desorption of SDP or IPN molecules on the FeS_2–water interface could be induced and accelerated by certain energy inputs (e.g., shaking, stirring and ultrasonic agitation) [44,45]. The dynamics of the adsorption-desorption equilibrium on the water–FeS_2 interface probably play an important role in this required time (12 h) to establish the apparent equilibrium states of the FeS_2-vesicle samples.

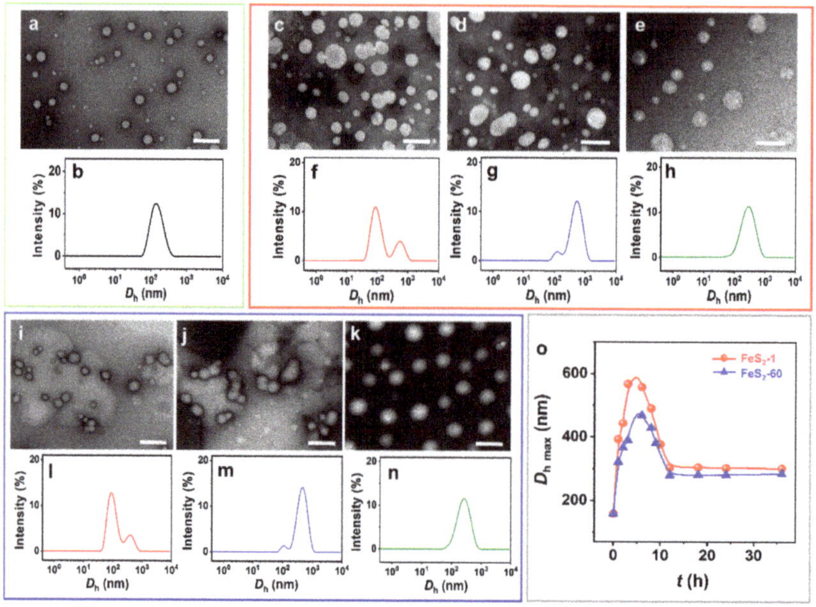

Figure 3. (**a**,**c**–**e**,**i**–**k**) NS-TEM images and (**b**,**f**–**h**,**l**–**n**) DLS size distributions of (**a**,**b**) vesicle, (**c**–**h**) FeS_2-1-vesicle and (**i**–**n**) FeS_2-60-vesicle solutions at different times. (**o**) DLS size distribution of FeS_2-vesicle solutions as a function of time. (**c**,**f**,**i**,**l**) 3 h; (**d**,**g**,**j**,**m**) 6 h; (**e**,**h**,**k**,**n**) 12 h. Scale bar: 500 nm.

Further, the increase in vesicle size after interaction with FeS_2-1 is more pronounced than the one with FeS_2-60 (Figure 3o), presumably due to the larger specific surface area of FeS_2-1. In addition, there is no significant change in the appearance, morphology and size of IPN@SDP vesicles after storage at 25.0 ± 5.0 °C for 6 months (Figure S3), which is similar with those solid interface-induced, simple single-chained amphiphilic molecule (SCA) vesicles in our previous study [12,42].

2.3. Influence of FeS_2 Particle Content

Varying amounts of FeS_2 particles are introduced into the vesicle solution to investigate the effect of FeS_2 particle content on the vesicles. As the FeS_2 content increases, the appearance of the FeS_2-vesicle solution (Figure 4a,b) varies from clear to turbid gradually. With the same content, the appearance of the vesicle solution with FeS_2-1 is more turbid than that of the one with FeS_2-60. Transmittance results (Figure 4c) show that the transmittance of the FeS_2-vesicle solution decreases with the increase in the content of FeS_2 particles, whereas there is no change in the transmittance of FeS_2-water under the same conditions, which is ~99.9%. This excludes the possibility of residual FeS_2 particles in the FeS_2-vesicle solution. With the same content, FeS_2-1 reduces the transmittance more strongly than FeS_2-60, which is consistent with the variation in appearance (Figure 4a,b). In addition, the decrease in the transmittance of the FeS_2-1-vesicle solution is 1.6–3.7 times that of the FeS_2-60-vesicle solution over a range of contents. This can be attributed to the fact that the specific surface area of FeS_2-1 is larger than that of FeS_2-60, and the interfacial adsorption and enrichment are more effective.

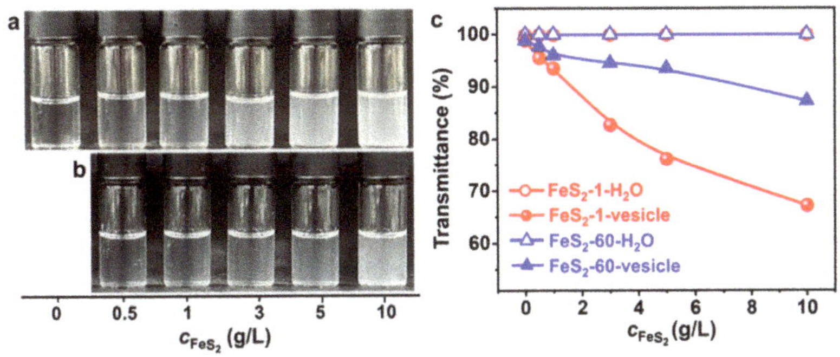

Figure 4. (**a**,**b**) Photographs of the appearance and (**c**) transmittance of FeS_2-vesicle solutions at different particle contents (24 h). Transmittance is the average of three measurements. (**a**) FeS_2-1-vesicle, (**b**) FeS_2-60-vesicle.

NS-TEM results show that the morphology of these vesicles does not change, but the size of the IPN@SDP vesicles grows gradually with the increase in FeS_2 particle content (Figure 5a–d,i–l). The DLS results indicate that the size of the vesicles in FeS_2-1-vesicle solution increases from ~158 nm to ~332 nm, while that in the FeS_2-60-vesicle solution increases to ~298 nm gradually (Figure 5e–h,m–p). This is consistent with the TEM results. As shown in Figure 6, the trend of FeS_2-1 in inducing a vesicle size increase is greater than that of FeS_2-60 particles, which is consistent with the variation in effects on the turbidity and the appearance (Figure 6).

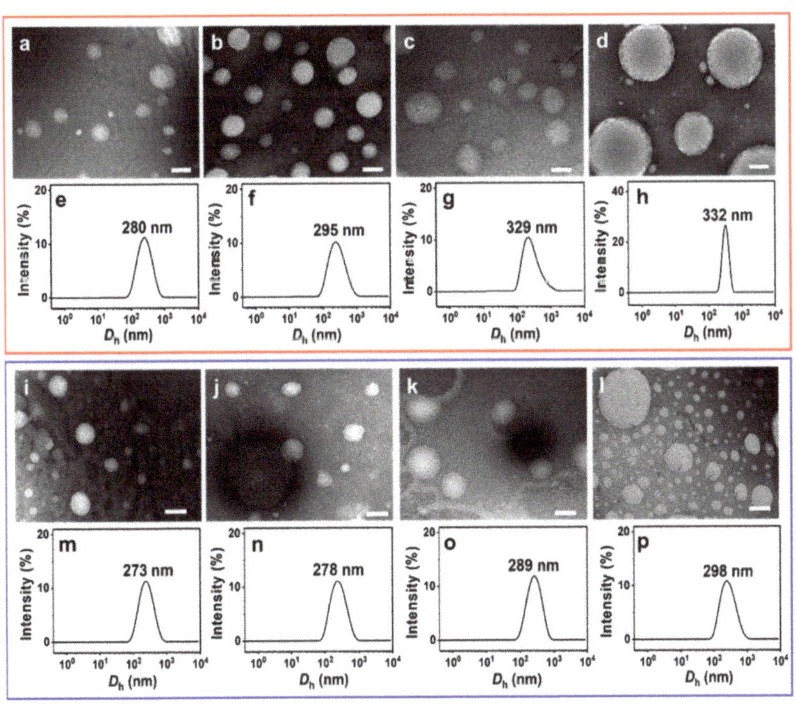

Figure 5. (**a–d**,**i**,**j**) NS-TEM images and (**e–h**,**m–p**) DLS size distributions of (**a–h**) FeS_2-1-vesicle and (**i–p**) FeS_2-60-vesicle solutions at different particle contents (24 h): (**a**,**e**,**i**,**m**) 0.5 g/L; (**b**,**f**,**j**,**n**) 1 g/L; (**c**,**g**,**k**,**o**) 5 g/L; (**d**,**h**,**l**,**p**) 10 g/L. Scale bar: 200 nm.

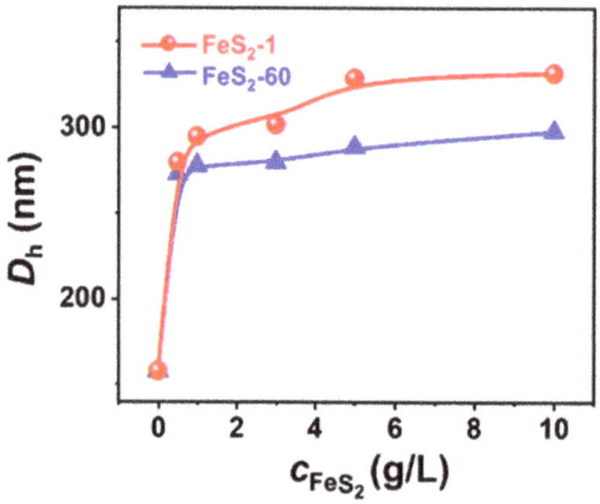

Figure 6. Variation in the DLS size distribution of FeS_2-vesicle solutions as a function of FeS_2 content.

In short, the total surface area of FeS_2 particles in the systems also gradually increases upon increasing FeS_2 content, which might lead to more remarkable bilayer adsorption and enrichment on the interface of FeS_2 particles; in other words, the matrix effect is enhanced. This may be the main reason why the vesicle size grows gradually with the increase in FeS_2 content.

2.4. Mechanism of FeS_2 Particle-Vesicle Solution Interaction

Optimized by Gaussian generalization theory, the length of the SDP molecule is 1.92 nm [46]. The small-angle X-ray scattering (SAXS) curves (Figure 7) of both the vesicle solution and FeS_2-vesicle solution samples show lamellar periodic diffraction peaks, indicating the presence of a vesicular bilayer structure. The thickness of the vesicle bilayer is about 3.70 nm (Figure S4), which is less than twice the length of the SDP molecule. This reveals that an interdigitated structure is adopted between the alkyl chains in the vesicular bilayers [46–48], with an interdigitated degree of 4.4% (Figure S4). The thicknesses of the vesicle bilayers shift to 3.74 nm and 3.76 nm through the interaction with FeS_2-1 (Figure 7a) and FeS_2-60 (Figure 7b), with interdigitated degrees of 3.1% and 2.5%, respectively. In fact, the thickness and interdigitated degrees of the vesicle membranes vary very slightly via the interaction with FeS_2 particles.

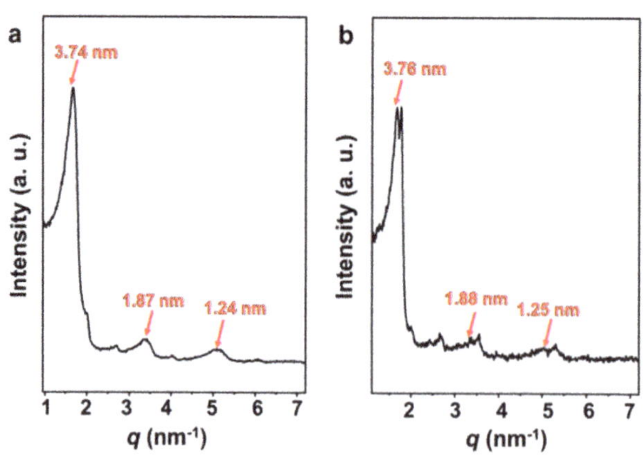

Figure 7. SAXS patterns of (**a**) FeS_2-1-vesicle and (**b**) FeS_2-60-vesicle samples.

Some simple single-chained amphiphilic molecules (SCAs) such as fatty acids [49,50], dodecylhydrogen sulfate [51] and monoalkyl phosphates [22,52,53] can form vesicles spontaneously in water close to their apparent pK_a. The structures of those SCA hydrogen bonding dimers play an important role in this vesicle formation process. As shown by the electrospray ionization mass spectrometry (ESI-MS) results (Figure 8), both FeS_2-1 and FeS_2-60 interact with vesicles and show peaks at m/z of ~289.15 ($C_{12}H_{25}OP(OH)O_2Na$), ~459.25 ($C_5H_9OH \cdots C_{12}H_{25}OPO_3^{2-}Na^+ \cdots C_5H_9OH$) and ~577.30 ($C_{12}H_{25}OP(OH)O_2Na \cdots C_{12}H_{25}OP(OH)O_2Na$). This indicates the presence of an SDP monomer, SDP-SDP dimer and IPN-SDP-IPN trimer (Table S1). Compared to the ESI-MS results of the vesicle solution (Figure S5), it is worth mentioning that the FeS_2-vesicle solution exhibits a new peak at m/z of 799.50, which indicates the generation of a new hydrogen-bonded trimer (SDP-SDP-SDP). It has been reported that minerals with prebiotic availability may have facilitated prebiotic chemistry by protecting organic molecules from UV radiation and thermal decomposition, concentrating them through adsorption and, finally, catalyzing polymerization reactions [3,54,55]. In this context, the presence of an SDP trimer reveals that pyrite mineral particles have the potential to induce the generation of new substances in simulated primitive Earth environments, which provides new possibilities for exploring the origin of life at the mineral–water interface.

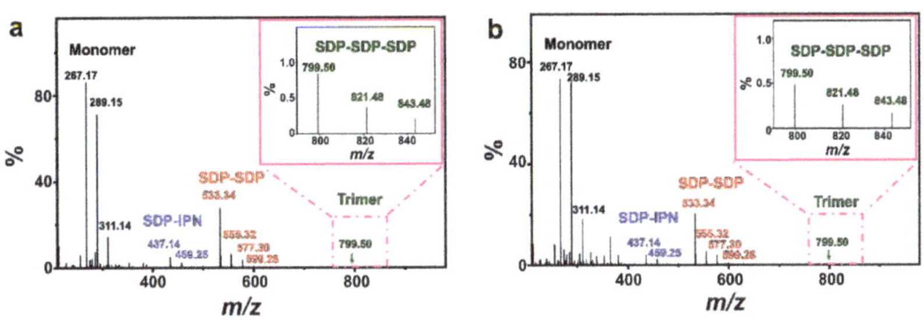

Figure 8. ESI-MS spectra of (**a**) FeS_2-1-vesicle and (**b**) FeS_2-60-vesicle samples (24 h). The insets in (**a**,**b**) are magnifications of the square part of the pink dashed line.

The self-assembly morphology of amphiphilic molecules is determined by their geometrical parameters, which are usually described by the molecular stacking parameter (P) defined as $P = v_0/a_s\, l_0$ [56,57], where v_0 is the volume of the hydrophobic chain of the amphiphilic molecule, l_0 is the length of the hydrophobic chain, and a_s is the area occupied by the polar headgroups on the surface of the aggregate. In general, spherical micelles are favored when $P \leq 1/3$, columnar micelles are favored when $1/3 \leq P \leq 1/2$ and bilayers or vesicles are favored only when $1/2 \leq P \leq 1$. Hydration and electrostatic repulsion between the headgroups can lead to large a_s values of amphiphiles, whereas the strong adsorption of amphiphiles on solid surfaces through electrostatic interactions, hydrogen bonding and van der Waals forces can significantly reduce the hydration of the headgroups, impede electrostatic repulsion between the headgroups and reduce the a_s values in amphiphilic molecular aggregates [12,58]. Based on the above background, we propose a possible mechanism of interaction of FeS_2 particles affecting the vesicle structure (Figure 9). The SDP headgroups are negatively charged, while the FeS_2 particles are positively charged. The electrostatic interactions allow the SDP monomers and bilayers of SDP/IPN vesicles to be adsorbed onto the surface of these FeS_2 particles, which reduces the interfacial energy of liquid–solid. This electrostatic adsorption behavior brings about two principle impacts: the fusion of bilayer membranes between vesicles and the growth of vesicles.

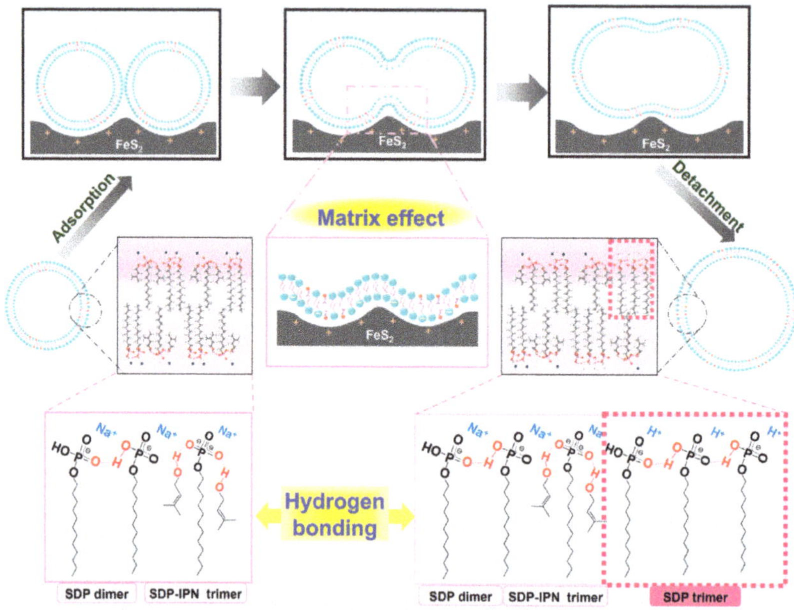

Figure 9. Schematic diagram of FeS_2-vesicle solution interaction mechanism.

(1) *Fusion of bilayer membranes between vesicles*. The electrostatic adsorption and the matrix effect [43] of the FeS_2–water interface increase the probability of contact between the vesicular bilayers. The "fast exchange" and "flip-flop" characteristics [59] of vesicles enable the contact or near enough area of the bilayer membranes to fuse with each other.

(2) *Growth of vesicles*. The FeS_2–water interface induces the dehydration of the SDP polar headgroups significantly and impedes electrostatic repulsion between the headgroups. It results in the diminution of the cross-sectional area (a_s) occupied by the SDP polar headgroups on the surface of the vesicle structures [12,58]. The reduced a_s decreases the curvature of the bilayer membranes, which leads to a lager radius of the vesicles. Cooperating with the fusion between vesicular bilayers, it results in the fusion–growth of vesicles.

As this process is repeated, the vesicles fuse and grow gradually until they reach the equilibrium of their transition (sphere–dumbbell–ellipsoid–larger sphere vesicles) (Figure 5). During this process, the FeS_2–water interface drives the formation of brand new hydrogen-bonded trimers (SDP-SDP-SDP) (Figure 8). Furthermore, the dumbbell-shaped and elliptical vesicle structures observed during this process (Figure 3) also corroborate this speculation.

In conclusion, based on the results of DLS, NS-TEM and ESI-MS (Figures 3, 5 and 8), the electrostatic adsorption between positively charged FeS_2 particles and negatively charged SDP molecules drives the vesicles to be adsorbed and enriched on the interface of FeS_2 particles, which further fuse and grow into vesicles with larger sizes. The vesicle fusion–growth process is influenced simultaneously by the electrostatic adsorption, matrix effects and the dehydration effects of the solid interface.

3. Materials and Methods

3.1. Materials

All chemicals used were of analytical reagent grade and utilized as received. Sodium monododecyl phosphate (SDP) was procured from TCI (Shanghai) Development Co., Ltd.,

Shanghai, China. Isopentenol (IPN) and natural pyrite (FeS_2) were sourced from Macklin Biochemical Technology Co., Ltd., Shanghai, China. Ultrapure water was obtained using a Hitech-Kflow water purification system (Hitech, Shanghai, China).

3.2. Size Regulation of FeS_2 Particles

Natural pyrite was ground into a granular form using an agate mortar and pestle. Pyrite particles with sizes ranging between the 200-mesh screen and the 290-mesh screen, corresponding to a size range of 60 ± 15 μm, were selected and designated as FeS_2-60. Additionally, pyrite was ground into a powder form, sieved through a 2800-mesh screen and characterized using a nanoparticle sizer and an optical microscope, revealing a size of approximately 1 μm, and it was recorded as FeS_2-1.

3.3. Preparation of Vesicle Solutions

Accurately weighed SDP was mixed with a designated amount of water, and a desired mass of IPN was added to the SDP/H_2O mixture. The sample was homogenized by shaking and vortexing and then kept at 25.0 ± 0.5 °C for 48 h to achieve equilibrium before measurements. The composition of the SDP/IPN/H_2O vesicular solution was fixed to be 0.30 wt% SDP (~10 mM) and 5.00 wt% IPN hereinafter if not otherwise specified.

3.4. Preparation of Particle–Solution Complexes

The FeS_2 particle was introduced into the SDP/IPN/H_2O solution and subjected to mixing by shaking and vortexing. The supernatant was investigated through centrifugation at 2000 rpm for 3 min using a high-speed centrifuge. If not otherwise indicated, FeS_2 was incorporated into the solution at a content of 3 g/L, and the vesicle solution was sampled with a composition of 0.30 wt% SDP and 5.00 wt% IPN. The resultant supernatant after interaction with the FeS_2 particle was noted as the FeS_2-vesicle solution. The FeS_2-vesicle solutions were kept at 25.0 ± 0.5 °C for at least 24 h after the addition of FeS_2 particles for the experiments in Section 2.3.

3.5. Characterization and Measurements

3.5.1. Scanning Electron Microscopy (SEM)

SEM images of samples were acquired using a Gemini 300 field-emission scanning electron microscope (Zeiss, Oberkochen, Germany) operated at an accelerating voltage of 3 kV. The sample solution was freeze-dried in a vacuum freeze dryer at approximately −40°. Subsequently, the freeze-dried samples were mounted on a 200-mesh C-coated grid and coated with a 5 nm thick layer of Pt using sputter coating.

3.5.2. Optical Microscope (OM)

An XPF-800C optical microscope (Tianxing, Shanghai, China) was employed to examine the morphology of the samples. Subsequently, the acquired images were processed using Nano Measurer 1.2 software to conduct size analysis and generate a histogram depicting the sample size distribution.

3.5.3. Dynamic Light Scattering (DLS) Measurements

The size distribution and average hydrodynamic diameter (D_h) of the aggregates were assessed using a Zetasizer Nano ZS90 dynamic light scattering instrument (Malvern, Worceterhire, UK) featuring a He-Ne laser (633 nm, 4 mW). Each sample underwent three measurements at 25.0 ± 0.5 °C.

3.5.4. Zeta Potential Measurement

A ZetaNano ZS-type zeta potential analyzer (Malvern, Worceterhire, UK) was employed to investigate the variation in zeta potential with pH in FeS_2 aqueous suspensions. The pH of the FeS_2 particle–water suspension was adjusted using diluted hydrochloric acid or sodium hydroxide, with the FeS_2 content set at 3 g/L. Temperature control achieved using a thermostatic water bath maintained at 25.0 ± 0.5 °C. Measurements were conducted three times concurrently, and the average values were calculated.

3.5.5. pH Measurement

A pH meter (Mettler Toledo, Shanghai, China) was utilized to measure the variation in pH of the sample solution. Measurements were performed three times concurrently, and the average value was calculated.

3.5.6. Contact Angle Meter

The water contact angle (θ_W) of the samples was determined by the sitting drop method using a DSA25 contact angle meter (KRÜSS, Hamburg, Germany). Firstly, the powder was pressed into a disc with a thickness of 1–2 mm under a pressure of 15 MPa, and then placed on the sample stage. The computer controlled the speed and volume of the droplets of the micro-syringe, and the droplets were dropped after stabilization, and the values of the droplet contact angle were photographed and recorded after equilibrium for 10 s. Each sample was tested three times, and the average value of the three tests was taken.

The FeS_2 particle–vesicle solution complexes were centrifuged at 2000 rpm for 3 min to remove the supernatant at 25.0 ± 0.5 °C. FeS_2 particles were washed 3 times by adding ultrapure water, and then air dried and recorded as FeS_2 mixed with vesicle solution. The dried FeS_2 samples were subjected to ultrasonic treatment in ultrapure water for 5 min, and then were centrifuged and air dried, noted as FeS_2 with ultrasonic treatment. The water contact angles of FeS_2 particles, FeS_2 mixed with vesicle solution and FeS_2 with ultrasonic treatment were measured separately.

3.5.7. Specific Surface Area and Pore Size Analyzer

The specific surface area and pore size distribution were obtained using an ASAP2460 specific surface area and pore size analyzer (Micromeritics, Norcross, USA) to determine the nitrogen adsorption–desorption isotherms of the samples under the condition of liquid nitrogen, and the specific surface area and pore size distribution were determined by using the BET method and BJH model. The samples were degassed under vacuum and 100 °C for 3 h before testing.

3.5.8. Ultraviolet Spectrophotometer (UV-Vis)

The transmittance of the samples was measured at a wavelength of 500 nm using a model 1800 UV-Vis spectrophotometer (Shimadzu, Kyoto, Japan) maintained at a constant temperature of 25.0 ± 0.5 °C. Quartz cuvettes with plastic caps were employed to prevent sample evaporation during analysis. Each liquid sample was placed in a quartz cuvette with a plastic lid to ensure sample integrity. Measurements were conducted three times concurrently, and the average value was calculated.

3.5.9. Negative Staining Transmission Electron Microscopy (NS-TEM)

The morphology of the aggregates was observed using uranyl acetate negative staining on a JEM-1011 transmission electron microscope (JEOL, Tokyo, Japan) operated at an accelerating voltage of 100 kV. A 10 μL aliquot of the sample was applied onto a carbon support film copper grid with a mesh size of 200 and allowed to settle for 2 min. Excess sample was blotted away using filter paper, followed by the addition of 7.5 μL of a 1.2% uranyl acetate ethanol solution. After 30 s, excess solution was removed by blotting with filter paper. Subsequently, the copper grid was dried under an infrared lamp for 30 min before the sample was transferred to a desiccator for electron microscope observation.

3.5.10. Small-Angle X-ray Scattering (SAXS)

The SAXS patterns were obtained using a SAXSess system (Anton-Paar, Graz, Austria) equipped with Cu Kα radiation and operated at 50 kV and 40 mA. Prior to measurements, the samples underwent freeze-drying under vacuum conditions.

3.5.11. Electrospray Ionization Mass Spectrometry (ESI-MS)

Mass spectrometry data were acquired in positive ion mode using a Bruker Impact M1 ultra-high performance liquid chromatography–quadrupole time-of-flight mass spectrometer (Bruker, Karlsruhe, Germany).

3.5.12. Molecular Dynamics Simulation

The molecular modeling and theoretical investigation for SDP dimers and trimers were carried out with LDA-DFT as implemented in the Dmol3 package provided by Materials Studio 2019. The local functional for the exchange correlation potential is LDA-PWC. The basis set is DND with unrestricted spin, minimum basis set, 3.5 basis files and fine cut-off [60]. The convergence tolerances for energy change, maximum force and maximum displacement between optimization cycles were set as 1.0×10^{-5} Ha, 0.002 Ha Å^{-1} and 0.005 Å, respectively [61].

4. Conclusions

In summary, we selected primitive relevant compounds (SDP, IPN and FeS_2) to construct a protocell model in water. The addition of FeS_2 particles enables the fusion and growth of IPN@SDP vesicles, induces the morphology transition of sphere–dumbbell–ellipsoid–larger sphere vesicles and reaches the kinetic equilibrium in ~12 h. The mediating behavior of pyrite solid–water interfaces increases the vesicle size from ~150 nm to ~300 nm, and the transmittance decreases gradually from ~99% to 94–83%. In particular, the pyrite solid–liquid interfacial mediating capacity of pyrite particles with small particle sizes (~1 μm, FeS_2-1) is more significant than the larger one (~60 μm, FeS_2-60), which may be due to the larger specific surface area of FeS_2-1. As the FeS_2 particle content increases, the total surface area increases, and the more pronounced the matrix effect at the solid–liquid interface is, inducing an increase in vesicle size. Briefly speaking, the matrix effect, intermolecular hydrogen bonding, and electrostatic interaction is demonstrated to be the main driving force for the fusion–growth behavior of the vesicle, the process in which the FeS_2–water interface drives the formation of the new substance (hydrogen-bonded trimers). This opens up the possibility for the generation of new substances in the prebiotic solid–liquid interface scenario. We expect this work to provide important insights into the effect of the solid–liquid interface on the self-assembly chemistry of SCAs and OSMs with primitive relevance in bulk solution, which, in the long run, may shed some light on the establishment of the model systems of early cell membranes for exploring the origin of life.

Supplementary Materials: The following supporting information can be downloaded at https://www.mdpi.com/article/10.3390/molecules29112664/s1: Figure S1: Vesicle phase diagram of the SDP/IPN/H_2O ternary system at 25.0 ± 0.5 °C; Figure S2: The water contact angles on the surface of (a,b) FeS_2 mixed with vesicle solution and (c,d) FeS_2 with ultrasonic treatment. (a,c) FeS_2-1, (b,d) FeS_2-60; Figure S3: (a,d) Photographs, (b,e) NS-TEM images and (c,f) DLS size distribution in (a–c) FeS_2-1-vesicle and (d–f) FeS_2-60-vesicle at 25.0 ± 0.5 °C over six months. Scale bar: 200 nm; Figure S4: SAXS patterns of vesicle samples; Table S1: Chemical structures corresponding to different m/z in ESI-MS spectra; Figure S5: ESI-MS spectra of vesicle samples (24 h).

Author Contributions: Conceptualization, N.D. and W.H.; methodology, D.G.; software, J.S. and H.Z.; validation, D.G.; investigation, D.G. and Z.Z.; resources, N.D.; data curation, D.G. and Z.Z.; writing—original draft preparation, D.G.; writing—review and editing, N.D. and W.H.; visualization, N.D.; supervision, N.D.; project administration, N.D.; funding acquisition, N.D. All authors have read and agreed to the published version of the manuscript.

Funding: This research was funded by the National Natural Science Foundation of China (No. 22272088, No. 22302165 and No. 21872082) and the Natural Science Foundation of Shandong Province, China (ZR2019MB025).

Institutional Review Board Statement: Not applicable.

Informed Consent Statement: Not applicable.

Data Availability Statement: The data presented in this study are available on request from the corresponding author.

Acknowledgments: The authors would like to acknowledge the technical support from Shandong University Structural Constituent and Physical Property Research Facilities. The authors would like to thank Shiling Yuan for his discussion and advice with respect to the work on computational chemistry and molecular dynamics simulations.

Conflicts of Interest: The authors declare no conflicts of interest.

References

1. Levine, A.G. *125 Questions: Exploration and Discovery*, 1st ed.; Science/AAAS Custom Publishing Office: New York, NY, USA, 2021; pp. 7–18.
2. Hanczyc, M.M.; Mansy, S.S.; Szostak, J.W. Mineral Surface Directed Membrane Assembly. *Orig. Life Evol. Biosph.* **2007**, *37*, 67–82. [CrossRef] [PubMed]
3. Dalai, P.; Sahai, N. Mineral–lipid interactions in the oigins of life. *Trends Biochem. Sci.* **2019**, *44*, 331–341. [CrossRef] [PubMed]
4. Wächtershäuser, G. Pyrite Formation, the First Energy Source for Life: A Hypothesis. *Syst. Appl. Microbiol.* **1988**, *10*, 207–210. [CrossRef]
5. Maurel, M.-C.; Décout, J.-L. Origins of life: Molecular foundations and new approaches. *Tetrahedron* **1999**, *55*, 3141–3182. [CrossRef]
6. Hartman, H. Speculations on the origin and evolution of metabolism. *J. Mol. Evol.* **1975**, *4*, 359–370. [CrossRef] [PubMed]
7. Costerton, J.W.; Cheng, K.J.; Geesey, G.G.; Ladd, T.I.; Nickel, J.C.; Dasgupta, M.; Marrie, T.J. Bacterial Biofilms in Nature and Disease. *Annu. Rev. Microbiol.* **1987**, *41*, 435–464. [CrossRef] [PubMed]
8. Nealson, K.H. Sediment bacteria: Who's there, what are they doing, and what's new? *Annu. Rev. Earth Planet. Sci.* **1997**, *25*, 403–434. [CrossRef] [PubMed]
9. Kleber, M.; Bourg, I.C.; Coward, E.K.; Hansel, C.M.; Myneni, S.C.B.; Nunan, N. Dynamic interactions at the mineral-organic matter interface. *Nat. Rev. Earth Environ.* **2021**, *2*, 402–421. [CrossRef]
10. Bernal, J.D. The physical basis of life. *Proc. Phys. Soc.* **1949**, *62*, 537–618. [CrossRef]
11. Wächtershäuser, G. Before enzymes and templates: Theory of surface metabolism. *Microbiol. Rev.* **1988**, *52*, 452–484. [CrossRef]
12. Du, N.; Song, R.; Li, H.; Song, S.; Zhang, R.; Hou, W. A Nonconventional Model of Protocell-like Vesicles: Anionic Clay Surface-Mediated Formation from a Single-Tailed Amphiphile. *Langmuir* **2015**, *31*, 12579–12586. [CrossRef] [PubMed]
13. Sahai, N.; Kaddour, H.; Dalai, P. The Transition from Geochemistry to Biogeochemistry. *Elements* **2016**, *12*, 389–394. [CrossRef]
14. Deamer, D.W. Boundary structures are formed by organic components of the Murchison carbonaceous chondrite. *Nature* **1985**, *317*, 792–794. [CrossRef]
15. Oba, Y.; Takano, Y.; Furukawa, Y.; Koga, T.; Glavin, D.P.; Dworkin, J.P.; Naraoka, H. Identifying the wide diversity of extraterrestrial purine and pyrimidine nucleobases in carbonaceous meteorites. *Nat. Commun.* **2022**, *13*, 2008. [CrossRef] [PubMed]
16. Ruiz-Mirazo, K.; Briones, C.; de la Escosura, A. Prebiotic Systems Chemistry: New Perspectives for the Origins of Life. *Chem. Rev.* **2014**, *114*, 285–366. [CrossRef] [PubMed]
17. Mansy, S.S.; Schrum, J.P.; Krishnamurthy, M.; Tobé, S.; Treco, D.A.; Szostak, J.W. Template-directed synthesis of a genetic polymer in a model protocell. *Nature* **2008**, *454*, 122–125. [CrossRef]
18. Zhang, S.J.; Lowe, L.A.; Anees, P.; Krishnan, Y.; Fai, T.G.; Szostak, J.W.; Wang, A. Passive endocytosis in model protocells. *Proc. Natl. Acad. Sci. USA* **2023**, *120*, e2221064120. [CrossRef] [PubMed]
19. Xu, H.; Du, N.; Song, Y.; Song, S.; Hou, W. Spontaneous vesicle formation and vesicle-to-micelle transition of sodium 2-ketooctanate in water. *J. Colloid Interface Sci.* **2018**, *509*, 265–274. [CrossRef] [PubMed]
20. Yao, Z.; Du, N.; Chen, N.; Liu, J.; Hou, W. Primitive nucleobases @ sodium 2-Ketooctanoate vesicles with high salt resistance. *J. Mol. Liq.* **2022**, *360*, 119516. [CrossRef]
21. Xu, H.; Wang, S.; Li, Y.; Liang, X.; He, H.; Du, N.; Hou, W. Aggregation behaviors of alkyl α-keto acids in water. *J. Mol. Liq.* **2024**, *394*, 123700. [CrossRef]
22. Gao, M.; Du, N.; Wang, Y.; Yuan, S.; Liu, L.; Liu, J.; Hou, W. Vesicles composed of the single-chain amphiphile sodium monododecylphosphate: A model of protocell compartment. *Colloids Surf. A Physicochem. Eng. Asp.* **2021**, *616*, 126374. [CrossRef]
23. Todd, Z.R.; Cohen, Z.R.; Catling, D.C.; Keller, S.L.; Black, R.A. Growth of Prebiotically Plausible Fatty Acid Vesicles Proceeds in the Presence of Prebiotic Amino Acids, Dipeptides, Sugars, and Nucleic Acid Components. *Langmuir* **2022**, *38*, 15106–15112. [CrossRef] [PubMed]

24. Xue, M.; Black, R.A.; Cornell, C.E.; Drobny, G.P.; Keller, S.L. A Step toward Molecular Evolution of RNA: Ribose Binds to Prebiotic Fatty Acid Membranes, and Nucleosides Bind Better than Individual Bases Do. *ChemBioChem* **2020**, *21*, 2764–2767. [CrossRef] [PubMed]
25. Ma, X.; Liang, H.; Pan, Q.; Prather, K.L.J.; Sinskey, A.J.; Stephanopoulos, G.; Zhou, K. Optimization of the Isopentenol Utilization Pathway for Isoprenoid Synthesis in *Escherichia coli*. *J. Agric. Food Chem.* **2022**, *70*, 3512–3520. [CrossRef] [PubMed]
26. Deamer, D.W.; Oro, J. Role of lipids in prebiotic structures. *Biosystems* **1980**, *12*, 167–175. [CrossRef] [PubMed]
27. Oró, J.; Miller, S.L.; Lazcano, A. The origin and early evolution of life on Earth. *Annu. Rev. Earth Planet. Sci.* **1990**, *18*, 317–356. [CrossRef] [PubMed]
28. Adamala, K.; Szostak, J.W. Nonenzymatic Template-Directed RNA Synthesis Inside Model Protocells. *Science* **2013**, *342*, 1098–1100. [CrossRef] [PubMed]
29. Imai, M.; Sakuma, Y.; Kurisu, M.; Walde, P. From vesicles toward protocells and minimal cells. *Soft Matter* **2022**, *18*, 4823–4849. [CrossRef] [PubMed]
30. Akbari, A.; Palsson, B.O. Positively charged mineral surfaces promoted the accumulation of organic intermediates at the origin of metabolism. *PLoS Comput. Biol.* **2022**, *18*, e1010377. [CrossRef]
31. Fisk, M.; Popa, R. Decorated Vesicles as Prebiont Systems (a Hypothesis). *Orig. Life Evol. Biosph.* **2023**, *53*, 187–203. [CrossRef]
32. Gaudu, N.; Farr, O.; Ona-Nguema, G.; Duval, S. Dissolved metal ions and mineral-liposome hybrid systems: Underlying interactions, synthesis, and characterization. *Biochimie* **2023**, *215*, 100–112. [CrossRef] [PubMed]
33. Holler, S.; Bartlett, S.; Löffler, R.J.G.; Casiraghi, F.; Diaz, C.I.S.; Cartwright, J.H.E.; Hanczyc, M.M. Hybrid organic–inorganic structures trigger the formation of primitive cell-like compartments. *Proc. Natl. Acad. Sci. USA* **2023**, *120*, e2300491120. [CrossRef] [PubMed]
34. Ferris, J.P. Mineral Catalysis and Prebiotic Synthesis: Montmorillonite-Catalyzed Formation of RNA. *Elements* **2005**, *1*, 145–149. [CrossRef]
35. Rode, B.M. Peptides and the origin of life. *Peptides* **1999**, *20*, 773–786. [CrossRef] [PubMed]
36. Hanczyc, M.M.; Fujikawa, S.M.; Szostak, J.W. Experimental Models of Primitive Cellular Compartments: Encapsulation, Growth, and Division. *Science* **2003**, *302*, 618–622. [CrossRef] [PubMed]
37. Sahai, N.; Kaddour, H.; Dalai, P.; Wang, Z.; Bass, G.; Gao, M. Mineral Surface Chemistry and Nanoparticle-aggregation Control Membrane Self-Assembly. *Sci. Rep.* **2017**, *7*, srep43418. [CrossRef] [PubMed]
38. Ilbert, M.; Bonnefoy, V. Insight into the evolution of the iron oxidation pathways. *Biochim. Biophys. Acta* **2013**, *1827*, 161–175. [CrossRef]
39. Liu, J.-S.; Wang, Z.-H.; Li, B.-M.; Zhang, Y.-H. Interaction between pyrite and cysteine. *Trans. Nonferrous Met. Soc. China* **2006**, *16*, 943–946. [CrossRef]
40. Liu, W.; Wang, X.; Wang, Z.; Miller, J. Flotation chemistry features in bastnaesite flotation with potassium lauryl phosphate. *Miner. Eng.* **2016**, *85*, 17–22. [CrossRef]
41. Chavez, P.; Ducker, W.; Israelachvili, J.; Maxwell, K. Adsorption of Dipolar (Zwitterionic) Surfactants to Dipolar Surfaces. *Langmuir* **1996**, *12*, 4111–4115. [CrossRef]
42. Song, R.; Du, N.; Zhu, X.; Li, H.; Song, S.; Hou, W. Rough Glass Surface-Mediated Transition of Micelle-to-Vesicle in Sodium Dodecylbenzenesulfonate Solutions. *J. Phys. Chem. B* **2015**, *119*, 3762–3767. [CrossRef] [PubMed]
43. Du, N.; Zhu, X.; Song, R.; Song, S.; Hou, W. Formation of simple single-tailed vesicles mediated by lipophilic solid surfaces. *Soft Matter* **2016**, *12*, 8574–8580. [CrossRef] [PubMed]
44. Zhu, X.; Du, N.; Song, R.; Hou, W.; Song, S.; Zhang, R. Rough Glass Surface-Mediated Formation of Vesicles from Lauryl Sulfobetaine Micellar Solutions. *Langmuir* **2014**, *30*, 11543–11551. [CrossRef] [PubMed]
45. Du, N.; Song, R.; Zhang, H.; Sun, J.; Yuan, S.; Zhang, R.; Hou, W. The formation and stability of sodium dodecylsulfate vesicles mediated by rough glass surface. *Colloids Surf. A Physicochem. Eng. Asp.* **2016**, *509*, 195–202. [CrossRef]
46. Gao, M.; Du, N.; Yao, Z.; Li, Y.; Chen, N.; Hou, W. Spontaneous vesicle formation and vesicle-to-α-gel transition in aqueous mixtures of sodium monododecylphosphate and guanidinium salts. *Soft Matter* **2021**, *17*, 4604–4614. [CrossRef] [PubMed]
47. Zhang, H.; Yuan, S.; Sun, J.; Liu, J.; Li, H.; Du, N.; Hou, W. Molecular dynamics simulation of sodium dodecylsulfate (SDS) bilayers. *J. Colloid Interface Sci.* **2017**, *506*, 227–235. [CrossRef]
48. Gao, M.; Du, N.; Yao, Z.; Li, Y.; Chen, N.; Hou, W. Vesicle formation of single-chain amphiphilic 4-dodecylbenzene sulfonic acid in water and micelle-to-vesicle transition induced by wet–dry cycles. *Soft Matter* **2021**, *17*, 2490–2499. [CrossRef] [PubMed]
49. Kanicky, J.R.; Poniatowski, A.F.; Mehta, N.R.; Shah, D.O. Cooperativity among Molecules at Interfaces in Relation to Various Technological Processes: Effect of Chain Length on the pK_a of Fatty Acid Salt Solutions. *Langmuir* **2000**, *16*, 172–177. [CrossRef]
50. Reijenga, J.; Van Hoof, A.; Van Loon, A.; Teunissen, B. Development of Methods for the Determination of pKa Values. *Anal. Chem. Insights* **2013**, *8*, 53–71. [CrossRef]
51. Liu, B.; Gao, M.; Li, H.; Liu, J.; Yuan, S.; Du, N.; Hou, W. Model of protocell compartments—Dodecyl hydrogen sulfate vesicles. *Phys. Chem. Chem. Phys.* **2018**, *20*, 1332–1336. [CrossRef]
52. Sakai, T.; Miyaki, M.; Tajima, H.; Shimizu, M. Precipitate Deposition around CMC and Vesicle-to-Micelle Transition of Monopotassium Monododecyl Phosphate in Water. *J. Phys. Chem. B* **2012**, *116*, 11225–11233. [CrossRef] [PubMed]
53. Gao, M.; Yao, Z.; Du, N.; Deng, Q.; Liu, L.; Liu, J.; Hou, W. Sodium Monododecylphosphate Vesicles Formed in Alcohol/Water Mixtures. *Chemnanomat* **2021**, *7*, 553–560. [CrossRef]

54. Ferris, J.P.; Hill, A.R.; Liu, R.; Orgel, L.E. Synthesis of long prebiotic oligomers on mineral surfaces. *Nature* **1996**, *381*, 59–61. [CrossRef] [PubMed]
55. Dalai, P.; Pleyer, H.L.; Strasdeit, H.; Fox, S. The Influence of Mineral Matrices on the Thermal Behavior of Glycine. *Orig. Life Evol. Biosph.* **2016**, *47*, 427–452. [CrossRef] [PubMed]
56. Israelachvili, J.N.; Mitchell, D.J.; Ninham, B.W. Theory of self-assembly of hydrocarbon amphiphiles into micelles and bilayers. *J. Chem. Soc.* **1976**, *72*, 1525–1568. [CrossRef]
57. Israelachvili, J.N.; Mitchell, D.; Ninham, B.W. Theory of self-assembly of lipid bilayers and vesicles. *Biochim. Biophys. Acta* **1977**, *470*, 185–201. [CrossRef] [PubMed]
58. Du, N.; Song, R.; Zhu, X.; Hou, W.; Li, H.; Zhang, R. Vesicles composed of one simple single-tailed surfactant. *Chem. Commun.* **2014**, *50*, 10573–10576. [CrossRef] [PubMed]
59. Mansy, S.S. Membrane Transport in Primitive Cells. *Cold Spring Harb. Perspect. Biol.* **2010**, *2*, a002188. [CrossRef] [PubMed]
60. Huang, T.H.; Yan, J.; Yang, H.; Du, H.M.; Zhang, M.H. Synthesis, structure, and spectroscopic properties of Cu^{+} complexes and its application to solar cells. *J. Coord. Chem.* **2015**, *68*, 1514–1527. [CrossRef]
61. Kumer, A.; Khan, M.W. The effect of alkyl chain and electronegative atoms in anion on biological activity of anilinium carboxylate bioactive ionic liquids and computational approaches by DFT functional and molecular docking. *Heliyon* **2021**, *7*, e07509. [CrossRef]

molecules

MDPI

Article

Sugar-Based Surfactants: Effects of Structural Features on the Physicochemical Properties of Sugar Esters and Their Comparison to Commercial Octyl Glycosides

Huiling Lu [1], Gwladys Pourceau [2,*], Benoit Briou [2], Anne Wadouachi [2], Théophile Gaudin [1], Isabelle Pezron [1] and Audrey Drelich [1,*]

1 Université de Technologie de Compiègne, ESCOM, TIMR (Transformations Intégrées de la Matière Renouvelable), Centre de Recherche Royallieu—CS 60 319, 60203 Compiègne Cedex, France; lhl@ujs.edu.cn (H.L.); isabelle.pezron@utc.fr (I.P.)

2 Laboratoire de Glycochimie, et des Agroressources d'Amiens (LG2A), UR 7378—Institut de Chimie de Picardie, Université de Picardie Jules Verne, 33 rue Saint Leu, 80039 Amiens Cedex, France; b.briou@orpiainnovation.com (B.B.); anne.wadouachi@u-picardie.fr (A.W.)

* Correspondence: gwladys.pourceau@u-picardie.fr (G.P.); audrey.drelich@utc.fr (A.D.)

Abstract: Two series of sugar esters with alkyl chain lengths varying from 5 to 12 carbon atoms, and with a head group consisting of glucose or galactose moieties, were synthesized. Equilibrium surface tension isotherms were measured, yielding critical micellar concentration (CMC) surface tensions at CMC (γcmc) and minimum areas at the air–water interface (Amin). In addition, Krafft temperatures (Tks) were measured to characterize the ability of molecules to dissolve in water, which is essential in numerous applications. As a comparison to widely used commercial sugar-based surfactants, those measurements were also carried out for four octyl D-glycosides. Impacts of the linkages between polar and lipophilic moieties, alkyl chain lengths, and the nature of the sugar head group on the measured properties were highlighted. Higher Tk and, thus, lower dissolution ability, were found for methyl 6-*O*-acyl-D-glucopyranosides. CMC and γcmc decreased with the alkyl chain lengths in both cases, but Amin did not appear to be influenced. Both γcmc and Amin appeared independent of the ester group orientation. Notably, alkyl (methyl α-D-glucopyranosid)uronates were found to result in noticeably lower CMC, possibly due to a closer distance between the carbonyl function and the head group.

Keywords: sugar esters; carbohydrate-based surfactants; glycolipids; physicochemical parameters

1. Introduction

Glycolipids are often considered safe alternatives to petroleum-based surfactants due to their renewable origin, high biodegradability, and low toxicity. Their surface-active properties make them highly interesting in a wide range of fields such as food, cosmetics, pharmaceuticals, and detergent industries [1–5]. These amphiphilic molecules comprise a saccharidic polar head group linked to a lipophilic chain. Depending on the linkage, its orientation and position, the length and degree of saturation of the alkyl chain, the sugar residue and its size, etc., a wide variety of sugar-based amphiphiles can be designed. Despite the fact that many studies have focused on relationships between their chemical structures and surface properties, important gaps remain, owing to the structural versatility of sugar-based surfactants [6–9]. Filling those gaps (regarding the knowledge of structure–property relationships) will help identify and anticipate trends, with the final aim of designing green bio-based surfactants for specific applications.

The structures of sugar-based surfactants are diverse (including glyco-amides, glyco-esters, glycosides, and other derivatives), and their properties also widely vary. For example, in previous work [10], we synthesized various glyco-amides, bearing an octyl chain, which differed by the position and the orientation of the amide linkage on the sugar head-group (Su-NHCO-R or Su-CONH-R), and measured significant differences for those molecules in physicochemical properties, such as solubility or CMC, as well as in cytotoxic effects. In particular, esterase enzymes found widely in nature can break down sugar esters, leading to high biodegradability for this specific surfactant family, and making it relevant whenever eco-compatibility is a concern [11]. Another advantage of sugar esters is the general ease of their synthesis, as the ester linkage can be obtained by simple chemical [12–14] or enzymatic routes [15–18]. However, their ability to dissolve in water appears to be lower than that of commercial alkyl glycosides, a difference observed across a diverse range of sugar residues combined with ester linkages (alkyl uronates, alkyl aldonates, acyl oses, acyl itols, etc.) [19–24]. Moreover, in most studies, mixtures of α and β isomers with an undetermined ratio amount are reported, even though this structural parameter influences dissolution and micellization [25–28]. As reported by Brown et al. [25], the solubility of octyl-β-D-glucoside (C8βGlc) is about seven times larger than that of octyl-α-D-glucoside (C8αGlc), due to the latter's more stable crystal structure. Consistently, glucosides with α-linkages are generally believed to have higher Krafft temperatures (Tks) than those with β-linkages [26]. Thus, it is essential to know exactly the α/β ratio or to protect the anomeric position to establish structure/function relationships. It is also necessary to keep in mind that the replacement of the anomeric -OH function with other groups should cause a decreased solubility of the corresponding esters at 25 °C [20]. In 2016, we reported on the synthesis and physicochemical behavior of one family of sugar esters with well-defined anomeric configurations, namely 1-*O*-methyl alkyl uronates (MeαGlc, MeβGal, and MeαMan derivatives) (Figure 1 left) [29]. This study revealed that, similar to corresponding alkyl polyglycosides or common polyoxyethylene nonionic surfactants, our newly synthesized sugar esters were able to significantly reduce the surface tensions of aqueous solutions, at low concentrations. Nevertheless, their solubility remained limited.

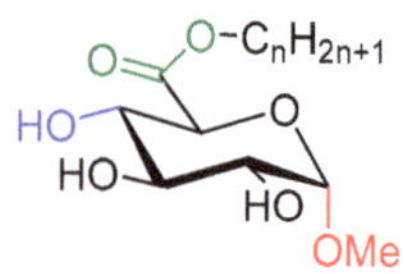

MeαGlc(C=O)OCn, n = 6, 8, 10, 12

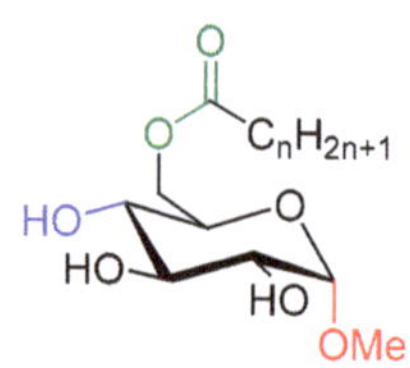

MeαGlcO(C=O)Cn, n = 5, 7, 9, 11

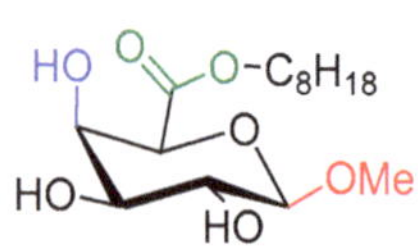

MeβGal(C=O)OC8

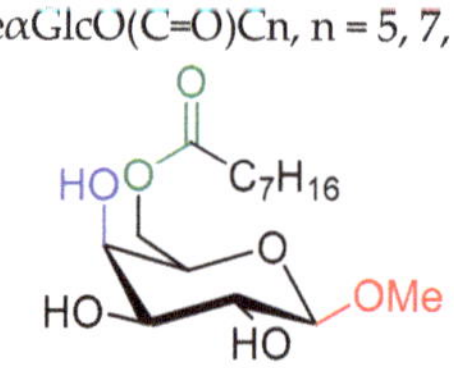

MeβGalO(C=O)C7

Figure 1. Structures of 1-*O*-methyl alkyl-D-glycuronates [29] (**left**) and methyl 6-*O*-acyl-D-glycopyranosides (**right**). Each molecule is abbreviated with the code annotated below.

Thus, we synthesized and characterized close, but distinct sugar esters with well-defined anomeric positions, derived from MeαGlc and MeβGal: methyl 6-*O*-acyl-D-glycopyranosides (Figure 1 right). Their physicochemical properties were compared to the corresponding 1-*O*-methyl alkyl uronates that we previously described. Thanks to the homology of both families (the same nature and position of the linkage, and the same anomeric configuration), the effects of other structural parameters, including the chain length (hexyl to dodecyl derivatives), the orientation of the ester groups, as well as the sugar head, can be directly compared.

Commercial alkyl α/β-glycosides have been, by far, the most extensively investigated sugar-based surfactants, making them relevant comparison models. Indeed, their structures are a straightforward combination of a typical sugar residue, such as glucose, with an alkyl chain, i.e., a prototypical surfactant that can be imagined with sugar as a polar head. Nevertheless, confidence in the available knowledge of their physicochemical behavior is limited by significant inconsistency and discrepancy for the same structure sometimes apparent within a large amount of published physicochemical measurements, as we highlighted previously [30]. For example, in 1997, Sakya et al. showed—for the first time—the phase diagram of octyl-β-D-galactoside (C8βGal), from which a Tk of 42 °C was deduced by Hato et al. [31]. In parallel, several researchers have measured the surface tension of C8βGal directly at 25 °C, assuming that the surfactant properly dissolves in water at this temperature [32,33]. To clarify the observed discrepancies in the literature regarding the structure–property relationships for alkyl glycosides, in the present study, properties for two pairs of α/β anomers, i.e., C8αGlc/C8βGlc and C8αGal/C8βGal, have also been investigated (Figure 2).

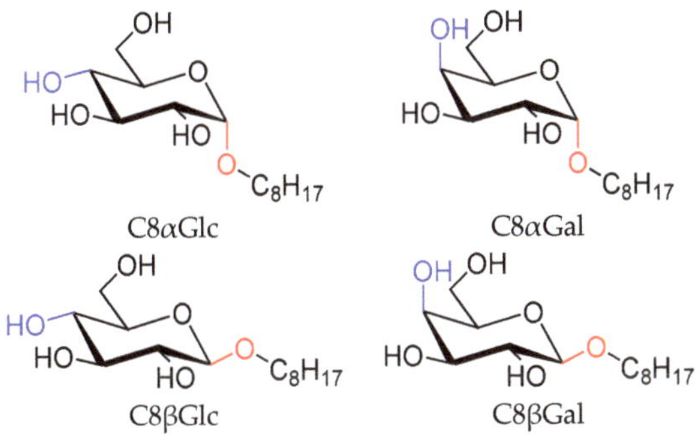

Figure 2. Structures of octyl-α/β-D-glucosides (**left**) and octyl-α/β-D-galactosides (**right**). Each molecule is abbreviated with the code annotated below.

2. Results and Discussion

2.1. Synthesis

Two families of sugar esters, alkyl (methyl D-glycopyranosid)uronates, and methyl 6-O-acyl-D-glycopyranosides, were synthesized. The uronates family was obtained using the two-step procedure that we previously described [29]. For the methyl 6-O-acyl-D-glycopyranosides, a known strategy ensuring regioselectivity on OH 6 [34] was successfully adapted and applied to methyl D-glycosides (Scheme 1). Firstly, α-methyl D-glucoside was per-silylated by treating with *N,O*-bis(trimethylsilyl)acetamide (BSA) and catalytic tetrabutylammonium fluoride (TBAF), and the more labile primary silyl ether was deprotected in a one-pot procedure during the work-up by the addition of K_2CO_3 to generate methyl 2,3,4-tri-*O*-trimethylsilyl-α-D-glucopyranoside **1** with an 80% yield. Methyl 2,3,4-tri-*O*-trimethylsilyl-β-D-galactopyranoside **2** was obtained with a 90% yield following the same protocol, starting from β-methyl D-galactoside. The TMS-protected derivative was then coupled to the fatty carboxylic acid by esterification mediated by 1-ethyl-3-(3-dimethylaminopropyl)carbodiimide (EDCI) and 4-(dimethylamino)pyridine (DMAP), leading to methyl 6-*O*-acyl-2,3,4-tri-*O*-trimethylsilyl-D-glycopyranosides **3–7** with good to excellent yields (70–88% yield, depending on the alkyl chain and the sugar head group). After quantitative deprotection of the TMS groups using Dowex-H^+ resin and purification on normal phase chromatography, methyl 6-*O*-acyl-α-D-glucopyranosides with C6, C8, C10 and C12 alkyl chains **8–11** and methyl 6-*O*-octanoyl-β-D-galactopyranoside **12** were obtained at the 2g-scale with good to excellent yields (71–83%). The high purity of each intermediate and final compound was confirmed by ^1H and ^{13}C NMR (reported in the Supplementary Materials) and HRMS.

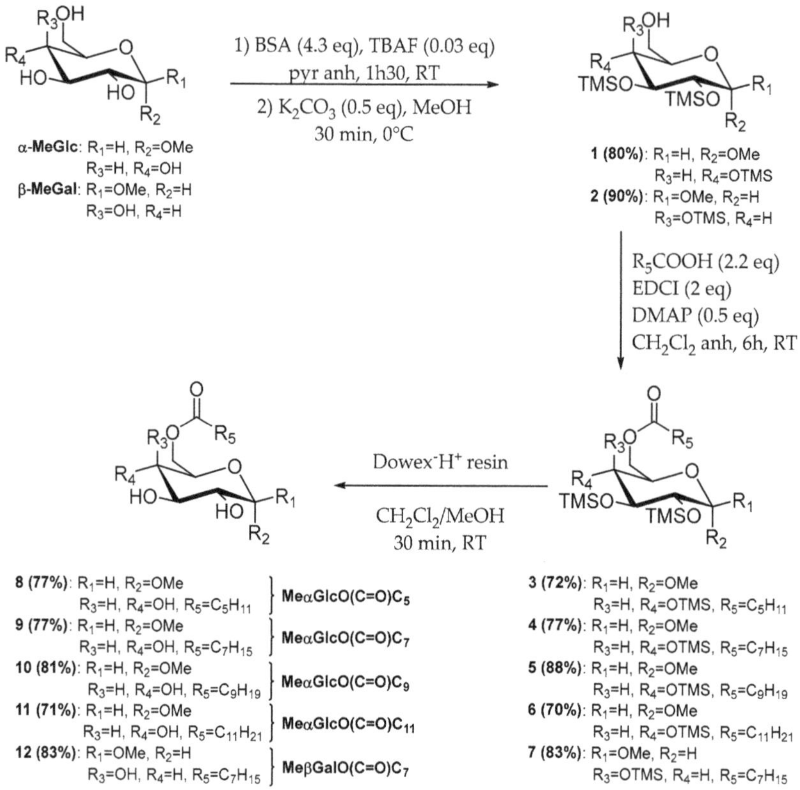

Scheme 1. Synthesis of methyl 6-*O*-acyl-D-glycopyranosides.

For the sake of brevity, from now on, methyl 6-*O*-acyl-α-D-glycopyranosides will be named MeαGlcO(C=O)Cn (n = 5, 7, 9, 11) and MeβGalO(C=O)C7, respectively, whereas alkyl (methyl D-glycopyranosid)uronates will be called MeαGlc(C=O)OCn (n = 6, 8, 10, 12) and MeβGal(C=O)OC8.

2.2. Solubility

Apart from the alkyl chain length, a cornerstone surfactant characteristic, it is generally believed that for sugar-based surfactants in particular, variations in physicochemical properties, including solubility, are often caused by differences in the molecular shape and intra/intermolecular hydrogen bonding [35]. Tk represents the temperature above which micelles can form. Below Tk, surfactants in hydrated solid form coexist with monomers at their solubility limit, which is visually determined at the transparency limit of the observed solutions. For example, for alkyl glucosides, the Tks of α-anomers are higher than those of the β-anomers and the β-anomers are much more soluble in water than the α-anomers [3]. Table 1 presents the results obtained from both the observation and DSC analysis. In Figure 3, Tk is plotted vs. the alkyl chain length for the various studied surfactant families. The results show that, although important, the alkyl chain length is far from the only relevant factor in Tk values for sugar-based surfactants. Indeed, the configuration at the anomeric center, as well as the relative orientation of one -OH of the polar head group can also have an important impact.

Table 1. Physical parameters related to the solubility properties (Krafft temperatures, solubility limits) for all studied molecules.

Family	Molecule	Tk (°C)	Solubility Limit (S/mM)	State
Alkyl glycosides	C8αGlc	39 ± 1	insoluble at RT; >100 at T > Tk	Powder
	C8βGlc	<4	>100 at RT	Powder
	C8αGal	<4	>100 at RT	Highly viscous liquid
	C8βGal	29.0 ± 0.5	10 < S < 20 at RT; >100 at T > Tk	Powder
Alkyl uronates	MeαGlc(C=O)OC6	<4	≤50 at RT	Wax
	MeαGlc(C=O)OC8	<4	15~20 at RT	Wax
	MeαGlc(C=O)OC10	<4	0.5~0.75 at RT	Powder
	MeαGlc(C=O)OC12	22.1 ± 0.2	0.02~0.05 at RT	Powder
	MeβGal(C=O)OC8	31.6 ± 0.9	insoluble at RT; >100 at T > Tk	Powder
Acyl glycosides	MeαGlcO(C=O)C5	<4	<40 at RT	Oil
	MeαGlcO(C=O)C7	<4	≤10 at RT	Powder
	MeαGlcO(C=O)C9	26.5 ± 0.2	insoluble at RT; ≤1 at T > Tk	Powder
	MeαGlcO(C=O)C11	42.9 ± 0.1	insoluble at RT; ≤0.1 at T > Tk	Powder
	MeβGalO(C=O)C7	<4	>100 at RT	Powder

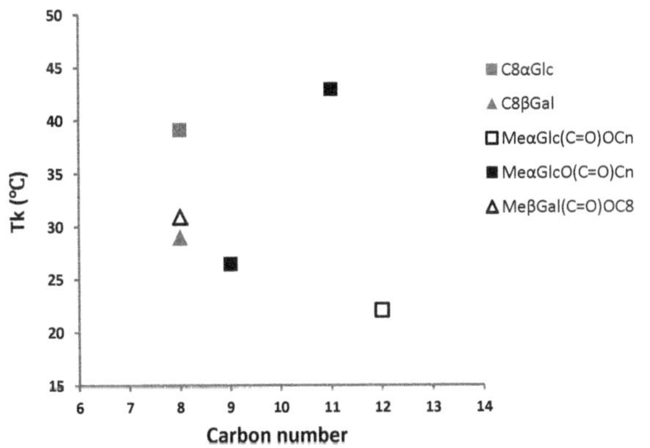

Figure 3. Tk as a function of alkyl chain length for studied molecules.

For commercial alkyl glucosides, the Tk decreases in the following sequence: C8αGlc > C8βGal >> C8βGlc ≈ C8αGal. All of them possess appreciable solubility limits above their Tks. Many experimental investigations have reported that C8βGlc is very soluble in water and, therefore, it has a Tk below room temperature, as is also evidenced by our results. The obtained Tk for C8αGlc is around 40 °C in accordance with previously published values [31,36]. However, as determined by the same experimental protocol, the Tk for C8βGal is about 29 °C, much lower than 42 °C derived from the phase diagram reported by Sakya et al. [37] A possible explanation is that Sakya et al. synthesized and purified C8βGal by themselves, whereas we carried our measurements on commercially purchased C8βGal. Interestingly, in contrast with C8αGlc, C8βGal displays a marked solubility at room temperature even if it is below its Tk. A solution containing 20 mM of C8βGal is nearly transparent at room temperature, with only a very low turbidity. For this reason, we assume that many researchers have been able to investigate their aqueous solutions at 25 °C, as mentioned in the introduction. In addition, when cooling the previously heated solutions with the same molecular concentration of 50 mM, we noted that C8αGlc precipitated much faster than C8βGal. The precipitation of C8αGlc took place within an hour, whereas no change was observed over one month for the C8βGal solution. This may, to some extent, be explained by a more favorable crystal structure for C8αGlc than C8βGal. With regard to C8αGal, which is presented in commercial samples as a highly viscous liquid, we expect it to have the weakest crystal structure and a very low Tk value. However, our experimental methods cannot give its exact Tk. Accordingly, one important conclusion can be drawn from our results: for the given type of octyl glycosides, a combination of "axial 4-OH and axial 1-alkyl chain" or "equatorial 4-OH and equatorial 1-alkyl chain", causes lower Tk values. While a combination like "axial 4-OH and equatorial 1-alkyl chain" or "equatorial 4-OH and axial 1-alkyl chain" can generate higher Tk values. This observation suggests that a different orientation for 4-OH and 1-alkyl chain promotes crystal packing.

After studying commercially available and well-known alkyl glycosides, we focused on the synthesized sugar esters. The observed dissolution phenomena are harder to interpret compared to commercial molecules. This can be a result of more structural contributions such as axial/equatorial 4-OH and 1-OMe, as well as two possible ester orientations. In the case of MeαGlc-derived esters, the solubility decreases, as expected, with the increase of the alkyl chain length. All molecules demonstrated the expected behavior of a lower solubility above their Tk, compared to commercial C8αGlc and C8βGlc. Moreover, both octyl glucuronate (MeαGlc(C=O)OC8) and octanoyl glycoside (MeαGlcO(C=O)C7) derivatives showed lower Tk than C8αGlc. A possible interpretation is that compared to the ether linker, the ester linker creates more disbalance toward hydrogen acceptor groups, which would not find hydrogen donor partners in the crystal structure but would find them in water, increasing aqueous solubility. As highlighted in Table 1, the alkyl (methyl α-D-glycopyranosid)uronates tend to show lower Tk values than their homologous methyl 6-*O*-acyl-α-D-glycopyranosides. This result is also in accordance with how the molecule presents when its pure form is exposed to some limited atmospheric water. Indeed, we could expect that a molecule showing up as a liquid or paste will likely have a lower Tk compared to a molecule presenting as a crystal in those circumstances. For instance, MeαGlc(C=O)OC8 presents as a semi-solid grease while MeαGlcO(C=O)C7 is a solid powder. No obvious structural interpretation arises regarding this difference. One could speculate that the C=O group in methyl 6-*O*-acyl-α-D-glycopyranosides is more available to form tight hydrogen bonding networks compared to alkyl (methyl α-D-glycopyranosid)uronates, where it lies closer to the hydrogen donors on the polar head.

For methyl 6-*O*-acyl-α-D-glucopyranosides, Tk is found to increase with the alkyl chain length, as is the case for alkyl α-glucopyranosides [38]. Moreover, according to Otto et al. [20], the 6-*O*-octanoyl-β-glucose, above its Tk, has a solubility limit of 25 mg/mL (>80 mM), which is much higher than that of our MeαGlcO(C=O)C7 (<10 mM). This indicates that replacing the anomeric -OH of the glucose with a methyl ether group increases hydrophobicity, making MeαGlcO(C=O)C7 less soluble in water.

In the case of MeβGal-derived esters, both methyl octyl galacturonate (MeβGal(C=O)OC8) and methyl 6-*O*-octanoyl galactoside (MeβGalO(C=O)C7) derivatives have high solubility limits above their Tks, not in line with MeαGlc-derived esters.

The measured Tk for MeβGalO(C=O)C7 is a little higher than that of C8βGal. Note that contrary to MeαGlc-derived C7/C8 esters, a carbonyl group closer to the sugar residue is associated with a lower ability to dissolve the molecule at room temperature. One hypothesis is that the polar alcohol group in axial orientation leads to a more compact polar head for Gal surfactants, shielding some polar surface area from contact with the solvent. To have a molecular perspective on this hypothesis, we carried out a comparison with the MeβGlcO(C=O)C8 equivalent. Electrostatic potential maps were obtained with Jmol software after a semi-empirical geometry optimization [39]. The results are displayed in Figure 4.

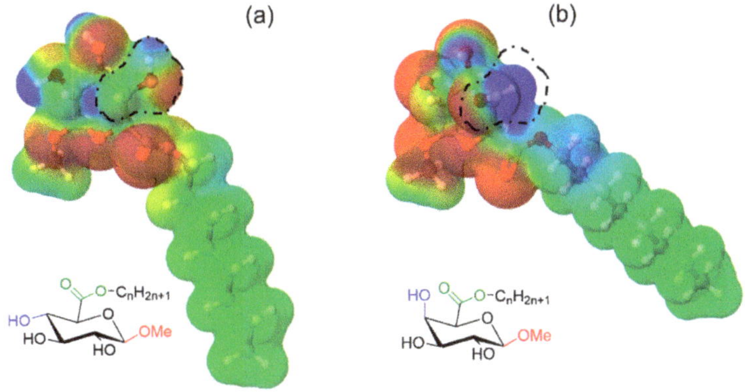

Figure 4. A 3D representation of (**a**) MeβGlcO(C=O)C8 and (**b**) MeβGalO(C=O)C8.

It can be seen that the polar head has a more compact shape; the axial OH is more restricted and has less potential for solvent exposure, supporting a slightly higher hydrophobicity.

2.3. Surface Activity

The aqueous tensiometric properties of all the studied molecules are listed in Table 2. CMCs are determined from the inflections in the plots of equilibrium surface tension against the logarithm of the concentration. For the interested reader, the Gibbs-free energy of micellization, ΔG_{mic}, the Gibbs-free energy of adsorption, ΔG_{ads}, and the efficiency of surfactant adsorption pC_{20} were calculated based on their definitions [40], as presented in Table 2.

Table 2. Surface tension data at T > Tk for different molecules.

Family	Molecule	CMC (mM)	γcmc (mN/m)	Amin (Å^2/molecule)	ΔG_{mic} (kcal/mol)	ΔG_{ads} (kcal/mol)	C20 (mM)	pC20	T (°C)
Alkyl glucosides	C8αGlc	15.5 ± 0.5	30.7 ± 0.3	43 ± 1.8	−4.7	−7.0	2.8	2.6	50
	C8βGlc	21.2 ± 0.8	31.0 ± 0.15	38.5 ± 1.9	−4.1	−6.4	2.7	2.6	25
	C8αGal	51.7 ± 1.5	29.5 ± 0.2	45.8 ± 0.9	−3.6	−6.4	3.8	2.4	25
	C8βGal	20 ± 1	28.5 ± 0.5	46.3 ± 1.7	−4.4	−7.1	3.2	2.5	40
Alkyl uronates	MeαGlc(C=O)OC6	55 ± 5	30.8 ± 0.8	39.2 ± 2.9	−3.6	−5.9	5.9	2.2	25
	MeαGlc(C=O)OC8	6 ± 0.5	29.2 ± 0.35	41 ± 1	−4.8	−7.3	0.5	3.3	25
	MeαGlc(C=O)OC10	0.65 ± 0.5	28.1 ± 0.1	42.4 ± 2	5.9	8.6	0.05	4.3	25
	MeαGlc(C=O)OC12	0.056 ± 0.004	28.3 ± 0.3	45.6 ± 5	−7.2	−10.1	0.004	5.4	25
	MeβGal(C=O)OC8	6.9 ± 0.4	29.4 ± 0.36	42.9 ± 3.9	−5.0	−7.4	1.2	2.9	40
Acyl glycosides	MeαGlcO(C=O)C5	105 ± 5	31 ± 0.1	40.1 ± 0.3	−3.3	−5.6	11.2	2.0	25
	MeαGlcO(C=O)C7	9.8 ± 0.76	29.5 ± 0.5	39 ± 2	−4.5	−6.9	1.0	3.0	25
	MeαGlcO(C=O)C9	0.92 ± 0.08	27.9 ± 0.05	41.8 ± 0.9	−5.8	−8.4	0.08	4.1	30
	MeαGlcO(C=O)C11	0.08 ± 0.005	26.7 ± 0.25	45.8 ± 2.8	−7.7	−10.4	0.007	5.1	50
	MeβGalO(C=O)C7	8.5 ± 0.5	28.3 ± 0.35	42.7 ± 2.4	−4.6	−7.3	0.7	3.1	25

Firstly, in the case of octyl D-glycosides, our measured CMC for C8βGlc, 21.2 mM, was consistent with the published values (18–26 mM), despite the numerous experimental methods used by the different authors [41–45]. However, there were discrepancies in the literature for C8αGlc, C8αGal, and C8βGal regarding both CMC and γcmc. Matsumura et al. [32] measured the static surface tension of C8αGlc and C8βGal by the Wilhelmy plate method at 25 °C. The authors have reported, respectively, CMC values of 12 mM and 16 mM. Both showed a γcmc of 30.5 mN/m. We measured the CMC values of 15.5 mM and 20 mM, respectively, higher than the published ones, as well as γcmc 30.7 mN/m and 28.5 mN/m, respectively, indicating a higher effectiveness of surface tension reduction for C8βGal. We questioned their results as the measurements were performed below their Tk

(39 °C and 29 °C, respectively, cf. Table 1). Schmidt-Lassen et al. [33] measured the CMC values for C8βGal and C8αGal by using isothermal titration calorimetry (ITC) and found that CMC values at 25 °C were 31.7 ± 0.7 mM for C8βGal and 30.2 ± 0.4 mM for C8αGal, showing no evident anomeric effect on the CMC. In contrast, we found CMCs of 20 mM for C8βGal and 51.7 mM for C8αGal, respectively, indicating that the anomeric orientation has a significant impact on the CMC here. Because the limited aqueous solubility of C8βGal was mentioned, again, we question the results published in their study.

After comparing our measurements with published ones, we focused on the structural interpretations arising from our own measurements. The order of their CMC for the 4 octyl glycosides was: C8αGal > C8βGlc ~ C8βGal > C8αGlc. Here, even at a constant alkyl chain length, a lower CMC correlates with a higher Tk. Apart from the alkyl chain length, the most obvious factor, it is known that polar head hydrophilicity can also significantly influence CMC, typically increasing it for more hydrophilic polar heads [6]. With all else equal, a higher CMC leads to a decrease in Tk due to the thermodynamic relationship between the two. Thus, the observed Tk could also reflect polar head hydrophilicity. As for γcmc, it was found that octyl D-galactosides can more effectively reduce the surface tension of pure water than octyl glucosides. Our Amin measurements suggest that the anomeric effect is more evident for C8α- and C8βGlc than for C8α- and C8βGal. In addition, the latter shows a greater Amin than the former, in agreement with Razafindralambo et al. [14]. It was said that an axial -OH at the 4-position (Gal) gives a larger minimum area occupied per molecule at the water surface. However, as the Amin depends strongly on the way of plotting the γ vs. log C curves, different authors have demonstrated different Amin values. Matsumura et al. [32] reported a range of 47–54 Å^2/molecule for C8–C12 alkyl D-glucosides and galactosides. Their study did not show any great difference between their α- and β-anomers but a slightly larger Amin for C8- and C10 βGal than for the C8- and C10αGlc. Kjellin et al. [23] showed an Amin of 38 Å^2/molecule for C8βGlc whereas an Amin of 42 Å^2/molecule was given for C8βGlc by Shinoda et al. [46]. Therefore, our results seem to be in good agreement with the literature.

The influence of the head saccharidic group of our new surfactants on the surface activity can also be analyzed. Figure 5 (top) represents the surface tension vs. log concentration plots for both families of MeαGlc-derived esters. Firstly, whatever the orientation of the ester group (whatever the family studied), the esters of glucose can efficiently reduce the surface tension of pure water with the obtained low γcmc values of 26–32 mN/m. It seems that the ester group orientation does not impact γcmc much. In general, it can be seen that γcmc slightly decreases with the alkyl chain length. Those results are in line with the identified trends [6] where γcmc appears to decrease asymptotically with the alkyl chain length. The suggested explanation is that as the alkyl chain length increases, alkyl groups get closer and closer to completely covering the surface, making it exhibit an alkane-like γcmc. The only exception comes with MeαGlc(C=O)OC12, which is very close to that of MeαGlc(C=O)OC10, very likely because it is challenging to measure a true γcmc for MeαGlc(C=O)OC12 as it precipitates easily. A similar phenomenon was observed by Otto et al. [20], who reported a higher γcmc for 6-*O*-decanoyl-β-D-glucopyranose (C10) than for 6-*O*-octanoyl-β-D-glucopyranose (C8).

A common conclusion in the literature is that the alkyl chain length is not an important factor in determining the Amin of n-alkyl glucosides at the air–water interface [3]. This observation can probably be extrapolated to sugar esters. Nevertheless, even though Amin is mainly decided by the hydrophilic group [47], the alkyl chain flexibility can contribute to whether molecules loosely or compactly pack at the interface [48]. It is reasonable to assume that given the higher number of rotational degrees of freedom, a longer saturated alkyl chain is more flexible, and based on this argument, we hypothesize that surfactants with longer alkyl chains at a constant polar head can exhibit the Amin increase. Our measurements are consistent with such a hypothesis, with a slight increase of Amin with alkyl chain lengths for both families. A similar trend was mentioned in the literature for other surfactant families, such as bolaform quaternary ammonium surfactants for which the Amin increases with the increase in the spacer carbon number (n < 10) [49]. Nevertheless, it must be noted that for some polyoxyethylene alkyl ethers, Amin decreases with an increasing carbon number but increases with the oxyethylene unit length [47,50,51]. Given that contrary to rigid polar rings of sugar-based surfactants, the polar heads of polyoxyethylene alkyl ethers are themselves flexible, their conformation may change with longer alkyl chains. Thus, the above analysis may not apply in their case.

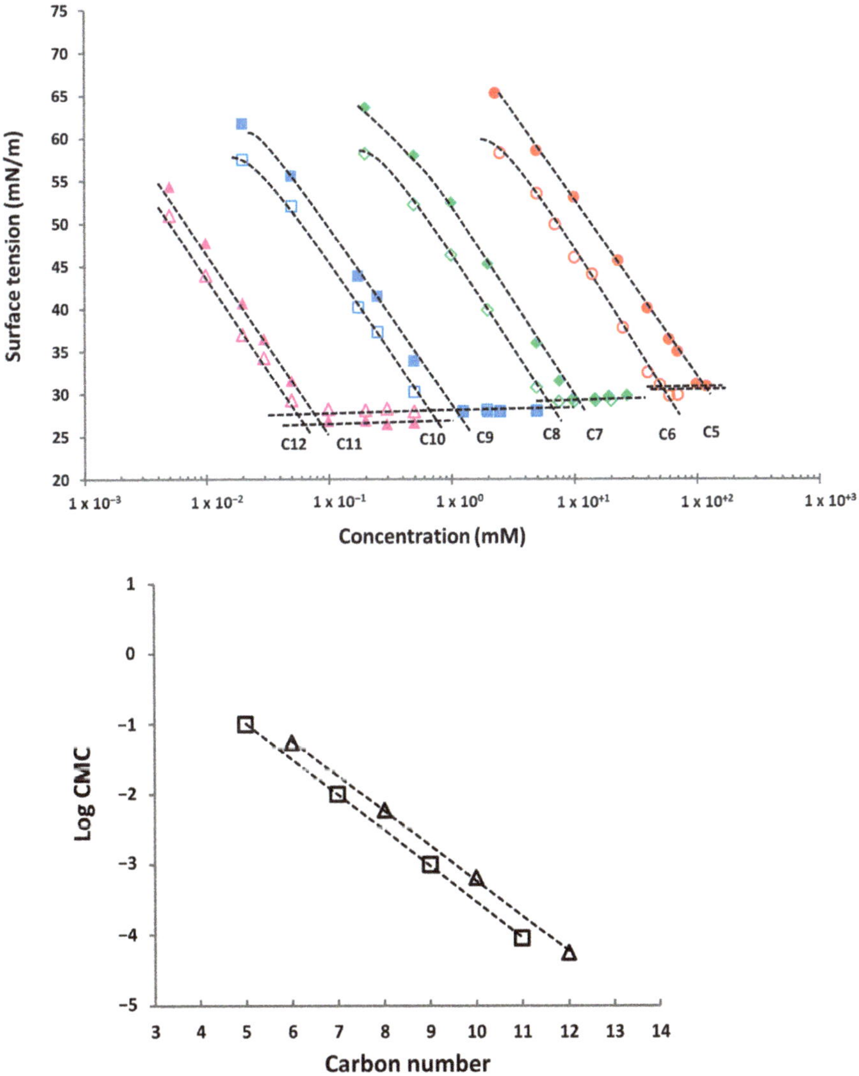

Figure 5. Surface tension as a function of concentration for methyl alkyl α-D-glucuronates at 25 °C (empty markers, MeαGlc(C=O)OCn, with n being an even number) and methyl 6-*O*-acyl-α-D-glucosides (filled markers, MeαGlcO(C=O)Cn, with n being an odd number) at 25 °C (C5 and C7), 30 °C (C9) and 50 °C (C11) (**top**) and the variation of the log CMC with the carbon numbers of the alkyl chain for alkyl methyl α-D-glucuronates (MeαGlc(C=O)OCn, △) and methyl 6-*O*-acyl-α-D-glucosides (MeαGlcO(C=O)Cn, □) (**bottom**). CMC is expressed in mol/L.

When it comes to CMC, the alkyl chain length, as well-known for other surfactants [6], is the most relevant structural parameter. The dependence of the log CMC on the number of carbon atoms in the alkyl chain for two families, generally, obeys the expected empirical equation [21]:

$$\mathrm{Log\,CMC} = \mathrm{A} - \mathrm{Bn} \tag{1}$$

where n represents the carbon numbers (the length) in the alkyl chain; A and B are empirical constants, which are, respectively, related to the contributions of the hydrophilic head group and a single hydrophobic methylene group to the change in free energy for micellization. Plots of log CMC values as functions of the alkyl chain length are depicted in Figure 5 (bottom).

The results show that a linear relationship between the alkyl chain length and the log CMC exists for both ester families. For alkyl D-glucuronates derivatives (MeαGlc(C=O)OCn), the values of A and B calculated are 0.5 and 1.7, while for acyl D-glucosides derivatives (MeαGlcO(C=O)Cn), A is 0.5 and B is 1.5. The slightly higher B value in the case of alkyl D-glucuronates suggests a more rapid decrease in CMC when one carbon is added. It is important to note that the MeαGlc(C=O)OCn series shows even-numbered carbon atoms while the MeαGlcO(C=O)Cn series has odd-numbered carbon atoms in the noted Cn part. The amphiphilic molecules can be, in a simplistic way, depicted as the combination of a polar head group with a lipophilic tail. In the case of both families, the ester linkage is attributed to the polar moiety, even if it is not a part of the sugar: for acyl glycosides, this functional group is not located on a carbon atom of the sugar, whereas for the alkyl D-glucuronate, the saccharidic C6 is oxidized, leading to a difference in the structure (and particularly in the number of carbon atoms) of the polar head. We assume that this slight difference has a negligible effect on the hydrophilicity of the head group. On the other hand, the alkyl D-glucuronates series (MeαGlc(C=O)OCn) exhibits a slightly lower CMC and γcmc compared to the acyl D-glucosides series (MeαGlcO(C=O)Cn), which can be partially associated with the fact that the former always has one more carbon in its lipophilic moiety than the latter.

Some published CMC values are also worth comparing to the measurements from the present work. Otto et al. [20] obtained a CMC of 18 mM for 6-*O*-octanoyl-D-glucopyranose, noticeably higher than the CMC of MeαGlcO(C=O)C7 (9.8 mM). In addition, Blecker et al. [17] also reported a CMC of 10.68 mM for octyl D-glucuronate compared to the 6 mM value we observed for the methyl octyl α-D-glucuronate (MeαGlc(C=O)OC8). Both comparisons indicate that the nonpolar axial methyl group at the 1-position brings additional hydrophobicity to the polar head.

For the first time, Razafindralambo et al. [52] studied the effect of the ester orientation (Su-O-CO-R or Su-CO-O-R) on the surface properties of similar molecules: D-glucosyl octanoate and octyl D-glucuronate. Their analysis was that octyl D-glucuronate exhibits a stronger hydrophobic character because its carbonyl group was closer to the head group, making the head more polar and resulting in stronger intermolecular interactions between the heads. The CMC values at 25 °C for octyl D-glucuronate and D-glucosyl octanoate were reported to be 10.7 and 19.1 mM, respectively, which were almost twice as large as the CMC values for MeαGlc(C=O)OC8 and MeαGlcO(C=O)C7. It is worth noting that the Razafindralambo molecules are a mixture of both α and β anomers, with the β-D-glucose derivative being more hydrophilic. In addition, they obtained a larger Amin for octyl D-glucuronate, which was supported by a computational approach. Based on those results, the authors suggested that a shorter distance between the carbonyl group and the sugar head introduces a tilt between the polar head and the alkyl chain, resulting in looser packing at the water surface. For the synthesized surfactants of this work, however, our measurements do not evidence a significant influence of the ester orientation on Amin. It is possible that the anomeric methyl group dilutes the influence of the mechanism proposed by Razafindralambo et al. Nevertheless, their proposed mechanism itself remains questionable as the surface is a dynamic environment where molecules not only take different conformations but also different orientations. By comparing our results with other literature work, we note that the location of the ester group at the head group also influenced the CMC: sugar esters with the carbonyl group located at the 3-position appear more hydrophobic than those with the carbonyl group located at the 6-position. For example, Gouéth et al. [19] reported that 3-*O*-octanoyl-D-glucopyranose, which is a mixture of α and β anomers, had a CMC of 0.61 mM while Savellie et al. [21] reported a CMC of 1.6 mM. Both studies evidenced a lower CMC than that for 6-*O*-octanoyl-β-D-glucopyranose and our octyl derivative esters, as discussed above. Moreover, these molecules (MeαGlc-(C=O)OC8 and MeαGlcO(C=O)C7) show much lower CMC values compared to the commercial C8αGlc and C8βGlc, as shown in Figure 6. The results indicate that both the methyl group at the 1-position and the transformation of 6-OH into the ester group increase the hydrophobicity but do not influence γcmc significantly.

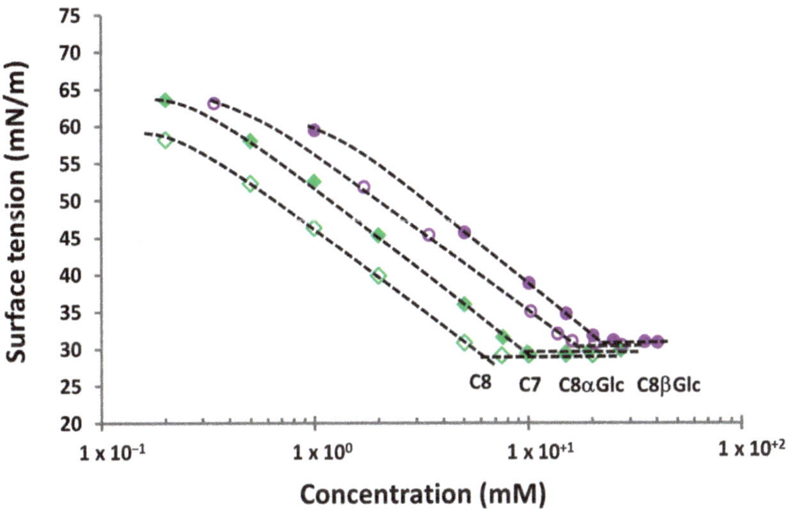

Figure 6. Surface tension as a function of the concentration for α-octyl glucoside (C8αGlc, ○) at 50 °C, β-octyl glucoside (C8βGlc, •) at 25 °C, octyl α-methyl glucuronate (MeαGlc(C=O)OC8, ◇) at 25 °C and 6-octanoyl-α-methyl glucoside (MeαGlcO(C=O)C7, ◆) at 25 °C.

Our Amin measurements (see Table 2) indicate a more compact packing for the synthesized esters at the interface of the air–aqueous solution. This might be related to the stronger intermolecular interactions between their sugar heads due to the presence of the ester group.

In the case of MeβGal-based esters derivatives (Figure 7), the octyl D-galacturonate derivative (MeβGal(C=O)OC8) also showed a lower CMC than its homologous octanoyl D-galactoside derivative (MeβGalO(C=O)C7), which is consistent with MeαGlc-derived C7/C8 esters. Given that the actual alkyl chain of MeβGal(C=O)OC8 contains 8 carbons vs. 7 for MeβGalO(C=O)C7, and that the alkyl chain length is the major parameter relevant to CMC, the observed CMC difference is in line with expectations. It can also be noticed that the CMCs observed for the esters (6–9.8 mM) are lower than those of corresponding octyl D-galactosides (20–51.7 mM). Given the higher observed Amin for octyl D-galactosides, the shape of the esters' polar heads may allow micellization more readily.

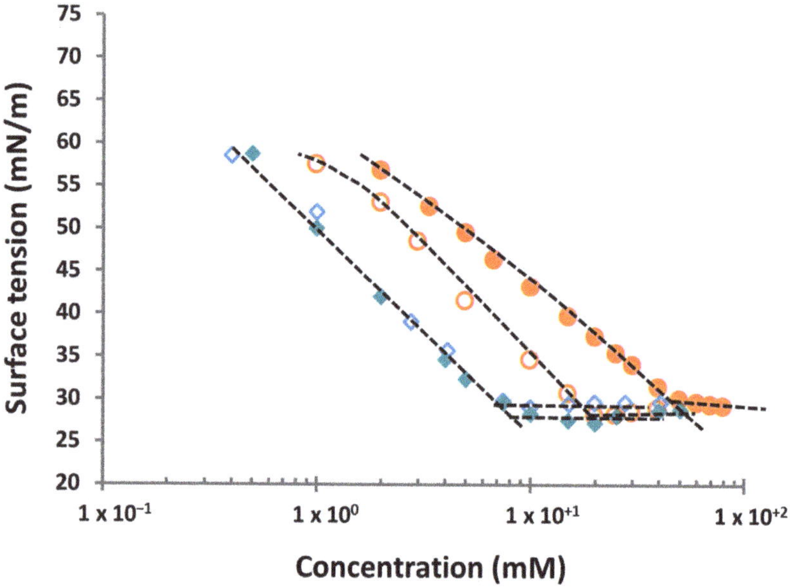

Figure 7. Surface tension as a function of concentration for β-octyl galactoside (C8βGal, ○) at 40 °C, α-octyl galactoside (C8αGal, •) at 25 °C, octyl β-methyl galacturonate (MeβGal(C=O)OC8, ◇) at 40 °C and 6-octanoyl-β-methyl galactoside (MeβGalO(C=O)C7, ◆) at 25 °C.

Overall, the main conclusions that arise by comparing the solubility and surface activity for all the synthesized C7/C8 esters are listed below:

- MeβGal-based esters show a much larger solubility limit than MeαGlc-based esters, suggesting that the solubility strongly depends on the head group configuration;
- Ester group orientation also influences the dissolution of MeβGal-based esters and MeαGlc-based esters;
- CMC is influenced by the ester orientation following the relationship: MeαGlc(C=O)OC8 < MeβGal(C=O)OC8 < MeβGalO(C=O)C7 < MeαGlcO(C=O)C7. The result suggests that the uronates derivatives exhibit lower CMC values than the acyl D-glycosides derivatives. The effect of ester group orientation on CMC was more evident for MeαGlc-derived esters than for MeβGal-derived esters;
- γ_{cmc} is very similar for all these esters, suggesting that the ester linkage orientation does not play an important role in the effectiveness of reducing surface tension.

3. Materials and Methods

3.1. General

Octyl α-D-glucopyranoside (>99%) and octyl β-D-glucopyranoside (>99%) were both purchased from Anatrace (Maumee, OH, USA). Octyl α-D-galactopyranoside and octyl β-D-galactopyranoside were purchased from Carbosynth (Compton, UK). They were all used without further purification. The deionized water with a resistivity of 18.2 Mcm was produced by a lab purification chain provided by Aquadem/Veolia Water STI (Saint-Maurice, France) and used to prepare all the solutions.

Alkyl (methyl D-glycopyranosid)uronates were synthesized using the two-step procedure developed in our lab, as previously described [29]. Briefly, the primary hydroxyl group of free methyl D-glycopyranosides was selectively and quantitatively oxidized by using 2,2,6,6-tetramethyl-1-piperidinyloxy free radical (TEMPO). Hydrophobic chains of different lengths were then introduced by acid-mediated esterification with fatty alcohols (hexyl to dodecyl alcohols) leading to the desired alkyl uronates with moderate to good yields (49–63%).

3.2. Synthesis of Methyl 6-O-acyl-D-glycopyranosides

3.2.1. General

All chemicals were purchased from Fisher Scientific (Illkirch, France), and Merck Sigma Aldrich (St Quentin Fallavier, France), and used as received. Mass analyses were performed on a Waters spectrometer (SINAPT TG2SI, Manchester, UK) using electrospray ionization (Z-Spray). The NMR analysis was performed on a BRUKER NMR spectrometer (Bruker, Wissembourg, France) operating at 400 MHz for ^{1}H and 100 MHz for ^{13}C. Samples of silylated products were prepared in deuterated chloroform ($CDCl_3$) while samples of free esters were dissolved in DMSO-*d6*.

3.2.2. General Procedure for One-Pot Protection/Selective Deprotection

N,*O*-bis(trimethylsilyl)acetamide (28.7 g, 4.3 eq.) and tetrabutylammonium fluoride (235 mg, 0.03 eq.) were added to a solution of methyl α-D-glucopyranoside (9.30 g, 32.9 mmol) in dry pyridine (7.5 mL). After 1.5 h of stirring at room temperature, the reaction was quenched by adding isopropanol (6.25 mL) and methanol (450 mL). After cooling to 0 °C, K_2CO_3 (2.28 g, 0.5 eq.) was added, the mixture was stirred for 30 min at 0 °C, and then neutralized with 2.5 mL of acetic acid. After concentration, the crude was extracted with Et_2O, washed with water, dried over Na_2SO_4, and concentrated to give methyl 2,3,4-tri-*O*-trimethylsilyl-α-D-glucopyranoside **1** as a white powder (10.94 g, yield 80%). The same procedure was applied to methyl β-D-galactopyranoside affording methyl 2,3,4-tri-*O*-trimethylsilyl-β-D-galactopyranoside **2** as a white powder (12.13 g, yield 90%).

3.2.3. General Procedure for Esterification

To a solution of methyl 2,3,4-tri-*O*-trimethylsilyl-α-D-glycopyranoside **1** or **2** (7.00 g, 17.1 mmol) in anhydrous dichloromethane (80 mL), 4-dimethylaminopyridine (1.04 g, 0.5 eq.), desired fatty acid (37.5 mmol, 2.2 eq.), and 1-ethyl-3-(3-dimethylaminopropyl)carbodiimide (5.29 g, 2 eq.) were successively added and the mixture was stirred at room temperature for 6 h. After concentration, the crude was purified by normal phase column chromatography on silica gel (mobile phase: cyclohexane/EtOAc 9:1) to give corresponding silylated sugar esters **3–7**.

3.2.4. General Procedure for Deprotection

To a solution of the as-obtained silylated ester glycoside, **3–7** (6 g) in CH_2Cl_2/MeOH 1:1 (100 mL), 3 g of Dowex-H^+ resin was added. The suspension was stirred for 30 min at room temperature, then the resin was eliminated by filtration, and the filtrate was concentrated. The residue was purified by column chromatography on silica gel (mobile phase: EtOAc/MeOH 98:2) to give pure corresponding methyl 6-*O*-acyl D-glycosides **8–12**.

3.3. Compound Characterization Data

Methyl 2,3,4-tri-O-trimethylsilyl-α-D-glucopyranoside **1**. White powder. Yield 80% from methyl α-D-glucopyranoside following the general procedure. ^{1}H and ^{13}C NMR spectra in accordance with the literature [53].

Methyl 2,3,4-tri-O-trimethylsilyl-β-D-galactopyranoside **2**. White powder. Yield 90% from methyl β-D-galactopyranoside following the general procedure. ^{1}H and ^{13}C NMR spectra in accordance with the literature [54].

Methyl 6-O-hexanoyl-2,3,4-tri-O-trimethylsilyl-α-D-glucopyranoside **3**. Colorless syrup. Yield 72% from the methyl 2,3,4-tri-*O*-trimethylsilyl-α-D-glucopyranoside **1** following the general procedure. ^{1}H NMR (400 MHz, $CDCl_3$), δ 0.11–0.16 (m, 27H, $\underline{(CH_3)_3}$-Si), 0.90 (t, $J_{CH3,CH2}$ = 6.9 Hz 3H, (CH_2)-$\underline{CH_3}$), 1.30–1.35 (m, 4H, CH_2-$\underline{(CH_2)_2}$-CH_3), 1.60–1.68 (m, 2H, CO-CH_2-$\underline{CH_2}$), 2.35 (o, J = 7.7 Hz, 2H, CO-$\underline{CH_2}$-CH_2), 3.35 (s, 3H, O-$\underline{CH_3}$), 3.42–3.49 (m, 2H, H-2,H-5), 3.68–3.76 (m, 2H, H-3, H-4), 4.03 (dd, 1H, $J_{H\text{-}6,H\text{-}6'}$ = 11.9 Hz, $J_{H\text{-}5,H\text{-}6'}$ = 5.4 Hz

H-6′), 4.40 (dd, 1H, $J_{H\text{-}6,H\text{-}6'}$ = 11.9 Hz, $J_{H\text{-}5,H\text{-}6}$ = 2.2 Hz, H-6), 4.70 (d, 1H, $J_{H\text{-}1,H\text{-}2}$ = 3.7 Hz, H-1). ^{13}C NMR (100 MHz, $CDCl_3$), δ 0.5–1.5 (3s, $(CH_3)_3$-Si), 13.9 (CH_2-CH_3), 22.3–31.3 (2s, CH_2-$(CH_2)_2$-CH_3), 24.5 (CO-CH_2-CH_2-$(CH_2)_2$), 34.2 (CO-CH_2-CH_2), 54.8 (O-CH_3), 63.3 (C-6), 69.4 (C-4), 72.5 (C-2), 73.7 (C-5), 75.1 (C-3), 99.8 (C-1), 173.6 (CO).

Methyl 6-O-octanoyl-2,3,4-tri-O-trimethylsilyl-α-D-glucopyranoside **4**. Colorless syrup. Yield 77% from the methyl 2,3,4-tri-*O*-trimethylsilyl-α-D-glucopyranoside **1** following the general procedure. ^{1}H NMR (400 MHz, $CDCl_3$) δ 0.10–0.15 (3 s, 27H, $(CH_3)_3$-Si), 0.85 (t, $J_{CH3,CH2}$ = 6.9 Hz, 3H, CH_2-CH_2-CH_3), 1.20–1.31 (m, 8H, CH_2-$(CH_2)_4$-CH_3), 1.58–1.63 (m, 2H, CO-CH_2-CH_2-CH_2), 2.32 (m, 2H,CO-CH_2-CH_2), 3.32 (s, 3H, O-CH_3), 3.42–3.48 (m, 2H, H-2,H-5), 3.67–3.75 (m, 2H, H-3, H-4), 4.02 (dd, 1H, $J_{H\text{-}6,H\text{-}6'}$ = 11.9 Hz, $J_{H\text{-}5,H\text{-}6'}$ = 5.4 Hz H-6′), 4.37 (d, 1H, $J_{H\text{-}6,H\text{-}6'}$ = 11.8 Hz), 4.62 (d, 1H, $J_{H\text{-}1,H\text{-}2}$ = 3.7 Hz, H-1). ^{13}C NMR (100 MHz, $CDCl_3$) δ 0–1.5 (3s, $(CH_3)_3$-Si), 14.0 (-CH_2-CH_3), 22.5–31.5 (4s, CH_2-$(CH_2)_4$-CH_3), 24.8 (CO-CH_2-CH_2-CH_2), 34.2 (CO-CH_2-CH_2), 54.7 (O-CH_3), 63.3 (C-6), 69.5 (C-4), 72.5 (C-2), 73.7 (C-5), 75.1 (C-3), 99.7 (C-1), 173.5 (CO).

Methyl 6-O-decanoyl-2,3,4-tri-O-trimethylsilyl-α-D-glucopyranoside **5**. Colorless syrup. Yield 88% from the methyl 2,3,4-tri-*O*-trimethylsilyl-α-D-glucopyranoside **1** following the general procedure. ^{1}H NMR (400 MHz, $CDCl_3$) δ 0.11–0.16 (3 s, 27H, $(CH_3)_3$-Si), 0.85 (t, $J_{CH3,CH2}$ = 6.9 Hz, 3H, CH_2-CH_3), 1.19–1.35 (m, 12H, CH_2-$(CH_2)_6$-CH_3), 1.55–1.65 (m, 2H, CO-CH_2-CH_2), 2.37 (td, 2H, CO-CH_2-CH_2), 3.35 (s, 3H, O-CH_3), 3.43–3.5 (m, 2H, H-2,H-5), 3.68–3.76 (m, 2H, H-3, H-4), 4.03 (dd, 1H, $J_{H\text{-}6,H\text{-}6'}$ = 11,8 Hz, $J_{H\text{-}5,H\text{-}6'}$ = 5.3 Hz, H-6′), 4.37 (dd, 1H, $J_{H\text{-}6,H\text{-}6'}$ = 11.8 Hz, $J_{H\text{-}5,H\text{-}6}$ = 2.3 Hz H-6), 4.60 (d, 1H, $J_{H\text{-}1,H\text{-}2}$ = 3.7 Hz, H-1). ^{13}C NMR (100 MHz, $CDCl_3$) δ 0–1.5 (3s, $(CH_3)_3$-Si), 14.0 (-CH_2-CH_3), 22.5–32 (6s, CH_2-$(CH_2)_6$-CH_3), 24,7 (CO-CH_2-CH_2-CH_2), 34.2 (CO-CH_2-CH_2), 55.0 (O-CH_3), 63.2 (C-6), 69.4 (C-4), 72.5 (C-2), 73.8 (C-5), 75.0 (C-3), 99.8 (C-1), 173.5 (CO).

Methyl 6-O-dodecanoyl-2,3,4-tri-O-trimethylsilyl-α-D-glucopyranoside **6**. Colorless syrup. Yield 70% from the methyl 2,3,4-tri-*O*-trimethylsilyl-α-D-glucopyranoside **1** following the general procedure. ^{1}H NMR (400 MHz, $CDCl_3$) δ 0.14–0.19 (3s, 27H, $(CH_3)_3$-Si), 0.75 (t, $J_{CH3,CH2}$ = 6.9 Hz, 3H, CH_2-CH_2-CH_3), 1.20–1.30 (m, 16H, CH_2-$(CH_2)_8$-CH_3), 1.61–1.64 (m, 2H,CO-CH_2-CH_2-CH_2), 2.35 (td, 2H,CO-CH_2), 3.35 (s, 3H, O-CH_3), 3.43–3.49 (m, 2H, H-2, H-5), 3.68–3.78 (m, 2H, H-3, H-4),4.05 (dd, 1H, $J_{H\text{-}6,H\text{-}6'}$ = 11.8 Hz, $J_{H\text{-}5,H\text{-}6'}$ = 5.4 Hz, H-6′), 4.39 (dd, 1H, $J_{H\text{-}6,H\text{-}6'}$ = 11.8 Hz, $J_{H\text{-}5,H\text{-}6}$ = 2.2 Hz, H-6), 4.62 (d, 1H, $J_{H\text{-}1,H\text{-}2}$ = 3.6 Hz, H-1). ^{13}C NMR (100 MHz, $CDCl_3$) δ 0–1.5 (3s, $(CH_3)_3$-Si), 14.0 (-CH_2-CH_3), 22.5–21.5 (8s, CH_2-$(CH_2)_8$-CH_3), 24.5 (CO-CH_2-CH_2), 34.3 (CO-CH_2), 55.0 (O-CH_3), 63.2 (C-6), 69.5 (C-4), 72.6 (C-2), 73.8 (C-5), 75.0 (C-3), 99.8 (C-1), 173.4 (CO).

Methyl 6-O-octanoyl-2,3,4-tri-O-trimethylsilyl-β-D-galactopyranoside **7**. Colorless syrup. Yield 83% from the methyl 2,3,4-tri-*O*-trimethylsilyl-β-D-galactopyranoside **2** following the general procedure. ^{1}H NMR (400 MHz, $CDCl_3$) δ 0.12 (3s, 27H $(CH_3)_3$-Si), 0.85 (t, $J_{CH3,CH2}$ = 6.9 Hz, 3H, CH_2-CH_2-CH_3), 1.20–1.31 (m, 8H, CH_2-$(CH_2)_4$-CH_3), 1.61 (p, 2H, J = 7.4 Hz CO-CH_2-CH_2-CH_2), 2.30 (t, 2H, J = 7.5 Hz, CO-CH_2-CH_2), 3.40 (dd, 1H, $J_{H\text{-}2,H\text{-}3}$ = 9.3 Hz, $J_{H\text{-}3,H\text{-}4}$ = 2.7 Hz, H-3), 3.47 (s, 3H, O-CH_3), 3.64 (m, 2H, H-2,H-5), 3.74 (d, 1H, $J_{H\text{-}3,H\text{-}4}$ = 2.3 Hz, H-4), 4.06 (d, 1H, $J_{H\text{-}1,H\text{-}2}$ = 7.5 Hz, H-1), 4.13 (dd, 1H, $J_{H\text{-}6,H\text{-}6'}$ = 11.0 Hz, $J_{H\text{-}5,H\text{-}6}$ = 6.5 Hz, H-6), 4.23 (dd, 1H, $J_{H\text{-}6,H\text{-}6'}$ = 11.0 Hz, $J_{H\text{-}5,H\text{-}6'}$ = 6.6 Hz, H-6′), ^{13}C NMR (100 MHz, $CDCl_3$) δ 0.6, 0.7, 0.8 (3s, $(CH_3)_3$-Si), 14.2 (-CH_2-CH_3), 22.7, 29.0, 29.2, 31.7 (4s, CH_2-$(CH_2)_4$-CH_3), 25.1 (CO-CH_2-CH_2-CH_2), 34.4 (CO-CH_2-CH_2), 57.2 (O-CH_3), 63.0 (C-6), 71.9 (C-4), 72.0 (C-2), 72.2 (C-5), 75.1 (C-3), 105.0 (C-1), 173.6 (CO).

Methyl 6-O-hexanoyl-α-D-glucopyranoside **8**. Colorless syrup. Yield 77% from the methyl 6-*O*-hexanoyl-2,3,4-tri-*O*-trimethylsilyl-α-D-glucopyranoside **3** following the general procedure for deprotection. ^{1}H NMR (400 MHz, DMSO-*d6*) δ 0.84 (t, 3H, $J_{CH3,CH2}$ = 6.9 Hz, CH_2-CH_3), 1.21–1.29 (m, 4H, -$(CH_2)_2$-CH_3), 1.44–1.57 (m, 2H, CO-CH_2-CH_2), 2.28 (t, 2H, $J_{CH2,CH2}$ = 7.3 Hz, CO-CH_2), 3.02–3.09 (m, 1H, H-4) 3.16–3.23 (m,1H, H-2) 3.25 (s, 3H, O-CH_3), 3.35–3.41 (m, 1H, H-3), 3.48–3.53 (m, 1H, H-5) 4.01 (dd, 1H, $J_{H\text{-}6,H\text{-}6'}$ = 11.6 Hz, $J_{H\text{-}5,H\text{-}6'}$ = 6.5 Hz, H-6′), 4.29 (dd, 1H, $J_{H\text{-}6,H\text{-}6'}$ = 11.6 Hz, $J_{H\text{-}5,H\text{-}6}$ = 2.1 Hz, H-6), 4.52 (d, 1H, $J_{H\text{-}1,H\text{-}2}$ = 3.7 Hz, H-1) 4.82 (d, 1H, $J_{OH,H\text{-}2'}$ = 6.4 Hz, OH-2), 4.90 (d, 1H, $J_{OH,H\text{-}3'}$ = 5.0 Hz, OH-3), 5.15 (d, 1H, $J_{OH,\,H\text{-}4'}$ = 5.8 Hz, OH-4). ^{13}C (100 MHz, DMSO-*d6*) δ 13.9 (-CH_2-CH_3),

22.9, 24.3, 30.7 (3 s, $-(\underline{CH_2})_3-CH_3$), 33.6 ($CO-\underline{CH_2}-$), 54.5 ($O-\underline{CH_3}$), 63.6 (C-6), 69.7 (C-5), 70.5 (C-4), 71.9 (C-2), 73.3 (C-3), 99.8 (C-1), 173.0 (CO). HRMS (ESI) m/z for $C_{13}H_{24}O_7Na^+$ $[M + Na]^+$ calcd 315.1420, found 315.1424.

Methyl 6-O-octanoyl-α-D-glucopyranoside **9**. White powder. Yield 77% from the methyl 6-*O*-octanoyl-2,3,4-tri-*O*-trimethylsilyl-α-D-glucopyranoside **4** following the general procedure for deprotection. ^{1}H NMR (400 MHz, DMSO-*d6*) δ 0.84 (t, 3H, $J_{CH3,CH2}$ = 7.3 Hz, $CH_2-\underline{CH_3}$), 1.17–1.30 (m, 8H, $-(\underline{CH_2})_4-CH_3$), 1.47–1.55 (m, 2H, $CO-CH_2-\underline{CH_2}$), 2.28 (t, 2H, $J_{CH2,CH2}$ = 7.3 Hz, $CO-\underline{CH_2}$), 3.02–3.09 (m, 1H, H-4) 3.16–3.23 (m, 1H, H-2) 3.25 (s, 3H, $O-\underline{CH_3}$), 3.35–3.41 (m, 1H, H-3), 3.50–3.56 (m, 1H, H-5) 4.01 (dd, 1H, $J_{H\text{-}6,H\text{-}6'}$ = 11.7 Hz, $J_{H\text{-}5,H\text{-}6'}$ = 6.7 Hz, H-6′), 4.28 (dd, 1H, $J_{H\text{-}6,H\text{-}6'}$ = 11.7 Hz, $J_{H\text{-}5,H\text{-}6}$ = 1.9 Hz, H-6), 4.52 (d, 1H, $J_{H\text{-}1,H\text{-}2}$ = 3.6 Hz, H-1) 4.82 (d, 1H, $J_{OH,H\text{-}2'}$ = 6.4 Hz, OH-2), 4.90 (d, 1H, $J_{OH,H\text{-}3'}$ = 4.9 Hz, OH-3), 5.15 (d, 1H, $J_{OH,H\text{-}4'}$ = 5.8 Hz, OH-4). ^{13}C (100 MHz, DMSO-*d6*) δ 14.1 ($-CH_2-\underline{CH_3}$), 22.2, 24.6, 28.5, 28.5, 31.2 (5 s, $(\underline{CH_2})_5-CH_3$), 33.6 ($CO-\underline{CH_2}-$), 54.4 ($O-\underline{CH_3}$), 63.6 (C-6), 69.7 (C-5), 70.5 (C-4), 71.9 (C-2), 73.3 (C-3), 99.8 (C-1), 173.0 (CO). HRMS (ESI) m/z for $C_{15}H_{28}O_7Na^+$ $[M + Na]^+$ calcd 343.1733, found 343.1726.

Methyl 6-O-decanoyl-α-D-glucopyranoside **10**. White powder. Yield 81% from the methyl 6-*O*-decanoyl-2,3,4-tri-*O*-trimethylsilyl-α-D-glucopyranoside **5** following the general procedure for deprotection. ^{1}H NMR (400 MHz, CD_3OD) δ 0.89 (t, 3H, $J_{CH3,CH2}$ = 7.3 Hz, $CH_2-\underline{CH_3}$), 1.24–1.39 (m, 12H, $-(\underline{CH_2})_6-CH_3$), 1.62 (p, 2H, $J_{CH2,CH2}$ = 7.2 Hz, $CO-CH_2-\underline{CH_2}$), 2.35 (t, 2H, $J_{CH2,CH2}$ = 7.4 Hz, $CO-\underline{CH_2}-CH_2$), 3.27 (dd, 1H, $J_{H\text{-}3,H\text{-}4}$ = 9.0 Hz, $J_{H\text{-}4,H\text{-}5}$ = 9.9 Hz, H-4) 3.37–3.42 (m, 4H, H-2, $O-\underline{CH_3}$), 3.61 (t, 1H, $J_{H\text{-}2,H\text{-}3} = J_{H\text{-}3,H\text{-}4}$ = 9.3 Hz, H-3), 3.70 (ddd, 1H, $J_{H\text{-}4,H\text{-}5}$ = 9.9 Hz, $J_{H\text{-}5,H\text{-}6}$ = 2.0 Hz, $J_{H\text{-}5,H\text{-}6'}$ = 6.0 Hz, H-5), 4.19 (dd, 1H, $J_{H\text{-}6,H\text{-}6'}$ = 11.8 Hz, $J_{H\text{-}5,H\text{-}6'}$ = 6.0 Hz, H-6′), 4.38 (dd, 1H, $J_{H\text{-}6,H\text{-}6'}$ = 11.8 Hz, $J_{H\text{-}5,H\text{-}6}$ = 2.1 Hz, H-6), 4.65 (d, 1H, $J_{H\text{-}1,H\text{-}2}$ = 3.7 Hz, H-1). ^{13}C (100 MHz, CD_3OD) δ 14.5 ($-CH_2-\underline{CH_3}$), 23.7, 26.1, 30.2, 30.4, 30.4, 30.6, 33.0 (7 s, $-(\underline{CH_2})_7-CH_3$), 35.0 ($CO-\underline{CH_2}-$), 55.6 ($O-CH_3$), 64.7 (C-6), 71.0 (C-5), 71.9 (C-4), 73.4 (C-2), 75.0 (C-3), 101.3 (C-1), 175.4 (CO). HRMS (ESI) m/z for $C_{17}H_{32}O_7Na^+$ $[M + Na]^+$ calcd 371.2046, found 371.2046.

Methyl 6-O-dodecanoyl-α-D-glucopyranoside **11**. White powder. Yield 71% from the methyl 6-*O*-dodecanoyl-2,3,4-tri-*O*-trimethylsilyl-α-D-glucopyranoside **6** following the general procedure for deprotection. ^{1}H NMR (400 MHz, CD_3OD) δ 0.89 (t, 3H, $J_{CH3,CH2}$ = 7.3 Hz, $CH_2-\underline{CH_3}$), 1.24–1.40 (m, 16H, $-(\underline{CH_2})_8-CH_3$), 1.62 (p, 2H, $J_{CH2,CH2}$ = 7.2 Hz, $-CO-CH_2-\underline{CH_2}$), 2.35 (t, 2H, $J_{CH2,CH2}$ = 7.4 Hz, $CO-\underline{CH_2}$), 3.27 (dd, 1H, $J_{H\text{-}3,H\text{-}4}$ = 9.0 Hz, $J_{H\text{-}4,H\text{-}5}$ = 10.0 Hz, H-4) 3.37–3.42 (m, 4H, H-2, $O-\underline{CH_3}$), 3.61 (t, 1H, $J_{H\text{-}2,H\text{-}3} = J_{H\text{-}3,H\text{-}4}$ = 9.3 Hz, H-3), 3.70 (ddd, 1H, $J_{H\text{-}4,H\text{-}5}$ = 9.9 Hz, $J_{H\text{-}5,H\text{-}6}$ = 2.0 Hz, $J_{H\text{-}5,H\text{-}6'}$ = 6.0 Hz, H-5), 4.19 (dd, 1H, $J_{H\text{-}6,H\text{-}6'}$ = 11.8 Hz, $J_{H\text{-}5,H\text{-}6'}$ = 6.0 Hz, H-6′), 4.38 (dd, 1H, $J_{H\text{-}6,H\text{-}6'}$ = 11.8 Hz, $J_{H\text{-}5,H\text{-}6}$ = 2.1 Hz, H-6), 4.65 (d, 1H, $J_{H\text{-}1,H\text{-}2}$ = 3.7 Hz, H-1). ^{13}C (100 MHz, CD_3OD) δ 14.5 ($-CH_2-\underline{CH_3}$), 23.8, 26.1, 30.2, 30.4, 30.5, 30.6, 30.8, 30.8, 33.1 (9 s, $-(\underline{CH_2})_9-CH_3$), 35.0 ($CO-CH_2-$), 55.6 ($O-CH_3$), 64.7 (C-6), 71.1 (C-5), 71.9 (C-4), 73.5 (C-2), 75.0 (C-3), 101.3 (C-1), 175.4 (CO). HRMS (ESI) m/z for $C_{19}H_{36}O_7Na^+$ $[M + Na]^+$ calcd 399.2359, found 399.2361.

Methyl 6-O-octanoyl-β-D-galactopyranoside **12**. White powder. Yield 83% from the methyl 6-*O*-octanoyl-2,3,4-tri-*O*-trimethylsilyl-β-D-galactopyranoside **7** following the general procedure for deprotection. ^{1}H NMR (400 MHz, CD_3OD) δ 0.89 (t, 3H, $J_{CH3,CH2}$ = 7.3 Hz, $CH_2-\underline{CH_3}$), 1.26–1.37 (m, 8H, $CH_2-(\underline{CH_2})_4-CH_3$), 1.57–1.65 (m, 2H, $CH_2-\underline{CH_2}-(CH_2)_4$), 2.34 (t, 2H, J = 7.4 Hz, $CO-\underline{CH_2}-CH_2$), 3.44–3.51 (m, 2H, H-2, H-3) 3.50 (s, 3H, $O-\underline{CH_3}$), 3.71 (ddd, 1H, $J_{H\text{-}5,H\text{-}6}$ = 7.3 Hz, $J_{H\text{-}5,H\text{-}6'}$ = 5.0 Hz, $J_{H\text{-}4,H\text{-}5}$ = 1.0 Hz, H-5), 3.79 (d, 1H, $J_{H\text{-}3,H\text{-}4}$ = 2.0 Hz, H-4), 4.12 (d, 1H, $J_{H\text{-}1,H\text{-}2}$ = 7.5 Hz, H-1), 4.21 (dd, 1H, $J_{H\text{-}6,H\text{-}6'}$ = 11.4 Hz, $J_{H\text{-}5,H\text{-}6}$ = 5.0 Hz, H-6), 4.30 (dd, 1H, $J_{H\text{-}6,H\text{-}6'}$ = 11.3 Hz, $J_{H\text{-}5,H\text{-}6'}$ = 7.4 Hz, H-6′), ^{13}C NMR (100 MHz, CD_3OD) δ 14.4 ($-CH_2-\underline{CH_3}$), 23.7, 26.1, 30.1, 30.2, 32.8 (5 s, $-(\underline{CH_2})_5-CH_3$), 35.0 ($CO-\underline{CH_2}-$), 57.2 ($O-\underline{CH_3}$), 64.6 (C-6), 70.3 (C-4), 72.3 (C-2), 73.9 (C-5), 74.7 (C-3), 105.9 (C-1), 175.3 (CO). HRMS (ESI) m/z for $C_{15}H_{28}O_7Na^+$ $[M + Na]^+$ calcd 343.1733, found 343.1743.

3.4. Physicochemical Characterizations

3.4.1. Solubility

Water solubility reflects the ability of a particular molecule to dissolve and its interaction with water. To characterize the water solubility of most amphiphile molecules, especially the ionic surfactants, the Krafft point (Tk) is often used, referring to the critical temperature above which micelles form or the melting point of the hydrated solid surfactant. Two general approaches were used to measure the Tks of the molecules investigated herein: (1) a visual examination of the dissolution phenomenon for binary molecule/water systems, and (2) a quantitative measurement by using differential scanning calorimetry (DSC) analysis. The visual observation procedure is detailed below. We start with adding the molecule into 5–50 mL water, in a concentration range between 10^{-2} mM and 10^3 mM (corresponding to a range of a surfactant mass concentration well below 20%), keeping the suspension under stirring for a certain time at room temperature (RT). Consequently, molecules can be classified into two major types: those soluble in water at RT and others insoluble at RT, with the solubility limit being called S (expressed in mM). The aspect of the sample, either transparent or turbid, is determined by direct visual examination. Tk determination with DSC is particularly applicable in the case of molecules insoluble at RT, for which a concentrated binary molecule/water mixture with a mass fraction of 20% of each molecule (C >> CMC in most cases) was prepared at 65 °C under stirring for 30 min to obtain a homogeneous system. These pre-homogenized aqueous systems were cooled and conserved in a refrigerator at 4 °C for a suitable period (>7 days), during which equilibrated hydrated crystals formed and precipitated. Then, for each molecule, about a 200–400 mg crystal sample was taken and put into the calorimeter cell. The first equilibrium phase was programmed at 4 °C for 1 h, followed by a slow heating stage from 4 °C to 65 °C at a rate of 0.5 °C/min. Tk was determined based on the location of the endothermic peak on the DSC curve, if any. All agitations were performed using a standard magnetic hotplate stirrer, MR 3001K (Heidolph Instruments, Schwabach, Germany). DSC curves were obtained from a commercial μDSC7 evo calorimeter (SETARAM Instrumentation, Caluire-et-Cuire, France). Tks are reported as the averages of at least two measurements as well as their standard deviations.

3.4.2. Surface Activity

The surface activity reflects the amphiphiles' adsorption behavior on the air–aqueous solution interface. The surface tension (γ) was obtained using the Wilhelmy plate method with a K100 Processor Tensiometer (KRUSS, Hamburg, Germany). Both the critical micellar concentration (CMC) and equilibrium surface tension at the CMC (γcmc) were derived from the typical plot of surface tension (γ) against the logarithm of concentration (Log C). The surface excess (Γmax) and minimal area per molecule at the air–aqueous solution interface (Amin) at different temperatures were calculated from the Gibbs adsorption equation [3,46]:

$$\Gamma_{\max} = -\frac{1}{2.303RT}\left(\frac{d\gamma}{dlogc}\right) \quad (2)$$

and

$$\mathrm{A}_{\min} = \frac{10^{20}}{\mathrm{N_A}\Gamma_{\max}} \quad (3)$$

where R is the ideal gas constant, 8.31 $\mathrm{J{\cdot}mol^{-1}{\cdot}K^{-1}}$; T is the measuring temperature, K; γ is the surface tension, N/m; c is the concentration, $\mathrm{mol.L^{-1}}$; $\mathrm{A_{min}}$ is the minimal area per molecule, $\mathrm{Å^2}$, and $\mathrm{N_A}$ is Avogadro's number, $6.022{\cdot}10^{23}\ \mathrm{mol^{-1}}$.

For each molecule, several solutions of a wide range of concentrations were prepared. Before measuring, each solution was finely homogenized with the help of a magnetic stirring bar. The stirring time was fixed at 30 min in all cases. For molecules with a Tk below RT, the preparation was performed at environmental conditions, whereas for molecules exhibiting a Tk higher than 25 °C, the aqueous mixtures were firstly heated and agitated at elevated temperatures (25–50 °C), depending on the corresponding Tk. Then, measurements were performed once for each concentration, at the solution preparation's temperature, with an uncertainty of ±1 °C.

4. Conclusions

Two series of methyl D-glycopyranoside-based esters, with different orientations of the ester group, were compared in terms of solubility and surface adsorption properties. This study, combined with a comparison of four well-studied commercial octyl glycosides (C8αGlc, C8βGlc, C8αGal, and C8βGal), allows for a careful exploration of how structural features affect their physicochemical properties.

Commercial octyl D-glycosides presented high solubility limits although their Krafft points were strongly influenced by the head group configuration. Only C8αGlc and C8βGal were found to have Tks above room temperature. Specific structural features causing differences in solubility were proposed for these molecules. For sugar esters families, the ester group orientation had an obvious impact on the dissolution behavior of the compounds.

With respect to the surface tension measurements, we observed, for both series of ester surfactants, a clear linear decrease in the log CMC with the alkyl chain length, in line with well-established surfactant science research. In addition, we could also evidence a slight decrease in γcmc and Amin with alkyl chain lengths, consistent with literature-based expectations for sugar-based surfactants. By comparing the commercial octyl D-glycosides and all the synthesized C7/C8 esters, we found that the latter exhibits a noticeably lower CMC. At a constant alkyl chain length, the various polar heads studied in this work did not seem to impact γcmc significantly.

Overall, this work offers a refined understanding of the various factors involved in the physicochemical properties of sugar-based amphiphiles. While being able to predict structure-amphiphilic property trends should allow the design of greener surfactants in a smarter way, it is obvious that predicting the solubility properties of new amphiphilic molecules is tricky and challenging, as sugar-based surfactant solubility arises from a delicate balance between three key factors—their crystal stability, hydrophilicity, and self-assembling tendency. Further research on the structure–property relationships of sugar-based surfactants are still needed to contribute to community knowledge and to provide significant hope regarding the replacement of petroleum-based substances.

Supplementary Materials: The following supporting information can be downloaded at https://www.mdpi.com/article/10.3390/molecules29102338/s1, ^{1}H and ^{13}C NMR spectra of final compounds (**8**–**12**).

Author Contributions: A.D., A.W., G.P. and I.P. conceived and designed the experiments; H.L. and B.B. performed the experiments; all authors analyzed the data; T.G. notably contributed to the structure–property relationships; G.P. and A.D. were the primary writers of the manuscript; all authors proofread the manuscript. All authors have read and agreed to the published version of the manuscript.

Funding: This work was supported by a grant awarded to B.B. and PhD funding for T.G., provided as part of the French Government's Investments for the Future, under the reference ANR-001-01.

Institutional Review Board Statement: Not Applicable.

Informed Consent Statement: Not Applicable.

Data Availability Statement: The original contributions presented in the study are included in the article/Supplementary Materials, further inquiries can be directed to the corresponding author/s.

Acknowledgments: H.L. acknowledges the Chinese Scholarship Council for the financial support of her PhD grant. This work was performed in partnership with SAS PIVERT, within the framework of the French Institute for the Energy Transition (Institut pour la Transition Energétique (ITE) P.I.V.E.R.T. (www.institutpivert.com)).

Conflicts of Interest: The authors declare no conflicts of interest.

References

1. Von Rybinski, W.; Hill, K. Alkyl polyglycosides—Properties and applications of a new class of surfactants. *Angew. Chem. Int. Ed.* **1998**, *37*, 1328–1345. [CrossRef]
2. Hill, K.; Rhode, O. Sugar-based surfactants for consumer products and technical applications. *Fett-Lipid* **1999**, *101*, 25–33. [CrossRef]
3. Ruiz, C.C. *Sugar-Based Surfactants: Fundamentals and Applications*; CRC Press: Boca Raton, FL, USA, 2008; pp. 245–306.
4. Rojas, O.J.; Stubenrauch, C.; Lucia, L.A.; Habibi, Y. Interfacial properties of sugar-based surfactants. In *Biobased Surfactants and Detergents: Synthesis, Properties, and Applications*; Hayes, D., Kitamoto, D., Solaiman, D., Ashby, R., Eds.; AOCS Press: Urbana, IL, USA, 2009; p. 457.
5. Pal, A.; Mondal, M.H.; Adhikari, A.; Bhattarai, A.; Saha, B. Scientific information about sugar-based emulsifiers: A comprehensive review. *RSC Adv.* **2021**, *11*, 33004–33016. [CrossRef] [PubMed]
6. Gaudin, T.; Lu, H.; Fayet, G.; Berthauld-Drelich, A.; Rotureau, P.; Pourceau, G.; Wadouachi, A.; Van Hecke, E.; Nesterenko, A.; Pezron, I. Impact of the chemical structure on amphiphilic properties of sugar-based surfactants: A literature overview. *Adv. Colloid Interface Sci.* **2019**, *270*, 87–100. [CrossRef] [PubMed]
7. Razafindralambo, H.; Blecker, C.; Paquot, M. Screening of basic properties of amphiphilic molecular structures for colloidal system formation and stability. In *Amphiphiles: Molecular Assembly and Applications*; American Chemical Society: Washington, DC, USA, 2011; Volume 1070, Chapter 4; pp. 53–66.
8. Razafindralambo, H.; Blecker, C.; Paquot, M. 8. Carbohydrate-based surfactants: Structure-activity relationships. In *Advances in Chemical Engineering*; Nawaz, Z., Ed.; InTech: Rijeka, Croatia, 2012.
9. Piispanen, P.S.; Persson, M.; Claesson, P.; Norin, T. Surface properties of surfactants derived from natural products. Part 1: Syntheses and structure/property relationships—Solubility and emulsification. *J. Surf. Det.* **2004**, *7*, 147–159. [CrossRef]
10. Lu, B.; Vayssade, M.; Miao, Y.; Chagnault, V.; Grand, E.; Wadouachi, A.; Postel, D.; Drelich, A.; Egles, C.; Pezron, I. Physico-chemical properties and cytotoxic effects of sugar-based surfactants: Impact of structural variations. *Colloids Surf. B* **2016**, *145*, 79–86. [CrossRef]
11. Baker, I.J.A.; Matthews, B.; Suares, H.; Krodkiewska, I.; Furlong, D.N.; Grieser, F.; Drummond, C.J. Sugar fatty acid ester surfactants: Structure and ultimate aerobic biodegradability. *J. Surf. Det.* **2000**, *3*, 1–11. [CrossRef]
12. Becerra, N.; Toro, C.; Zanocco, A.L.; Lemp, E.; Günther, G. Characterization of micelles formed by sucrose 6-O-monoesters. *Colloids Surf. A* **2008**, *327*, 134–139. [CrossRef]
13. Cook, A.G.; Wardell, J.L.; Imrie, C.T. Carbohydrate liquid crystals: Synthesis and characterisation of the methyl-6-O-(n-acyl)-α-D-glucopyranosides. *Chem. Phys. Lipids* **2011**, *164*, 118–124. [CrossRef]
14. Razafindralambo, H.; Richel, A.; Wathelet, B.; Blecker, C.; Wathelet, J.P.; Brasseur, R.; Lins, L.; Minones, J.; Paquot, M. Monolayer properties of uronic acid bicatenary derivatives at the air-water interface: Effect of hydroxyl group stereochemistry evidenced by experimental and computational approaches. *Phys. Chem. Chem. Phys.* **2011**, *13*, 15291–15298. [CrossRef]
15. Ducret, A.; Giroux, A.; Trani, M.; Lortie, R. Enzymatic preparation of biosurfactants from sugars or sugar alcohols and fatty acids in organic media under reduced pressure. *Biotechnol. Bioeng.* **1995**, *48*, 214–221. [CrossRef] [PubMed]
16. Skagerlind, P.; Larsson, K.; Barfoed, M.; Hult, K. Glucoside ester synthesis in microemulsions catalyzed by Candida antartica component B lipase. *J. Am. Oil Chem. Soc.* **1997**, *74*, 39–42. [CrossRef]
17. Blecker, C.; Piccicuto, S.; Lognay, G.; Deroanne, C.; Marlier, M.; Paquot, M. Enzymatically prepared n-alkyl esters of glucuronic acid: The effect of hydrophobic chain length on surface properties. *J. Colloid Int. Sci.* **2002**, *247*, 424–428. [CrossRef] [PubMed]
18. Moreau, B.; Lognay, G.C.; Blecker, C.; Brohée, J.C.; Chéry, F.; Rollin, P.; Paquot, M.; Marlier, M. Synthesis of novel D-glucuronic acid fatty esters using Candida antarctica lipase in tert-butanol. *Biotechnol. Lett.* **2004**, *26*, 419–424. [CrossRef] [PubMed]
19. Gouéth, P.Y.; Gogalis, P.; Bikanga, R.; Godé, P.; Postel, D.; Ronco, G.; Villa, P. Synthesis of Monoesters as Surfactants and Drugs from D-Glucose. *J. Carbohydr. Chem.* **1994**, *13*, 249–272. [CrossRef]
20. Otto, R.T.; Bornscheuer, U.T.; Syldatk, C.; Schmid, R.D. Lipase-catalyzed synthesis of arylaliphatic esters of β-D(+)-glucose, n-alkyl- and arylglucosides and characterization of their surfactant properties. *J. Biotechnol.* **1998**, *64*, 231–237. [CrossRef]
21. Savelli, M.P.; Van Roekeghem, P.; Douillet, O.; Cavé, G.; Godé, P.; Ronco, G.; Villa, P. Effects of tail alkyl chain length (n), head group structure and junction (Z) on amphiphilic properties of 1-Z-R-D,L-xylitol compounds (R=C_nH_{2n+1}). *Int. J. Pharm.* **1999**, *182*, 221–236. [CrossRef] [PubMed]
22. Garofalakis, G.; Murray, B.S.; Sarney, D.B. Surface Activity and Critical Aggregation Concentration of Pure Sugar Esters with Different Sugar Headgroups. *J. Colloid Interface Sci.* **2000**, *229*, 391–398. [CrossRef] [PubMed]
23. Kjellin, U.R.M.; Claesson, P.M.; Vulfson, E.N. Studies of N-Dodecyllactobionamide, Maltose 6′-O-Dodecanoate, and Octyl-β-glucoside with Surface Tension, Surface Force, and Wetting Techniques. *Langmuir* **2001**, *17*, 1941–1949. [CrossRef]
24. Soultani, S.; Ognier, S.; Engasser, J.M.; Ghoul, M. Comparative study of some surface-active properties of fructose esters and commercial sucrose esters. *Colloids Surf. A* **2003**, *227*, 35–44. [CrossRef]
25. Brown, G.M.; Dubreuil, P.; Ichhaporia, F.M.; Desnoyers, J.E. Synthesis and properties of some α-D-alkyl glucosides and mannosides: Apparent molal volumes and solubilization of nitrobenzene in water at 25 °C. *Can. J. Chem.* **1970**, *48*, 2525–2531. [CrossRef]
26. Nilsson, F.; Söderman, O.; Johansson, I. Physical–Chemical Properties of the n-Octyl β-D-Glucoside/Water System. A Phase Diagram, Self-Diffusion NMR, and SAXS Study. *Langmuir* **1996**, *12*, 902–908. [CrossRef]

27. Dupuy, C.; Auvray, X.; Petipas, C.; Rico-Lattes, I.; Lattes, A. Anomeric effects on the structure of micelles of alkyl maltosides in water. *Langmuir* **1997**, *13*, 3965–3967. [CrossRef]
28. Fukada, K.; Kawasaki, M.; Seimiya, T.; Abe, Y.; Fujiwara, M.; Ohbu, K. Stereochemical aspects of micellar properties of esterified glucoside surfactants in water: Apparent molar volume, adiabatic compressibility, and aggregation number. *Colloid Polym. Sci.* **2000**, *278*, 576–580. [CrossRef]
29. Lu, H.; Drelich, A.; Omri, M.; Pezron, I.; Wadouachi, A.; Pourceau, G. Catalytic synthesis of a new series of alkyl uronates and evaluation of their physicochemical properties. *Molecules* **2016**, *21*, 1301. [CrossRef] [PubMed]
30. Lu, H.; Pezron, I.; Gaudin, T.; Drelich, A. Non-equilibrium micelles formed by sugar-based surfactants under their Krafft temperature. *Colloids Surf. A* **2018**, *540*, 167–176. [CrossRef]
31. Hato, M. Synthetic glycolipid/water systems. *Curr. Opin. Colloid Interface Sci.* **2001**, *6*, 268–276. [CrossRef]
32. Matsumura, S.; Imai, K.; Yoshikawa, S.; Kawada, K.; Uchibor, T. Surface activities, biodegradability and antimicrobial properties of n-alkyl glucosides, mannosides and galactosides. *J. Am. Oil Chem. Soc.* **1990**, *67*, 996–1001. [CrossRef]
33. Schmidt-Lassen, J.; Lindhorst, T.K. Exploring the meaning of sugar configuration in a supramolecular environment: Comparison of six octyl glycoside micelles by ITC and NMR spectroscopy. *MedChemComm* **2014**, *5*, 1218–1226. [CrossRef]
34. Khan, A.A.; Chee, S.H.; McLaughlin, R.J.; Harper, J.L.; Kamena, F.; Timmer, M.S.M.; Stocker, B.L. Long-Chain Lipids Are Required for the Innate Immune Recognition of Trehalose Diesters by Macrophages. *ChemBioChem* **2011**, *12*, 2572–2576. [CrossRef]
35. Kotena, Z.M.; Behjatmanesh-Ardakani, R.; Hashim, R. AIM and NBO analyses on hydrogen bonds formation in sugar-based surfactants (α/β-D-mannose and n-octyl-α/β-D-mannopyranoside): A density functional theory study. *Liq. Cryst.* **2014**, *41*, 784–792. [CrossRef]
36. Straathof, A.J.J.; Van Bekkum, H.; Kieboom, A.P.G. Solid State and Solution Properties of Octyl D-Glucopyranosides. *Starch* **1988**, *40*, 438–440. [CrossRef]
37. Sakya, P.; Seddon, J.M. Thermotropic and lyotropic phase behaviour of monoalkyl glycosides. *Liq. Cryst.* **1997**, *23*, 409–424. [CrossRef]
38. Boyd, B.J.; Drummond, C.J.; Krodkiewska, I.; Grieser, F. How Chain Length, Headgroup Polymerization, and Anomeric Configuration Govern the Thermotropic and Lyotropic Liquid Crystalline Phase Behavior and the Air–Water Interfacial Adsorption of Glucose-Based Surfactants. *Langmuir* **2000**, *16*, 7359–7367. [CrossRef]
39. Jmol Development Team. Jmol. 2016. Available online: http://jmol.sourceforge.net/ (accessed on 22 April 2024).
40. Rosen, M.J.; Kunjappu, J.T. *Surfactants and Interfacial Phenomena*, 4th ed.; Wiley: Hoboken, NJ, USA, 2006; ISBN 978-0-470-54194-4.
41. Capalbi, A.; Gente, G.; La Mesa, C. Solution properties of alkyl glucosides, alkyl thioglucosides and alkyl maltosides. *Colloids Surf. A* **2004**, *246*, 99–108. [CrossRef]
42. Lainez, A.; Del Burgo, P.; Junquera, E.; Aicart, E. Mixed Micelles Formed by n-Octyl-β-D-glucopyranoside and Tetradecyltrimethylammonium Bromide in Aqueous Media. *Langmuir* **2004**, *20*, 5745–5752. [CrossRef] [PubMed]
43. Mańko, D.; Zdziennicka, A.; Jańczuk, B. Thermodynamic properties of adsorption and micellization of n-octyl-β-D-glucopyranoside. *Colloids Surf. B* **2014**, *114*, 170–176. [CrossRef] [PubMed]
44. Silva, F.V.; Goulart, M.; Justino, J.; Neves, A.; Santos, F.; Caio, J.; Lucas, S.; Newton, A.; Sacoto, D.; Barbosa, E.; et al. Alkyl deoxy-arabino-hexopyranosides: Synthesis, surface properties, and biological activities. *Bioorganic Med. Chem.* **2008**, *16*, 4083–4092. [CrossRef]
45. Frindi, M.; Michels, B.; Zana, R. Ultrasonic absorption studies of surfactant exchange between micelles and bulk phase in aqueous micellar solutions of nonionic surfactants with a short alkyl chain. 3. Surfactants with a sugar head group. *J. Phys. Chem.* **1992**, *96*, 8137–8141. [CrossRef]
46. Shinoda, K.; Yamaguchi, T.; Hori, R. The Surface Tension and the Critical Micelle Concentration in Aqueous Solution of β-D-Alkyl Glucosides and their Mixtures. *Bull. Chem. Soc. Jpn.* **1961**, *34*, 237–241. [CrossRef]
47. Sulthana, S.B.; Rao, P.V.C.; Bhat, S.G.T.; Nakano, T.Y.; Sugihara, G.; Rakshit, A.K. Solution Properties of Nonionic Surfactants and Their Mixtures: Polyoxyethylene (10) Alkyl Ether [CnE10] and MEGA-10. *Langmuir* **2000**, *16*, 980–987. [CrossRef]
48. Wang, X.; Yan, F.; Li, Z.; Zhang, L.; Zhao, S.; An, J.; Yu, J. Synthesis and surface properties of several nonionic–anionic surfactants with straight chain alkyl-benzyl hydrophobic group. *Colloids Surf. A* **2007**, *302*, 532–539. [CrossRef]
49. Alami, E.; Beinert, G.; Marie, P.; Zana, R. Alkanediyl-.alpha.,.omega.-bis(dimethylalkylammonium bromide) surfactants. 3. Behavior at the air-water interface. *Langmuir* **1993**, *9*, 1465–1467. [CrossRef]
50. Ueno, M.; Takasawa, Y.; Miyashige, H.; Tabata, Y.; Meguro, K. Effects of alkyl chain length on surface and micellar properties of octaethyleneglycol-n alkyl ethers. *Colloid Polym. Sci.* **1981**, *259*, 761–766. [CrossRef]
51. Zhou, T.; Yang, H.; Xu, X.; Wang, X.; Wang, J.; Dong, G. Synthesis, surface and aggregation properties of nonionic poly(ethylene oxide) gemini surfactants. *Colloids Surf. A* **2008**, *317*, 339–343. [CrossRef]
52. Razafindralambo, H.; Blecker, C.; Mezdour, S.; Deroanne, C.; Crowet, J.M.; Brasseur, R.; Lins, L.; Paquot, M. Impacts of the Carbonyl Group Location of Ester Bond on Interfacial Properties of Sugar-Based Surfactants: Experimental and Computational Evidences. *J. Phys. Chem. B* **2009**, *113*, 8872–8877. [CrossRef] [PubMed]

53. Li, G.-L.; Kung, K.K.-Y.; Wong, M.-K. Gold-catalyzed amide synthesis from aldehydes and amines in aqueous medium. *Chem. Comm.* **2012**, *48*, 4112–4114. [CrossRef]
54. Wang, H.; Cui, Y.; Zou, R.; Cheng, Z.; Yao, W.; Mao, Y.; Zhang, Y. Synthesis of oligosaccharides using per-O-trimethylsilyl-glycosyl iodides as glycosyl donors. *Carbohydr. Res.* **2016**, *427*, 1–5. [CrossRef]

molecules

MDPI

Article

Co-Assembled Supramolecular Organohydrogels of Amphiphilic Zwitterion and Polyoxometalate with Controlled Microstructures

Peilin Wei [1,†], Yu Duan [1,†], Chen Wang [1], Panpan Sun [2,*] and Na Sun [1,*]

[1] College of Pharmacy, Shandong Second Medical University, Weifang 261053, China; 19510200661@163.com (P.W.); miaoyintian@163.com (Y.D.); wangchchw@163.com (C.W.)

[2] School of Bioscience and Technology, Shandong Second Medical University, Weifang 261053, China

* Correspondence: sunpanpan@sdsmu.edu.cn (P.S.); sunna@sdsmu.edu.cn (N.S.)

† These authors contributed equally to this work.

Abstract: The organization of modifiable and functional building components into various superstructures is of great interest due to their broad applications. Supramolecular self-assembly, based on rationally designed building blocks and appropriately utilized driving forces, is a promising and widely used strategy for constructing superstructures with well-defined nanostructures and diverse morphologies across multiple length scales. In this study, two homogeneous organohydrogels with distinct appearances were constructed by simply mixing polyoxometalate (phosphomolybdic acid, HPMo) and a double-tailed zwitterionic quaternary ammonium amphiphile in a binary solvent of water and dimethyl sulfoxide (DMSO). The delicate balance between electrostatic attraction and repulsion of anionic HPMo clusters and zwitterionic structures drove them to co-assemble into homogeneous organohydrogels with diverse microstructures. Notably, the morphologies of the organohydrogels, including unilamellar vesicles, onion-like vesicles, and spherical aggregates, can be controlled by adjusting the ionic interactions between the zwitterionic amphiphiles and phosphomolybdic acid clusters. Furthermore, we observed an organohydrogel fabricated with densely stacked onion-like structures (multilamellar vesicles) consisting of more than a dozen layers at certain proportions. Additionally, the relationships between the self-assembled architectures and the intermolecular interactions among the polyoxometalate, zwitterionic amphiphile, and solvent molecules were elucidated. This study offers valuable insights into the mechanisms of polyoxometalate-zwitterionic amphiphile co-assembly, which are essential for the development of materials with specific structures and emerging functionalities.

Keywords: organohydrogels; supramolecular co-assembly; zwitterionic amphiphiles; polyoxometalates

1. Introduction

Molecular self-assembly represents a spontaneous natural process and a relatively common strategy for creating highly ordered nano- to micro-architectures, fostering the development of diverse functional materials with enhanced complexity, synergetic and dynamic properties. Among these materials, supramolecular gels, resulting from the self-assembly of low-molecular-weight gelators (LMWGs) through an array of supramolecular noncovalent interactions, have garnered substantial research attention due to their emerging application prospects such as biomaterials, sensing, stimulus-responsive and self-healing materials [1–5]. Self-assembly represents an efficient and low-energy pathway for the spontaneous generation of programmable and versatile nanostructures. By manipulating the distinct moieties, such as hydrophilicity and hydrophobicity of building blocks, along with the types of interactions between them, and controlling assembly conditions (solvent polarity, pH, light, and temperature etc.), devisable and multitudinous assemblies can be achieved, including micelles, vesicles, liquid crystals, and fibers [6–9]. Currently, the engineering of innovative and high-tech applications of conventional hydrogels has garnered significant interest. Specifically, organohydrogels, prepared in the water-organic solvent mixture, have attracted broad attention due to their intriguing properties, such as superior freezing-resistance, anti-drying capacity, and enhanced mechanical performance [10–12]. Therefore, there is an urgent need to develop supramolecular organohydrogels based on novel building blocks, which provide a straightforward strategy and fresh perspective for constructing diverse microstructures with additional and specific functionalities.

Among the myriad building blocks and strategies employed in constructing functional supramolecular gels, inorganic-organic co-assembly, merging the diversity of organic building blocks with the versatility of inorganic assemblies, has been considered as a promising approach. Moreover, by tailoring the specific properties of individual components, the intrinsic capabilities of resulting hybrid co-assembled systems can be precisely adjusted. Furthermore, the co-assembling process serves as a valid strategy for organizing both inorganic and organic building blocks into well-defined supramolecular nanostructures, which greatly improves properties through forming local concentrated environment, such as photochromism, luminescence, catalytic performance and antibacterial ability [13]. Various inorganic materials including polyoxometalates (POMs), quantum dots (QDs), polyhedral oligomeric silsesquioxane (POSS), have been widely employed as inorganic components to create well-defined inorganic-organic hybrid materials [14–17]. Particularly, POMs, which are anionic nanoscale inorganic clusters composed of early transition metal oxides in their highest oxidation states, exhibit diverse compositions and structures as well as versatile and tunable chemical properties, making them excellent candidates as inorganic building blocks for the construction of inorganic-organic co-assembled materials [18,19]. Due to the anionic characteristic of POMs in polar solvents, cationic amphiphiles or polymers are commonly used to co-assemble with POMs through electrostatic interactions. Wu et al. reported that nanodisks, nanocones, nanotubes, and onionlike hybrid nanostructures could be formed in co-assembling systems of POM nanoclusters and cationic amphiphiles [20,21]. Hao et al. studied that the co-assembly of double-tailed cationic surfactant possessing magnetic properties and magnetic POM clusters, which led to the formation of supramolecular magnetic aggregate structures comprised of layered structures and virus-like particles [22,23]. However, because of strong electrostatic interactions, the aforementioned co-assembled complexes are nearly insoluble in water and challenging to process into functional materials. Zwitterions, characterized by a covalently connected cation and anion, can interact with various ionic species and co-assemble into diverse aggregates via electrostatic forces. Recently, the delicate balance between electrostatic attraction and repulsion of zwitterionic amphiphiles and POM clusters has been discovered, leading to the successful construction of homogenous supramolecular hydrogel systems. In our previous studies, single-tailed imidazolium-type zwitterions were commonly employed to co-assemble with POM clusters, resulting in the observation that these systems exhibited a distinct tendency to form well-organized wormlike micelles [24,25]. Consequently, appropriately designing the structure of building blocks to regulate driving forces represents a rational strategy for generating diverse assemblies and investigating the assembly mechanism.

In this work, we designed and synthesized a double-tailed zwitterionic quaternary ammonium amphiphile (dioctylmethylammonium propanesulfonate, $(C8)_2MeAS$) for co-assembly with Keggin-type POM clusters (phosphomolybdic acid, HPMo), followed by a systematic investigation of their co-assembled characteristics. Leveraging the suitable ionic interaction between zwitterionic amphiphiles and anionic POM clusters, $(C8)_2MeAS$ and HPMo clusters were able to co-assemble into supramolecular organohydrogels with tunable nanostructures in a binary solvent of water and dimethyl sulfoxide (DMSO) (4:1 v/v). The morphologies of co-assemblies inside the supramolecular organohydrogels transformed from onion-like vesicles to spherical aggregates as the molar ratio of $(C8)_2MeAS$ to HPMo varied from 10:1 to 6:1. The gelation behavior, gelation mechanism, and rheological properties were detailed studied to reveal a theoretical comprehension of zwitterionic amphiphiles-POMs co-assembly in the binary solvent. Zwitterionic amphiphiles significantly broaden the scope of POM clusters' applications as building blocks in the fabrication of versatile co-assembled nanostructures and soft materials.

2. Results and Discussion

2.1. Phase Diagram and Morphologies of the Organohydrogels

The zwitterionic compound $(C8)_2$MeAS exhibits limited solubility in water because of its two hydrophobic alkyl chains. To enhance the solubility of $(C8)_2$MeAS and induce more complex co-assembly behavior, DMSO was used as a cosolvent [26,27]. To investigate the co-assembling behavior of zwitterionic $(C8)_2$MeAS and anionic HPMo clusters in a water-DMSO binary solvent, the phase diagram of $(C8)_2$MeAS/HPMo system at various $(C8)_2$MeAS concentrations and varying $(C8)_2$MeAS/HPMo molar ratios in water-DMSO binary solvent (4:1 v/v) was primarily examined. As illustrated in Figure 1, the phase boundary was primarily determined through visual observation and the tube-inversion method. $(C8)_2$MeAS cannot dissolve on its own in a mixture of water-DMSO mixed solvent with the volume ratio of 4:1. The addition of hydrophilic HPMo clusters significantly increases the solubility of $(C8)_2$MeAS, resulting in the formation of a homogenous solution. With the successive increase of HPMo concentration, planar lamellar phase (L_α), gel phase and precipitate region was observed. The insets of Figure 1 demonstrate that gels with $(C8)_2$MeAS/HPMo molar ratio from 7:1 to 12:1 exhibit good transparency, while gels with $(C8)_2$MeAS/HPMo molar ratio of 5:1 and 6:1 are optically opaque. For a more detailed examination of gel phase, representative samples with a fixed $(C8)_2$MeAS concentration of 200 mM were initially chosen for specific analysis (depicted by orange points in Figure 1). Optical photographs of organohydrogels, observed under crossed polarizers, reveal a birefringent texture in samples with $(C8)_2$MeAS/HPMo molar ratios of 12:1, 10:1 and 8:1 (Figure S1a). Additionally, the phase transition of the samples at a constant $(C8)_2$MeAS/HPMo molar ratio of 10:1 was investigated (indicated by the pink points in Figure 1). When the concentration of $(C8)_2$MeAS is lower than 120 mM, a homogeneous and transparent solution with no birefringence was obtained (Figure S1b). Optical photographs of organohydrogels with (C8)2MeAS concentration of 130 mM and 150 mM observed under crossed polarizers show birefringent textures (Figure S1b). This suggests that a transition from a micellar solution to a vesicle-based gel phase occurs at a $(C8)_2$MeAS concentration of 130 mM. Upon increasing the concentration of $(C8)_2$MeAS to 270 mM, the optically transparent gel sample turned to a viscous, translucent, and flowing substance exhibiting an oil-like texture under polarized optical microscope (POM) (Figure S3). These results demonstrate that a phase transition from vesicles to planar lamellar structures occurs as the concentration of $(C8)_2$MeAS increases to higher values.

To further visualize and identify the inner microstructure of assembled organohydrogels, cryogenic transmission electron microscopy (cryo-TEM) and freeze fracture-transmission electron microscope (FF-TEM) were conducted. Figure 2a,b display the cryo-TEM and FF-TEM images of the organohydrogel with a $(C8)_2$MeAS/HPMo molar ratio of 10:1. Densely stacked onion-like vesicles (multilamellar vesicles) and accompanying unilamellar vesicles are clearly observed. The onion-like vesicles have diameters ranging from 380 to 700 nm and interlamellar distances of approximately 34 nm. The unilamellar vesicles have smaller diameters ranging from 54 to 110 nm. As shown in Figure 2c,d, the organohydrogel with a $(C8)_2$MeAS/HPMo molar ratio of 8:1 also consists of large multilamellar vesicles with the diameters ranging from 850 nm to 1.4 μm, along with polydispersed unilamellar vesicles with diameters ranging from 50 to 500 nm. Multilamellar vesicles exhibit a smaller bilayer distance of about 30 nm compared to those in the organohydrogel of a 10:1 ratio. Moreover, slight agglomeration of unilamellar vesicles occurs in both organohydrogels with 10:1 and 8:1 ratios, promoting gel formation [28,29]. Increasing the $(C8)_2$MeAS/HPMo molar ratio to 6:1 resulted in a morphological transformation from vesicles to spherical aggregates. As shown in Figure 2e,f, the organohydrogel with a 6:1 ratio consists of polydisperse and well-defined spherical aggregates ranging in diameter from about 30 nm to near 1 μm, corresponding to the opaque appearance of organohydrogels [30]. Scanning electron microscopy (SEM) images reveal the adhesion of spherical aggregates, leading to network formation and subsequent gelation [31]. Similar spherical aggregates can be observed for the organohydrogel with a 5:1 ratio (Figure S2). As mentioned earlier, the morphological transformation with the increase in HPMo content is further confirmed by zeta potential measurements, typically used to monitor the surface potential of aggregates in colloidal systems. Figure 3 shows the zeta potentials of mixtures with a constant $(C8)_2$MeAS concentration and varying $(C8)_2$MeAS/HPMo molar ratios. The negative zeta potential of the mixtures demonstrates that anionic POM clusters reside the surface of the aggregates [17]. The zeta potential becomes increasingly negative as the $(C8)_2$MeAS/HPMo molar ratio varies from 12:1 and 7:1. The interaction between POM clusters and the covalently connected cation and anion of zwitterionic amphiphiles via electrostatic forces results in the bilayers becoming more negatively charged upon incorporation of more anionic POM clusters [24,32]. However, the zeta potential becomes less negative as the HPMo content increases to molar ratios of 6:1 and 5:1, corresponding with the observed morphological transformation. The slight increase in zeta potential might be attributed to the accumulation of counterions (H^+) on the aggregate surface upon addition of excess HPMo [33].

Small-angle X-ray scattering (SAXS) measurement is a potent technique for examining co-assembled microstructures in samples. Firstly, SAXS spectra were detected for samples with different molar ratios of $(C8)_2$MeAS/HPMo and fixed $(C8)_2$MeAS concentration of 200 mM (Figure 4a). Two distinctive scattering peaks, corresponding to the (100) and (200) reflections of lamellar structures, were observed with a relative ratio of 1:2. This observation concurs with the presence of multilamellar vesicles as depicted in TEM images. The interlayer spacing (d) can be determined using the equation $d = 2\pi/q$, derived from the first scattering factor (q_1). As the HPMo content increases from 12:1 to 5:1, the scattering peaks gradually shift towards higher q values, indicating a reduction in interlayer spacing from 31.08 to 26.04 nm. This trend suggests a denser packing of bilayers [28]. The introduction of HPMo molecules on the bilayer surface enhances interactions between adjacent bilayers, resulting in a more tightly packed bilayer structure. For organohydrogels with the $(C8)_2$MeAS/HPMo molar ratios of 10:1 and 8:1, the calculated d-spacing is 31.08 and 28.34 nm, respectively. Then, the SAXS patterns were measured for samples with a constant $(C8)_2$MeAS/HPMo molar ratio of 10:1 and varying $(C8)_2$MeAS concentrations, as shown in Figure 4b. In the case of the sample with a $(C8)_2$MeAS concentration of 110 mM, no significant scattering peak is observed. However, with an increase in $(C8)_2$MeAS concentration to 130 mM, scattering peaks corresponding to lamellar structures emerge. Notably, these scattering peaks become sharper with increasing $(C8)_2$MeAS concentration, indicating the more regular arrangement of bilayers in the samples. Increasing the $(C8)_2$MeAS concentration to 300 mM results in evident peaks at q values in a 1:2:3 ratio, indicating more ordered planar lamellar structures. With the $(C8)_2$MeAS concentration rising from 130 mM to 300 mM, the d-spacing reduces from 43.80 nm to 21.41 nm. This can be attributed to the more closely packed molecules with increasing $(C8)_2$MeAS concentration [28,34]. Thus, the augmentation of both HPMo and $(C8)_2$MeAS concentration ineluctably leads to a reduction in interlayer spacing of bilayers.

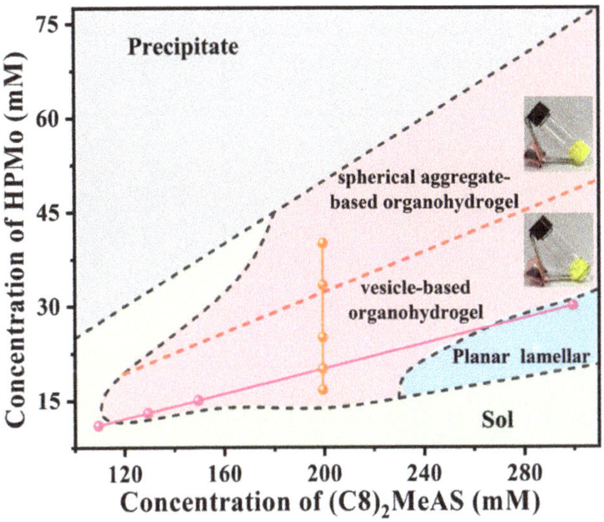

Figure 1. Phase diagram of mixtures of $(C8)_2$MeAS and HPMo at different molar ratios in water-DMSO binary solvent (4:1 v/v); the orange points represent samples with a fixed $(C8)_2$MeAS concentration of 200 mM; the pink points represent samples with a fixed $(C8)_2$MeAS/HPMo molar ratio of 10:1; the insets are photos of organohydrogels with the molar ratios of 10:1 and 6:1, respectively (at a fixed $(C8)_2$MeAS concentration of 200 mM).

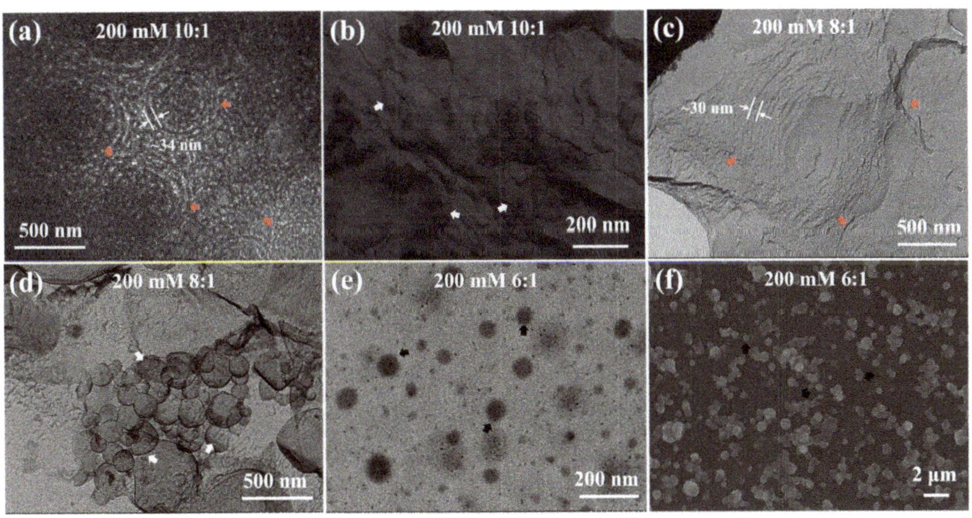

Figure 2. (**a**,**b**) Cryo-TEM and FF-TEM images of organohydrogel with the $(C8)_2$MeAS/HPMo molar ratio of 10:1; (**c**,**d**) FF-TEM and cryo-TEM images of organohydrogel with the $(C8)_2$MeAS/HPMo molar ratio of 8:1; (**e**,**f**) TEM and SEM images of organohydrogel with the $(C8)_2$MeAS/HPMo molar ratio of 6:1. The concentration of $(C8)_2$MeAS was fixed at 200 mM. The unilamellar vesicles, multilamellar vesicles and spherical aggregates in the images have been depicted by white, red and black arrows, respectively.

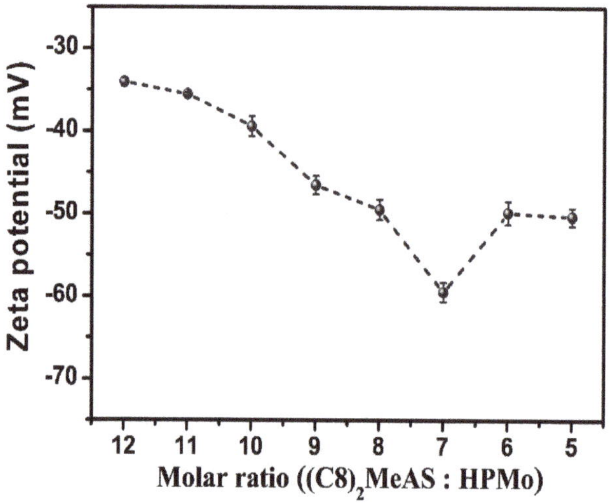

Figure 3. Zeta potentials for mixtures at $(C8)_2MeAS$ concentration of 90 mM with different $(C8)_2MeAS$/ HPMo molar ratios.

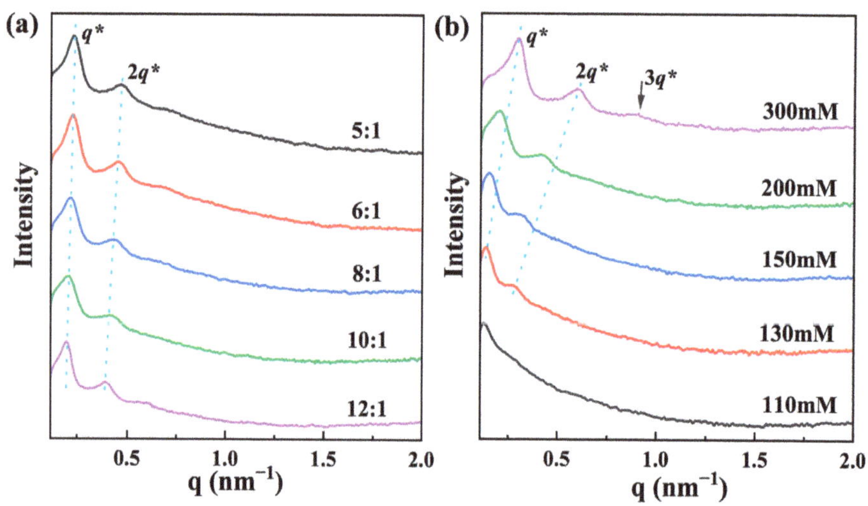

Figure 4. (**a**) SAXS spectra of samples with different $(C8)_2MeAS$/HPMo molar ratios and fixed $(C8)_2MeAS$ concentration of 200 mM; (**b**) SAXS data of samples with various $(C8)_2MeAS$ concentrations and the fixed $(C8)_2MeAS$/HPMo molar ratio of 10:1.

2.2. Rheological Properties of the Organohydrogels

Rheological measurements are generally used to investigate the macroscopic properties of samples. Figure 5a shows frequency sweep results for organohydrogels with different $(C8)_2MeAS$/HPMo molar ratios. The elastic moduli (G') of all samples exceed the viscous moduli (G'') by approximately one order of magnitude. Furthermore, the G' and G'' curves run almost parallel and remain frequency-independent, further certifying the good gel character of samples [35]. With the $(C8)_2MeAS$/HPMo molar ratio ranging from 10:1 to 5:1, the G' value, a significant parameter for assessing gel mechanical strength, gradually decreases, implying reduced ability to resist mechanical disturbance [24]. The decrease of mechanical strength arises from two factors: the microstructures within the gels and excessive crosslinking by HPMo clusters. Obviously, the G′ values of vesicle-based organohydrogels (8:1 and 10:1) higher than those of spherical aggregate-based organohydrogels (5:1 and 6:1). Vesicle-based gels feature densely packed multilamellar and unilamellar vesicles with minimal inter-vesicle spacing, forming a notably stiffer network compared to gels composed of adhesive spherical aggregates. Moreover, because of the hydrophilicity of POMs clusters, HPMo clusters situated on the outer surface of adjacent aggregates can interact with each other and serve as cross-linkers, facilitating the binding of neighboring aggregates and promoting gel formation [24]. However, an excessive addition of HPMo clusters may result in an increase in electrostatic repulsive forces among neighboring aggregates, significantly weakening the crosslinking of HPMo clusters. In addition, we also explored the frequency-dependent oscillatory profiles of organohydrogels with different $(C8)_2$MeAS concentrations but a fixed $(C8)_2$MeAS/HPMo molar ratio of 10:1 (Figure 5b). Both G′ and G″ remain independent of frequency. Moreover, all samples possess higher G′ values, indicating their elasticity dominant property, a characteristic feature of gel materials. The mechanical strength of organohydrogels gradually enhances with the rise in $(C8)_2$MeAS content. Summarily, the mechanical strength of gels can be facilely tuned by altering the content of $(C8)_2$MeAS and HPMo clusters.

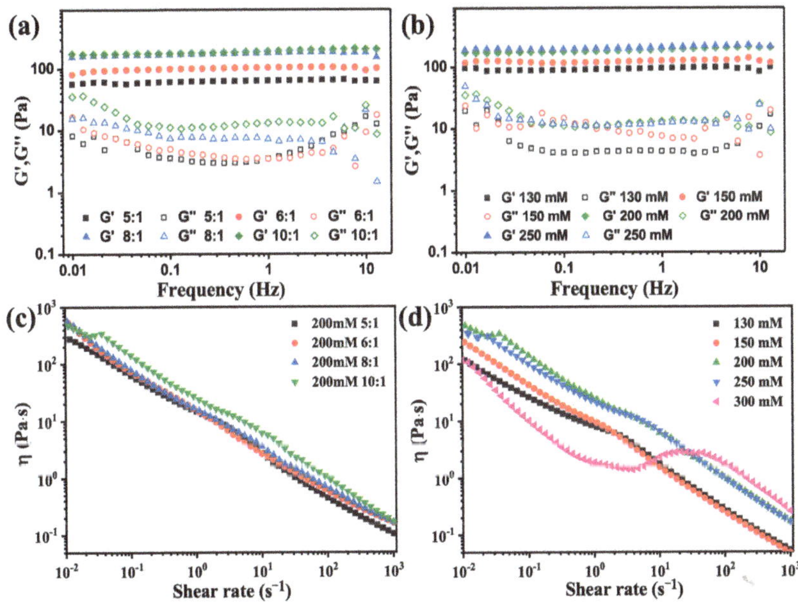

Figure 5. The rheological properties of the prepared samples: (**a**,**c**) frequency-dependent oscillatory profiles and shear viscosity as a function of shear rate for organohydrogels with different $(C8)_2$MeAS/HPMo molar ratios (The concentration of $(C8)_2$MeAS was fixed at 200 mM); (**b**,**d**) frequency sweep plots and shear rate dependence of the viscosity for organohydrogels with varying $(C8)_2$MeAS concentration ($(C8)_2$MeAS/HPMo = 10:1).

Additionally, the microstructure transformation of self-assembled systems could be distinguished from rheological measurements. Steady-shear measurements for samples with various components were further conducted. In Figure 5c, the viscosity of organohydrogels, with a fixed $(C8)_2$MeAS concentration of 200 mM, decreased gradually as the shear rate increased, implying the good shear-thinning property of gels. Therefore, these organohydrogels have potentially application as injectable materials. In Figure 5d, for organohydrogels with fixed $(C8)_2$MeAS/HPMo molar ratios of 10:1, the viscosity of gels rises with increasing $(C8)_2$MeAS concentration. This occurs because the increasing number of molecules results in tighter packing of vesicles within the organohydrogel. At a $(C8)_2$MeAS content of 300 mM, the sample transitioned into a viscous fluid with a planar lamellar structure, resulting in a significant decreased in viscosity. Notably, a partial shear thickening behavior is observed within the shear rate range of 4 s^{-1} to 47 s^{-1}, potentially associated with the transition from planar bilayers to onion-like vesicles. This transition from lamellar to onion-like structures is a universal phenomenon in certain surfactant/water system [36–38]. The subsequent shear-thinning region is attributed to the deformation of vesicle bilayers as the shear rate increases.

2.3. *Co-Assembly Mechanism of the Formation of Organohydrogels*

During the co-assembly process of the $(C8)_2$MeAS/HPMo system, an essential prerequisite is the matching of hydrophobicity of the zwitterionic amphiphiles with the hydrophilicity of the POMs. When the HPMo clusters co-assembled with either significantly more hydrophilic dihexylmethylammonium propanesulfonate ($(C6)_2$MeAS) or notably more hydrophobic didecylmethylammonium propanesulfonate ($(C10)_2$MeAS), dilute aqueous solution and precipitates were obtained at equivalent concentrations, respectively (Figure S4).

Another crucial factor contributing to the abundant phase behavior is the delicate interaction between zwitterions and POM clusters. Isothermal titration calorimetry (ITC) technique was employed to explore the thermodynamic behavior in the amphiphile-POMs mixing process. In the case of the cationic amphiphile ($(C8)_2$MeABr) and the anionic HPMo system (Figure 6a), an exothermic process with a distinct enthalpy change occurred during titration, confirming the strong electrostatic attraction between $(C8)_2$MeABr and HPMo clusters. This strong electrostatic attraction between the cationic $(C8)_2$MeABr and the anionic HPMo induces the formation of light-yellow precipitation (Figure S4). Conversely, as depicted in Figure 6b, enthalpy value of the endothermic process is low and remains relatively constant during the binding process of zwitterionic $(C8)_2$MeAs and HPMo, resulting from the covalently connected cation and anion of the zwitterion. For zwitterionic $(C8)_2$MeAS, along with the electrostatic attraction between the positively charged quaternary ammonium cation and the $[PMo_{12}O_{40}]^{3-}$ anion, the electrostatic repulsion also exists between the sulfonic anion and the $[PMo_{12}O_{40}]^{3-}$ anion, which is crucial for maintaining the electrostatic balance, as shown in Figure S5. The appropriate balance between electrostatic attraction and repulsion results in small enthalpy values in the ITC measurement and facilitates the formation of diverse phase behaviors and homogenous gel systems.

The intermolecular noncovalent forces involved in the co-assembly process were further investigated. The interaction between zwitterionic amphiphiles and HPMo clusters during co-assembly was initially studied using ^{1}H NMR spectra (Figure 7a). Protons adjacent to the charge (H_a~H_d) of $(C8)_2$MeAS molecules in non-co-assembled state (the black line) exhibit much sharper peaks compared to the protons of $(C8)_2$MeAS molecules in the co-assembled state. Furthermore, the protons of $(C8)_2$MeAS exhibit a noticeable downfield shift after mixing with the HPMo cluster, ascribing to the altered dielectric constant of the solvent and the conformational effects induced by molecular aggregation in the presence of HPMo [39,40]. Moreover, with the continuous increase in HPMo content, the proton of the methyl on the quaternary ammonium cation further shifts slightly downfield, attributed to the electrostatic interaction between $(C8)_2$MeAS and HPMo clusters. Fourier transform infrared spectroscopy (FT-IR) was also carried out to further elucidate the interaction mechanism between $(C8)_2$MeAS and HPMo clusters. FT-IR spectra of $(C8)_2$MeAS, HPMo, and xerogels with different $(C8)_2$MeAS/ HPMo molar ratios are shown in Figure 7b. The fundamental absorptions of HPMo clusters, including the asymmetric $P-O_a$ stretch at 1057 cm^{-1} and the Mo=O terminal stretch ($Mo=O_t$) at 955 cm^{-1}, are observed [41]. After co-assembling with $(C8)_2$MeAS, the vibrational characteristics of HPMo are still preserved. In comparison with the pure HPMo clusters, the absorptions of $P-O_a$ and $Mo=O_t$ shift to 1061 cm^{-1} and 993 cm^{-1} respectively, which can be ascribed to the intermolecular electrostatic interactions between HPMo and $(C8)_2$MeAS [24]. The results of ^{1}H NMR and

FT-IR experiments above reveal that intermolecular interaction plays an essential role in the co-assembly process.

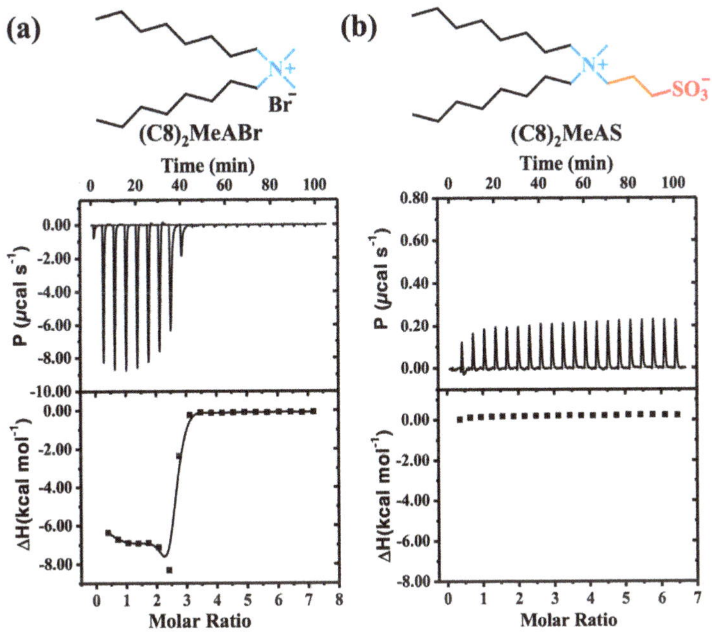

Figure 6. ITC titration data of (**a**) ammonium-type cationic amphiphile ((C8)$_2$MeABr, 3 mM) and (**b**) zwitterionic amphiphile ((C8)$_2$MeAS, 3 mM) titrated into HPMo solution (0.1 mM) at 25.0 °C. The water-DMSO binary solvent with the volume ration of 4:1 was used to prepare above solutions.

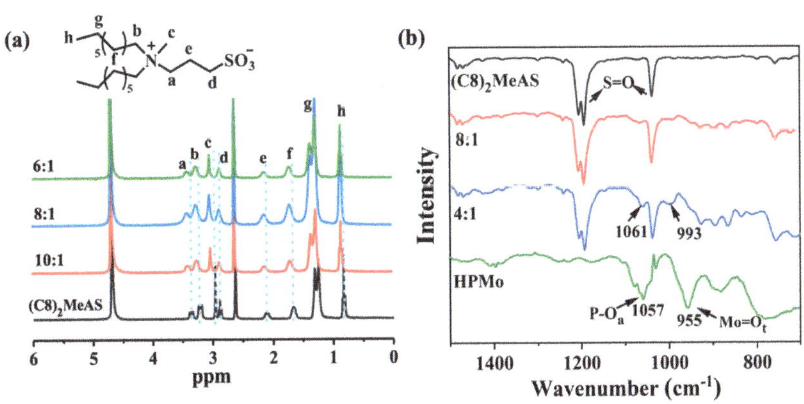

Figure 7. (**a**) ^{1}H NMR of mixtures of (C8)$_2$MeAS and HPMo in D$_2$O/DMSO-d_6 with various (C8)$_2$MeAS/HPMo molar ratios (the concentration of (C8)$_2$MeAS is fixed at 5 mM, the volume ratio of D$_2$O/DMSO-d_6 is 4:1); (**b**) FT-IR spectra of (C8)$_2$MeAS, HPMo and xerogels with the (C8)$_2$MeAS/HPMo molar ratios of 4:1 and 8:1.

The results above unequivocally demonstrate the critical role of electrostatic interaction between (C8)$_2$MeAS molecules and HPMo in the formation of organohydrogels with diverse microstructures. For the case of molar ratio at 10:1 and 8:1, HPMo cluster primarily increase the solubility of zwitterionic amphiphiles. Consequently, the morphology of co-assemblies mainly dictated by the geometric properties of $(C8)_2$MeAS amphiphile. The topological structure of $(C8)_2$MeAS is similar to the conventional double-tailed amphiphiles, which prefers to form bilayers in solution [23,42]. In the solvent system of water-DMSO, the charged zwitterionic headgroups reside outside of the vesicles, while hydrophilic HPMo clusters anchor at the interface through electrostatic interactions. Furthermore, HPMo clusters located on the outer surface of vesicles can interact with HPMo clusters on adjacent vesicles by virtue of hydrogen-bonding, van der Waals forces and electrostatic interactions. These interactions lead to the adhesion and rearrangement of bilayers, facilitating the formation of multilamellar vesicles. With the increasing concentrations of $(C8)_2$MeAS and HPMo, a sturdy network of multilamellar and unilamellar vesicles is formed, retaining the solvent and promoting gel formation. The probably mechanism for the transformation of aggregates is proposed, as shown in Scheme 1. As the molar ratio changes to 6:1 and 5:1, the transition in morphology can be explained by the surface occupation model [18,20]. At these molar ratios, the occupancy of zwitterionic amphiphiles on the POM core is less than 1/3, favoring a linear-shaped geometry wherein $(C8)_2$MeAS resides predominantly on one side of POM core, thus, promoting bilayer formation. On account of the excessive interactions between HPMo clusters on adjacent bilayers, the bilayer structure transformed into well-defined spherical assemblies with tidily and closely packed bilayers. Such HPMo cluster-cluster interactions among adjacent spherical particles lead to a network of these spherical aggregates, consequently resulting in the formation of organohydrogels [43].

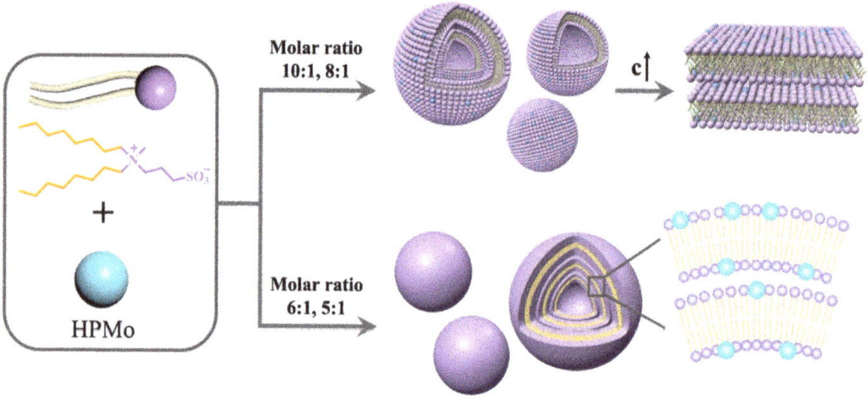

Scheme 1. The proposed mechanism for the formation of different organohydrogels.

The aggregation behavior of amphiphilic molecules usually can be manipulated by the presence of a cosolvent [27,29]. Thus, we explored the effect of DMSO content on the co-assembly process of $(C8)_2$MeAS and HPMo. ^{1}H NMR spectra of mixtures of $(C8)_2$MeAS and HPMo in a binary solvent with different D_2O/DMSO-d_6 volume ratios were recorded, as shown in Figure 8a. Despite varying the D_2O/DMSO-d_6 volume ratio from 3:1 to 6:1, the protons located on $(C8)_2$MeAS did not show obvious shift, indicating that the altered polarity of the binary solvent with increasing DMSO-d_6 volume has minimal impact on the electrostatic interaction between $(C8)_2$MeAS and HPMo [44]. SAXS spectra of samples prepared in water-DMSO mixed solvent with various volume ratios were also measured (Figure 8b). All SAXS spectra exhibit typical scatting peaks corresponding to lamellar structures with a q value ratio of 1:2. Relative to samples prepared in pure water, those in a water-DMSO mixed solvent (9:1) exhibit an increase in d-spacing from 28.38 nm to 31.08 nm, suggesting the looser bilayer structure in the presence of DMSO. The amphiphilic $(C8)_2$MeAS molecule is completely soluble in pure DMSO. Some of the added DMSO molecules solubilize into the hydrophobic layer of aggregates, causing the swelling of bilayers, and eventually resulting in an increase in d-spacing [45–47]. Further augmenting the water-DMSO ratio from 9:1 to 7:3, there is no significant variation of d-spacing, revealing that alterations in solution polarity do not affect the internal morphology. Noticeably, with the increasing DMSO content, the peaks become much sharper, indicating the lamellar structure turns to more ordered.

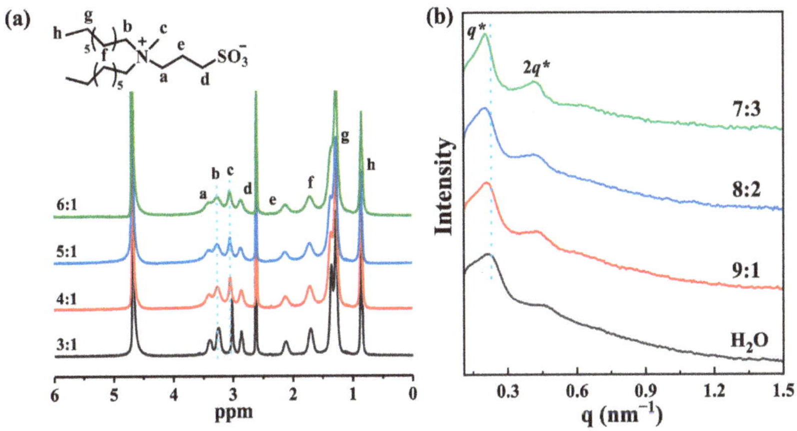

Figure 8. (**a**) ^{1}H NMR spectra of mixtures of $(C8)_2$MeAS and HPMo in the binary solvent with different D_2O/DMSO-d_6 volume ratios; (**b**) SAXS spectra of samples prepared in water-DMSO mixed solvent with various volume ratios (the concentration of $(C8)_2$MeAS is fixed at 200 mM; $(C8)_2$MeAS/HPMo = 10:1).

3. Materials and Methods

3.1. Materials

The compounds N-methyl-N-octyl-1-octanamine (98%), 1, 3-propanesultone (99%) and dimethyldioctylammonium bromide ($(C8)_2$MeBr, 98%) were purchased from J&K Scientific Co., Ltd. (Beijin, China). Phosphomolybdic acid (HPMo, 99%) was acquired from Aladdin Co., Ltd. (Shanghai, China). Dimethyl sulfoxide (DMSO) was obtained from Sinopharm Chemical Reagent Co., Ltd. (Shanghai, China). All materials were used as received. Deionized water was used throughout all experiments.

3.2. Synthesis of Dioctylmethylammonium Propanesulfonate ($(C8)_2$MeAS)

$(C8)_2$MeAS was synthesized according to the literature method [48]. More specifically, to a solution of N-methyl-N-octyl-1-octanamine in acetonitrile placed in a round bottom flask, the equimolar amount of 1,3-propanesultone was slowly added under a nitrogen atmosphere. The resulting mixture was refluxed at 80 °C for 5 days. After the solvent was removed by filtration, the residue was washed at least three times with diethyl ether and ethyl acetate respectively. Then, the white podwer was dried in vacuum oven at room temperature for 24 h. The ^{1}H NMR spectrum of $(C8)_2$MeAS in $CDCl_3$ is depicted in Figure S6. ^{1}H NMR (400 MHz, $CDCl_3$, δ/ppm): 3.716–3.770 (2H, m), 3.217–3.248 (4H, m), 3.151 (3H, s), 2.907–2.950 (2H, t), 2.209–2.260 (2H, m), 1.689 (4H, s), 1.274–1.348 (20H, m), 0.864–0.908 (6H, t).

3.3. Sample Preparation

Specific molar concentrations of $(C8)_2$MeAS and HPMo were weighted into stoppered glass vials. Water-DMSO mixture with volume ratio of 4:1 was served as solvent. The samples were dissolved and homogenized at 60 °C, and then, equilibrated at 25 °C for at least two weeks.

3.4. Characterization

^{1}H NMR Measurements. ^{1}H NMR spectroscopy was measured on a Bruker Avance II 400 MHz NMR spectrometer equipped with a pulse field gradient module (Z-axis) using a 5 mm BBO probe at 25 °C.

ITC Measurements. ITC were performed using an MicroCal VP-ITC apparatus at 25 °C. The 3 mM water-DMSO solution of $(C8)_2$MeAS and $(C8)_2$MeBr were titrated into a 0.1 mM water-DMSO solution of HPMo (1.4 mL) by using a 300 μL syringe, respectively. The total injection was 29 drops.

FT-IR Measurements. FT-IR spectra of $(C8)_2$MeAS, HPMo and freeze-dried hydrogels were measured by using an FT-IR spectrometer (PerkinElmer Spectrum Two) in the range 4000–450 cm^{-1}.

Cryo-TEM and FF-TEM Observations. The preparation of samples for testing followed established methods in the literature [42,49]. For cryo-TEM, samples were prepared using a controlled environment vitrification system (CEVS) maintained at 25 °C with 95% relative humidity. A micropipette was utilized to apply 5 μL of organohydrogels onto a TEM copper grid, which was subsequently blotted with two pieces of filter paper, resulting in the formation of thin films suspended across the mesh holes. After approximately 10 s, the samples were swiftly plunged into a reservoir of liquid ethane (cooled by nitrogen) at −165 °C. The vitrified samples were preserved in liquid nitrogen until transferred to a cryogenic sample holder (Gatan 626). The preparation procedure for FF-TEM is described below. A small amount of the organohydrogel was placed on a 0.1 mm thick copper disc, covered with a second copper disc, and then, plunged into liquid propane cooled by liquid nitrogen. Fracturing and replication were conducted using Balzers BAF-400D equipment at −150 °C. Pt/C was deposited at an angle of 45°. Cryo-TEM and FF-TEM images of the organohydrogels were acquired using a transmission electron microscope (JEOL JEM-1400, 120 kV).

TEM and SEM Observations. The sample preparation involved loading about 5 μL of organohydrogels onto a carbon-coated copper grid and removing the excess organohydrogels with filter paper after 5 min. Subsequently, samples were rapidly dehydrated via freeze-drying. TEM (JEOL JEM-1400, 120 kV) was used to characterize the microcosmic structures. SEM images were observed by scanning electron microscopy (SEM, JEOL JSM-7600F).

Zeta potential Measurements. Zeta potentials were measured at 25 °C by using a Brookhaven ZetaPALS instrument. The concentration of $(C8)_2$MeAS in the solutions was fixed at 100 mm. The molar ratio of $(C8)_2$MeAS /HPMo was varied from 5:1 to 10:1.

SAXS Measurements. The spectra of SAXS were recorded on the SAXSess mc2 X-ray scattering system (Anton Paar) with Cu K_α radiation (0.154 nm) operating at 50 kV and 40 mA. The distance between the sample and detector was about 264.5 mm and the wavelength of X-rays was 1.542 Å. The exposure time was 600 s for all samples.

Rheological Measurements. The rheological measurements were conducted by using a ThermoHaake RS300 rheometer with a parallel plate on a Peltier plate at 25 °C. Dynamic stress sweep spectra were recorded at a frequency of 1 Hz. Dynamic frequency sweep was carried out in the linear viscoelastic region determined from dynamic stress sweep measurements. The steady-state shear measurements were conducted with the shear rate increasing from 0.01 to 1000 s^{-1}.

4. Conclusions

In summary, we have successfully fabricated a series of supramolecular organohydrogels through the co-assembly of double-tailed zwitterionic quaternary ammonium amphiphiles and POMs in a water-DMSO binary solvent. The organohydrogels, featuring unilamellar vesicles, onion-like vesicles, and spherical aggregates with a layered structure, were formed by controlling the ionic interaction between zwitterionic amphiphiles and HPMo clusters. The addition of DMSO barely affected the electrostatic interaction between zwitterionic amphiphiles and POM clusters. Notably, DMSO molecules are able to increase the solubility of the hydrophobic layer of aggregates and induce bilayer swelling, resulting in a more relaxed bilayer structure. The inspiration drawn from the combination of zwitterionic amphiphiles and POMs in this work provides a vivid strategy and fresh perspective for designing functional hybrid materials incorporating nanoclusters or nanoparticles with rich potentials in addition to POMs.

Supplementary Materials: The following supporting information can be downloaded at: https://www.mdpi.com/article/10.3390/molecules29102286/s1, Figure S1: Optical photographs of organohydrogels observed with crossed polarizers; Figure S2: TEM and SEM images of the organohydrogel with $(C8)_2$MeAS/HPMo molar ratio of 5:1; Figure S3: POM images of samples with different $(C8)_2$MeAS concentration of 280 mM and 300 mM; Figure S4: Photos of mixtures composed of different amphiphiles and HPMo; Figure S5: Schematic illustration of electrostatic interaction for $(C8)_2$MeABr/HPMo and $(C8)_2$MeAS/HPMo systems.

Author Contributions: P.W.: Data curation, Formal analysis, Investigation, Writing—original draft, Writing—review & editing. Y.D.: Conceptualization, Methodology, Project administration, Resources, Software, Writing—original draft. C.W.: Data curation, Formal analysis, Investigation. N.S.: Writing—review & editing, Visualization, Supervision, Project administration, Funding acquisition. P.S.: Validation, Funding acquisition. All authors have read and agreed to the published version of the manuscript.

Funding: N. Sun was financially supported by the National Natural Science Foundation of China (No. 22102118). P. Sun was funded by the Natural Science Foundation of Shandong Province (No. ZR2023QB052). This work is funded by the domestic visiting project of Shandong Second Medical University.

Institutional Review Board Statement: Not applicable.

Informed Consent Statement: Not applicable.

Data Availability Statement: Data are contained within the article and Supplementary Materials.

Conflicts of Interest: The authors declare no conflicts of interest.

References

1. Gao, D.; Thangavel, G.; Lee, J.; Lv, J.; Li, Y.; Ciou, J.-H.; Xiong, J.; Park, T.; Lee, P.S. A Supramolecular Gel-Elastomer System for Soft Iontronic Adhesives. *Nat. Commun.* **2023**, *14*, 1990. [CrossRef] [PubMed]
2. Perera, M.M.; Ayres, N. Dynamic Covalent Bonds in Self-Healing, Shape Memory, and Controllable Stiffness Hydrogels. *Polym. Chem.* **2020**, *11*, 1410–1423. [CrossRef]
3. Phillip, R.A.; Chivers, D.K.S. Shaping and Structuring Supramolecular Gels. *Nat. Rev.* **2019**, *4*, 463–478.
4. Panja, S. Stimuli Responsive Dynamic Transformations in Supramolecular Gels. *Chem. Soc. Rev.* **2021**, *50*, 5165–5200. [CrossRef] [PubMed]
5. Wang, Z.; Zhang, Y.; Yin, Y.; Liu, J.; Li, P.; Zhao, Y.; Bai, D.; Zhao, H.; Han, X.; Chen, Q. High-Strength and Injectable Supramolecular Hydrogel Self-Assembled by Monomeric Nucleoside for Tooth-Extraction Wound Healing. *Adv. Mater.* **2022**, *34*, 2108300. [CrossRef] [PubMed]
6. Uchida, J.; Soberats, B.; Gupta, M.; Kato, T. Advanced Functional Liquid Crystals. *Adv. Mater.* **2022**, *34*, 2109063. [CrossRef] [PubMed]
7. Yin, C.; Jiang, F.; Li, B.; Wu, L. Multiple Modulations for Supramolecular Hydrogels of Bola-Form Surfactants Bearing Rigid and Flexible Groups. *Soft Matter* **2019**, *15*, 5034–5041. [CrossRef] [PubMed]
8. Liu, J.; Liu, P.; Du, J.; Wang, Q.; Chen, X.; Zhao, L. Review on High-Temperature-Resistant Viscoelastic Surfactant Fracturing Fluids: State-of-the-Art and Perspectives. *Energy Fuels* **2023**, *37*, 9790–9821. [CrossRef]
9. Tabet, A.; Forster, R.A.; Parkins, C.C.; Wu, G.; Scherman, O.A. Modulating Stiffness with Photo-Switchable Supramolecular Hydrogels. *Polym. Chem.* **2019**, *10*, 467–472. [CrossRef]
10. Liu, J.; Chen, Z.; Chen, Y.; Rehman, H.U.; Guo, Y.; Li, H.; Liu, H. Ionic Conductive Organohydrogels with Dynamic Pattern Behavior and Multi-Environmental Stability. *Adv. Funct. Mater.* **2021**, *31*, 2101464. [CrossRef]
11. Helgeson, M.E.; Moran, S.E.; An, H.Z.; Doyle, P.S. Mesoporous Organohydrogels from Thermogelling Photocrosslinkable Nanoemulsions. *Nat. Mater.* **2012**, *11*, 344–352. [CrossRef] [PubMed]
12. Qin, Z.; Dong, D.; Yao, M.; Yu, Q.; Sun, X.; Guo, Q.; Zhang, H.; Yao, F.; Li, J. Freezing-Tolerant Supramolecular Organohydrogel with High Toughness, Thermoplasticity, and Healable and Adhesive Properties. *ACS Appl. Mater. Interfaces* **2019**, *11*, 21184–21193. [CrossRef] [PubMed]

13. Cameron, J.M.; Guillemot, G.; Galambos, T.; Amin, S.S.; Hampson, E.; Mall Haidaraly, K.; Newton, G.N.; Izzet, G. Supramolecular Assemblies of Organo-Functionalised Hybrid Polyoxometalates: From Functional Building Blocks to Hierarchical Nanomaterials. *Chem. Soc. Rev.* **2022**, *51*, 293–328. [CrossRef] [PubMed]
14. Huang, S.-C.; Xia, X.-X.; Fan, R.-X.; Qian, Z.-G. Programmable Electrostatic Interactions Expand the Landscape of Dynamic Functional Hydrogels. *Chem. Mater.* **2020**, *32*, 1937–1945. [CrossRef]
15. Pruksawan, S.; Lim, J.W.R.; Lee, Y.L.; Lin, Z.; Chee, H.L.; Chong, Y.T.; Chi, H.; Wang, F. Enhancing Hydrogel Toughness by Uniform Cross-Linking Using Modified Polyhedral Oligomeric Silsesquioxane. *Commun. Mater.* **2023**, *4*, 75. [CrossRef]
16. Sun, X.; Yao, F.; Li, J. Nanocomposite Hydrogel-Based Strain and Pressure Sensors: A Review. *J. Mater. Chem. A* **2020**, *8*, 18605–18623. [CrossRef]
17. Zhou, W.; Hu, Z.; Wei, J.; Dai, H.; Chen, Y.; Liu, S.; Duan, Z.; Xie, F.; Zhang, W.; Guo, R. Quantum Dots-Hydrogel Composites for Biomedical Applications. *Chin. Chem. Lett.* **2022**, *33*, 1245–1253. [CrossRef]
18. Li, B.; Li, W.; Li, H.; Wu, L. Ionic Complexes of Metal Oxide Clusters for Versatile Self-Assemblies. *Acc. Chem. Res.* **2017**, *50*, 1391–1399. [CrossRef] [PubMed]
19. Luo, J.; Liu, T. Competition and Cooperation among Different Attractive Forces in Solutions of Inorganic–Organic Hybrids Containing Macroionic Clusters. *Langmuir* **2019**, *35*, 7603–7616. [CrossRef]
20. Nisar, A.; Zhuang, J.; Wang, X. Cluster-Based Self-Assembly: Reversible Formation of Polyoxometalate Nanocones and Nanotubes. *Chem. Mater.* **2009**, *21*, 3745–3751. [CrossRef]
21. Li, H.; Sun, H.; Qi, W.; Xu, M.; Wu, L. Onionlike Hybrid Assemblies Based on Surfactant-Encapsulated Polyoxometalates. *Angew. Chem. Int. Ed.* **2007**, *46*, 1300–1303. [CrossRef] [PubMed]
22. Zhao, W.; Sun, H.; Wang, Y.; Eastoe, J.; Dong, S.; Hao, J. Self-Assembled Magnetic Viruslike Particles for Encapsulation and Delivery of Deoxyribonucleic Acid. *Langmuir* **2018**, *34*, 7171–7179. [CrossRef] [PubMed]
23. Zhao, W.; Cui, J.; Hao, J.; Van Horn, J.D. Co-Assemblies of Polyoxometalate $\{Mo_{72}Fe_{30}\}$/Double-Tailed Magnetic-Surfactant for Magnetic-Driven Anchorage and Enrichment of Protein. *J. Colloid Interface Sci.* **2019**, *536*, 88–97. [CrossRef] [PubMed]
24. Sun, N.; Wu, A.; Yu, Y.; Gao, X.; Zheng, L. Polyoxometalate-Based Photochromic Supramolecular Hydrogels with Highly Ordered Spherical and Cylindrical Micellar Nanostructures. *Chem. Eur. J.* **2019**, *25*, 6203–6211. [CrossRef] [PubMed]
25. Wu, A.; Sun, P.; Sun, N.; Yu, Y.; Zheng, L. Coassembly of a Polyoxometalate and a Zwitterionic Amphiphile into a Luminescent Hydrogel with Excellent Stimuli Responsiveness. *Chem. Eur. J.* **2018**, *24*, 16857–16864. [CrossRef] [PubMed]
26. Oh, K.; Rajesh, K.; Stanton, J.F.; Baiz, C.R. Quantifying Hydrogen-Bond Populations in Dimethyl Sulfoxide/Water Mixtures. *Angew. Chem. Int. Ed.* **2017**, *56*, 11375–11379. [CrossRef]
27. Das, S.; Mondal, S.; Ghosh, S. Physicochemical Studies on the Micellization of Cationic, Anionic, and Nonionic Surfactants in Water–Polar Organic Solvent Mixtures. *J. Chem. Eng. Data* **2013**, *58*, 2586–2595. [CrossRef]
28. Dong, R.; Zhong, Z.; Hao, J. Self-Assembly of Onion-like Vesicles Induced by Charge and Rheological Properties in Anionic–Nonionic Surfactant Solutions. *Soft Matter* **2012**, *8*, 7812. [CrossRef]
29. Liu, X.; Gitsov, I. Nonionic Amphiphilic Linear Dendritic Block Copolymers. Solvent-Induced Self-Assembly and Morphology Tuning. *Macromolecules* **2019**, *52*, 5563–5573. [CrossRef]
30. Nieh, M.-P.; Dolinar, P.; Kučerka, N.; Kline, S.R.; Debeer-Schmitt, L.M.; Littrell, K.C.; Katsaras, J. Formation of Kinetically Trapped Nanoscopic Unilamellar Vesicles from Metastable Nanodiscs. *Langmuir* **2011**, *27*, 14308–14316. [CrossRef]
31. Cao, X.; Gao, A.; Hou, J.; Yi, T. Fluorescent Supramolecular Self-Assembly Gels and Their Application as Sensors: A Review. Coord. *Chem. Rev.* **2021**, *434*, 213792. [CrossRef]
32. Zhang, T.; Li, H.-W.; Wu, Y.; Wang, Y.; Wu, L. Self-Assembly of an Europium-Containing Polyoxometalate and the Arginine/Lysine-Rich Peptides from Human Papillomavirus Capsid Protein L1 in Forming Luminescence-Enhanced Hybrid Nanoparticles. *J. Phys. Chem. C* **2015**, *119*, 8321–8328. [CrossRef]
33. Cheng, C.-Y.; Wang, T.-Y.; Tung, S.-H. Biological Hydrogels Formed by Swollen Multilamellar Liposomes. *Langmuir* **2015**, *31*, 13312–13320. [CrossRef] [PubMed]
34. Song, Z.; Kim, H.; Ba, X.; Baumgartner, R.; Lee, J.S.; Tang, H.; Leal, C.; Cheng, J. Polypeptide Vesicles with Densely Packed Multilayer Membranes. *Soft Matter* **2015**, *11*, 4091–4098. [CrossRef] [PubMed]
35. Himmelein, S.; Lewe, V.; Stuart, M.C.A.; Ravoo, B.J. A Carbohydrate-Based Hydrogel Containing Vesicles as Responsive Non-Covalent Cross-Linkers. *Chem. Sci.* **2014**, *5*, 1054–1058. [CrossRef]
36. Briceño-Ahumada, Z.; Soltero, A.; Maldonado, A.; Perez, J.; Langevin, D.; Impéror-Clerc, M. On the Use of Shear Rheology to Formulate Stable Foams. Example of a Lyotropic Lamellar Phase. *Colloids Surf. Physicochem. Eng. Asp.* **2016**, *507*, 110–117. [CrossRef]
37. Sato, D.; Obara, K.; Kawabata, Y.; Iwahashi, M.; Kato, T. Re-Entrant Lamellar/Onion Transition with Varying Temperature under Shear Flow. *Langmuir* **2013**, *29*, 121–132. [PubMed]
38. Fujii, S.; Richtering, W. Shear Quench-Induced Disintegration of a Nonionic Surfactant C10E3 Onion Phase. *Soft Matter* **2013**, *9*, 5391. [CrossRef]
39. Wu, A.; Gao, X.; Liang, L.; Sun, N.; Zheng, L. Interaction among Worm-like Micelles in Polyoxometalate-Based Supramolecular Hydrogel. *Langmuir* **2019**, *35*, 6137–6144. [CrossRef]
40. Huang, X.; Han, Y.; Wang, Y.; Wang, Y. Aggregation Behavior of Nitrophenoxy-Tailed Quaternary Ammonium Surfactants. *J. Phys. Chem. B* **2007**, *111*, 12439–12446. [CrossRef]

41. Gunaratne, K.D.D.; Johnson, G.E.; Andersen, A.; Du, D.; Zhang, W.; Prabhakaran, V.; Lin, Y.; Laskin, J. Controlling the Charge State and Redox Properties of Supported Polyoxometalates via Soft Landing of Mass-Selected Ions. *J. Phys. Chem. C* **2014**, *118*, 27611–27622. [CrossRef]
42. Fong, C.; Le, T.; Drummond, C.J. Lyotropic Liquid Crystal Engineering–Ordered Nanostructured Small Molecule Amphiphileself-Assembly Materials by Design. *Chem. Soc. Rev.* **2012**, *41*, 1297–1322. [CrossRef]
43. Wang, Y.; Li, W.; Wu, L. Organic—Inorganic Hybrid Supramolecular Gels of Surfactant-Encapsulated Polyoxometalates. *Langmuir* **2009**, *25*, 13194–13200. [CrossRef] [PubMed]
44. Zhang, J.; Chen, X.; Li, W.; Li, B.; Wu, L. Solvent Dielectricity-Modulated Helical Assembly and Morphologic Transformation of Achiral Surfactant-Inorganic Cluster Ionic Complexes. *Langmuir* **2017**, *33*, 12750–12758. [CrossRef] [PubMed]
45. Gao, X.; Lu, F.; Liu, Y.; Sun, N.; Zheng, L. The Facile Construction of an Anion Exchange Membrane with 3D Interconnected Ionic Nano-Channels. *Chem. Commun.* **2017**, *53*, 767–770. [CrossRef] [PubMed]
46. Li, B.-Y.; Li, Y.-C.; Lu, Z.-Y. The Important Role of Cosolvent in the Amphiphilic Diblock Copolymer Self-Assembly Process. *Polymer* **2019**, *171*, 1–7. [CrossRef]
47. Hao, L.-S.; Wu, J.; Peng, Y.-R.; Wang, Y.; Xiao, K.; Hu, Y.; Nan, Y.-Q. Short-Chain n -Alcohol-Induced Changes in Phase Behaviors of Aqueous Mixed Cationic/Anionic Surfactant System. *Langmuir* **2018**, *34*, 7319–7333. [CrossRef] [PubMed]
48. Ferreira, A.M.; Passos, H.; Okafuji, A.; Tavares, A.P.M.; Ohno, H.; Freire, M.G.; Coutinho, J.A.P. An Integrated Process for Enzymatic Catalysis Allowing Product Recovery and Enzyme Reuse by Applying Thermoreversible Aqueous Biphasic Systems. *Green Chem.* **2018**, *20*, 1218–1223. [CrossRef]
49. Sun, N.; Shi, L.; Lu, F.; Xie, S.; Sun, P.; Zheng, L. Spontaneous Vesicle Phase Formation by Linear Pseudo-Oligomeric Surfactant in Aqueous Solutions. *Langmuir* **2015**, *31*, 2281–2287. [CrossRef]

molecules

MDPI

Article

Quantifying the Hydrophobic Effect per CF_2 Moiety from Adsorption of Fluorinated Alcohols at the Water/Oil Interface

Boyan Peychev [1], Dimitrinka Arabadzhieva [2], Ivan L. Minkov [2,3], Elena Mileva [2] and Radomir I. Slavchov [1,*]

[1] School of Engineering and Materials Science, Queen Mary University of London, Mile End Road, London E1 4NS, UK; b.peychev@qmul.ac.uk

[2] Rostislaw Kaischew Institute of Physical Chemistry, Bulgarian Academy of Sciences, Acad. G. Bonchev Str., bl. 11, 1113 Sofia, Bulgaria; dimi@ipc.bas.bg (D.A.); minkov.ivan@gmail.com (I.L.M.); mileva@ipc.bas.bg (E.M.)

[3] Department of Chemistry, Biochemistry, Physiology, and Pathophysiology, Faculty of Medicine, Sofia University, 1 Koziak Str., 1407 Sofia, Bulgaria

* Correspondence: r.slavchov@qmul.ac.uk

Abstract: Amphiphilic fluorocarbon substances are a trending topic of research due to their wide range of applications accompanied by an alarming environmental and health impact. In order to predict their fate in the environment, use them more economically, develop new water treatment methods, etc., a better understanding of their physicochemical behavior is required. Their hydrophobicity in water/oil systems is particularly sensitive to one key thermodynamic parameter: the free energy of transfer of a perfluoromethylene group from oil to water. However, for the $–CF_2–$ moiety, the transfer energy values reported in the literature vary by more than ±25%. Due to the exponential relationship between this energy and the adsorption constants or the partition coefficients, such an uncertainty can lead to orders of magnitude error in the predicted distribution of fluorinated species. We address this problem by presenting an experimental determination of the hydrophobic effect of a $–CF_2–$ moiety with a greater certainty than currently available. The transfer energy is determined by measuring the interfacial tension of water|hexane for aqueous solutions of short-chained fluorotelomer alcohols. The obtained results for the free energy of transfer of a $–CF_2–$ moiety from oil to water are $1.68 \pm 0.02 \times RT_0$, $1.75 \pm 0.02 \times RT_0$, and $1.88 \pm 0.02 \times RT_0$ at 288.15 K, 293.15 K, and 303.15 K, respectively.

Keywords: hydrophobic effect; fluorotelomer alcohol; adsorption model; water-soluble perfluoroalkyl substances; tensiometer

1. Introduction

Fluorocarbons and their derivatives have a number of unique properties that make them valuable in many industrial branches. In fact, in a recent study, Glüge et al. identified almost 300 unique uses for perfluoroalkyl substances [1]. In more than a third of these applications, the perfluoroalkyl substances are used because of their superior surface activity. Due to the larger size of the fluorine atom compared to the hydrogen atom, fluorocarbons are more hydrophobic than hydrocarbons, making fluorosurfactants more effective at reducing the interfacial energy compared to conventional surfactants of similar molecular size [2]. Furthermore, due to the extreme electronegativity and low polarizability of the fluorine atom, fluorocarbons are also oleophobic, allowing for coatings that are simultaneously water- and oil-repellent [2,3]. On the other hand, fluorine-containing organic substances have received much media attention under the umbrella term "forever chemicals" because their chemical-, thermal-, and photo-stability leads to accumulation in the environment. The shorter water-soluble fluorosurfactants, often surface-active products of chemical degradation, have been transported by the water cycle all over the world [4–6]. Furthermore, perfluoroalkyl substances have been associated with numerous health hazards [7–9]. Thus, there is regulatory action being taken to limit the use of fluorocarbon-based substances [7,10] and a great interest in remediation technologies [7,11,12]. A better understanding of the physicochemical properties of fluorosurfactants could allow for their more efficient use (i.e., use of smaller quantities); could allow for more accurate modeling of their fate in the environment and in the human body or of their interaction with lipid structures [13]; and could aid in the design of separation methods.

In general, fluorocarbon substances are much less studied than their hydrocarbon analogues. For instance, even though separation methods for fluorocarbon substances are a trending topic [7,11,12], their partitioning behavior has not been extensively investigated yet [12]. For perfluoroalkylated substances, the bulk partitioning and the adsorption out of aqueous solutions are driven by the hydrophobic effect, as quantified by the free energy $\Delta\mu_{CF_2}$ for transferring the nonpolar moiety –CF_2– from the hydrophobic phase to the aqueous phase. The values of its aliphatic counterpart, $\Delta\mu_{CH_2}$, are well-known for both air and oil phases under a range of conditions [14–22] and prove to be very useful for calculating the thermodynamic parameters of homologous series of aliphatic substances, such as partition coefficients, adsorption constants, critical micelle concentrations, etc. In contrast, for the –CF_2– fragment, the value of the transfer energy between oil and water has not been established with certainty. The reported results for $\Delta\mu_{CF_2}$ from oil to water are based on data for the change in the length of the hydrophobic tail (C_nF_{2n+1}–) of the partition coefficient [23] and on the adsorption constant of fluorinated ionic surfactants [24] (see the discussion in Peychev and Slavchov [25]), and they range from $\Delta\mu_{CF_2} = 1.6$ to $2.2 \times RT_0$ (R is the universal gas constant, $T_0 = 298.15$ K). This high uncertainty results in at least an order of magnitude uncertainty in the predicted adsorption constants and the oil/water partition coefficients of fluorinated substances.

The main reason for this status quo is that the fluorinated substances that are in common use have long chains and negligible solubility in water. This makes the adsorption out of aqueous solution impossible and the analysis of their partitioning behavior difficult. In contrast, the typical fluorocarbon bioaccumulating pollutants are the shorter water-soluble species, which can be circulated in the water cycle. Furthermore, at least in some cases, the shorter species are more ecotoxic [26]. However, the properties of the water-soluble fluorosurfactants, in particular the nonionic ones, have been scarcely studied. Another reason for the uncertainty in the $\Delta\mu_{CF_2}$ estimates is that, unlike their fatty analogues, unbranched perfluorinated chains have two preferred conformations and can transition between the two, as demonstrated by structural studies [27–30], making $\Delta\mu_{CF_2}$ length-dependent.

Perhaps the most precise method of determining $\Delta\mu_{CF_2}$ is through the adsorption constant K_a for fluorosurfactant adsorbing from a water solution to the water|oil interface. The adsorption constant has an exponential dependence on $\Delta\mu_{CF_2}$, i.e., each unit of –CF_2– added to the hydrophobic tail increases the surface activity of the surfactant by a factor of $\exp(\Delta\mu_{CF_2}/RT)$, which is on the order of 5–6, e.g., for ideal solutions, one –CF_2– decreases sixfold the concentration needed to achieve the same interfacial tension, a variant of Traube's rule [31]. This dependence allows the determination of $\Delta\mu_{CF_2}$ from tensiometric data. Accordingly, in this work, we aim to do exactly that—determine $\Delta\mu_{CF_2}$ via tensiometry for homologous series of water-soluble fluorinated nonionic surfactants adsorbing on the water|oil interface. The practical limitation of this approach is that perfluoroalkylated nonionic compounds are very hydrophobic. Perfluorinated chains with as few as five carbon atoms already partition almost completely to the oil phase and, moreover, have exceedingly low solubility in water. On the other hand, the shortest members of a homologous series are known to deviate from Traube's rule. Thus, no more than a few homologues can be used for the determination, and the result has to be corrected for the expected deviations.

One solution to the low solubility problem is the use of fluorinated ionic surfactants, as done by Mukerjee and Handa [24], who used salts of perfluorinated carboxylic acids. However, using ionic surfactants complicates the interpretation of the results due to the change in the surface charge, the ionic strength effects, and the possibility for hydrolysis at the surface. Moreover, specifically in Mukerjee and Handa's [24] experiments, only two homologues were studied and only in the infinite dilution region, where impurities, adsorption kinetics, and possible depletion effects may complicate the determination of the equilibrium interfacial tension. These factors increase the uncertainty of the obtained $\Delta\mu_{CF_2}$. Instead, here we focus on nonionic surfactants, which both eliminates potential sources of error due to surface charging and makes the analysis simpler.

To achieve the set goal, we collected tensiometric data for the three shortest water-soluble fluorotelomer alcohols of the type $F(CF_2)_nCH_2OH$, with $n = 1, 2, 3$, adsorbing on the water|hexane interface (see Table 1). The particular choice of alcohols as surfactants and hexane as the oil phase was based on the large amount of previous work done for similar systems, usually with longer oil-soluble homologues; this includes experimental [32–38], theoretical [25], and simulation studies [39]. Furthermore, we also investigated the temperature dependence of $\Delta\mu_{CF_2}$ in the range 288.15–303.15 K.

Table 1. The investigated fluoroalcohols.

n	Formula	Name
1	CF_3CH_2OH	2,2,2-trifluoro-1-ethanol
2	$CF_3CF_2CH_2OH$	2,2,3,3,3-pentafluoro-1-propanol
3	$CF_3CF_2CF_2CH_2OH$	2,2,3,3,4,4,4-heptafluoro-1-butanol

For the analysis of the tensiometric data, we used the sticky disk adsorption model [40,41] and the molecular thermodynamic model for K_a of Ivanov et al. [21,22,41], both of which were shown previously to work well for longer oil-soluble fluorotelomer alcohols [25]. The explicit model for K_a accounts for the deviations from Traube's rule for short tail-lengths, which allows us to determine a more accurate value of $\Delta\mu_{CF_2}$.

2. Theory

2.1. SD Model

The sticky disk (SD) adsorption model was developed to describe fluid monolayers (i.e., nonlocalized adsorption) on soft matter interfaces [40,41]. At its core, it is a hard-disk fluid model based on the scaled particle theory, wherein surfactant molecules occupy a certain hard-disk area α and can move freely on the surface without overlapping [42]. Ivanov's group also added a 1D sticky potential correction [40] to account for the lateral attraction. This results in an additional attraction parameter β, making the SD model a three-parametric model, i.e., α, β, and the adsorption constant K_a. Furthermore, it was shown that β implicitly accounts for the depletion attraction effect from the solvent present in the monolayer [43]. The SD model works well for noncohesive and weakly cohesive monolayers, with lateral attraction parameter $\beta << 38$, but fails when strong attraction is present $\beta \geq 38$ [21,22]. A comparison of the SD model to other, more popular three-parametric adsorption models, such as those of Frumkin and Van der Waals, shows that the SD model is superior in the sense that all of its parameters correspond to their theoretical definition and can be predicted from independent data, while for the others, they are more or less empirical [22]. We have previously shown that the SD model works very well for fluorotelomer alcohols with 8–12 carbon atoms [25] that are structurally similar but more cohesive than the ones studied here.

The SD equation of state reads:

$$\Pi = RT\frac{R_\beta - 1}{2\alpha\beta(1-\alpha\Gamma)}, \quad \text{where} \quad R_\beta = \sqrt{1 + 4\beta\frac{\alpha\Gamma}{1-\alpha\Gamma}}, \tag{1}$$

and where $\Pi = \gamma_0 - \gamma$ is interfacial pressure, γ is interfacial tension, γ_0 is the interfacial tension of the surfactant-free interface, Γ is surfactant adsorption, α is the hard-disk area of the surfactant per mole (i.e., parking area; the reader is referred to [44] for a discussion about the different types of molecular areas), and T is temperature. The associated surface activity coefficient f_a of the surfactant in the monolayer is:

$$f_{\mathrm{a}}=\frac{1}{1-\alpha\Gamma}\left(\frac{2}{1+R_{\beta}}\right)^{2+1/\beta}\exp\left[\frac{\alpha\Gamma(4-3\alpha\Gamma)}{(1-\alpha\Gamma)^{2}}\times\frac{2}{1+R_{\beta}}\right]. \quad (2)$$

2.2. *Hard-Disk Area α*

For the SD model, the area parameter corresponds exactly to the cross-sectional area of the amphiphile molecule standing upright. Since the –OH group and the hydrocarbon chain are smaller than the perfluorocarbon chain, α is determined by the cross-sectional area of a perfluorocarbon chain, i.e., $\alpha = N_A\alpha_{CF_2}$. The fluorocarbon chain has two preferential conformations: distorted antiperiplanar and helical [27,28], with different cross-sectional areas. Assuming a single constant hard-disk area for all studied surfactants, we previously found that $\alpha_{CF_2} = 24.5$ Å^2, consistent with a helical conformation of the F-block, which agrees well with the tensiometric data for oil-soluble surfactants [25]. We will use this constant value throughout this paper as well, as an approximation. In reality, the cross-sectional area is a function of both the length of the blocks [29,30] and the temperature [28], and it has to be appreciated that the low molecular weight surfactants studied here may approach the area of the antiperiplanar configuration, $\alpha_{CF_2} = 21.6$ Å^2, as the fluorocarbon chain length decreases. However, these 3 Å^2 of difference result in a relatively small (approx. 5%) change in the transfer energy calculated below.

2.3. *Attraction Parameter β*

As we did previously, we calculate β based on a combination of osmotic attraction and effective Van der Waals interaction between the F-blocks through hexane [43]:

$$\beta=\beta_{\mathrm{osm}}+\frac{1}{R_{\mathrm{CF_2}}^{2}}\int_{2R_{\mathrm{CF_2}}}^{\infty}\left\{\exp\left[\frac{nL_{\mathrm{CF_2}}}{4RTl_{\mathrm{CF_2}}r^{5}}\left(\frac{nl_{\mathrm{CF_2}}r}{r^{2}+n^{2}l_{\mathrm{CF_2}}^{2}}+3\arctan\frac{nl_{\mathrm{CF_2}}}{r}\right)\right]-1\right\}r\mathrm{d}r, \quad (3)$$

where $l_{CF_2} = 1.306$ Å [45] is the height of the –CF$_2$– fragment, and $R_{CF_2} = \sqrt{\alpha_{CF_2}/\pi} = 2.79$ Å is the effective radius of the –CF$_2$– fragment (see Figure 1). The presence of solvent molecules in the monolayer results in the depletion contribution, which was found to be $\beta_{osm} = 0.17$ for perfluoroalkylated alcohols at water|hexane [25]. The parameter $L_{CF_2} = 6.32 \times 10^{-54}$ J·m^6/mol is an effective interaction constant for –CF$_2$– groups trough hexane [25]. Equation 3 represents well the dependence of β on the number n of perfluorinated carbon atoms for longer, oil-soluble homologues [25]. However, the dependence of β on temperature extracted from the experimental data is somewhat steeper than the theoretical one, probably due to the neglected effect of the temperature on the hard-disk area α. For the studied surfactants of $n = 1, 2, 3$ at 293 K, Equation (3) predicts the values $\beta = 0.26, 0.51$, and 0.94, respectively.

2.4. *Adsorption Constant K_a.*

In general, the adsorption isotherm of the surfactant reads:

$$K_{\mathrm{a}}C=f_{\mathrm{a}}\Gamma, \quad (4)$$

assuming that the aqueous surfactant solution is ideal (that the surfactant's bulk activity coefficient is unity). The surface activity coefficient f_a is controlled by the intralayer interactions, through the area per molecule α (a measure of the repulsive interaction) and the attraction parameter β (a measure of the attractive interactions in the monolayer); see the SD model's Equation (2). It should be noted that f_a varies significantly with the adsorption model. In particular, it is different for theories with localized (site) and nonlocalized (two-dimensional fluid) adsorption. The third adsorption parameter appearing in the adsorption isotherm—the adsorption constant K_a—is a characteristic of the adsorption of a single molecule on the neat interface; therefore, the value of K_a is independent of the adsorption model.

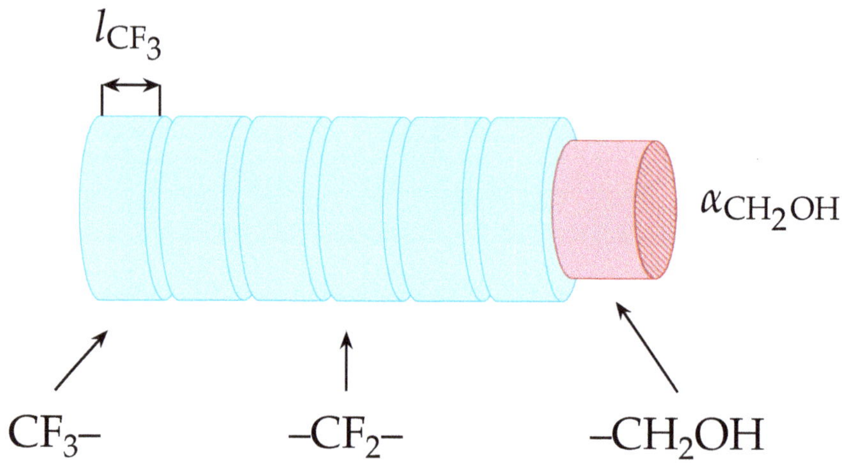

Figure 1. A fluorotelomer alcohol molecule modeled as a structure of connected cylindrical segments, as used in Ivanov's K_a model. Each segment is characterized by a height l_i, cross-sectional area α_i, and a free energy of transfer of the segment $\Delta\mu_i$ from oil to water.

For long monoblock surfactants, the adsorption constant is an exponential function of the so-called adsorption free energy, $K_a = \delta_a \exp(-E_a/RT)$, where the adsorption energy E_a corresponds to the minimum of the free energy of the surfactant molecule at the surface [14] and is a linear function of the chain length n. The pre-exponential factor δ_a is known as the adsorption length (due to its dimensions), and it is common to assume that it is equal to the length of the surfactant molecule, as suggested by Davies [14]. However, Ivanov et al. [41] showed that the actual value of δ_a is about an order of magnitude lower than the size of the surfactant [22,41] and does not depend on the chain length. The theory of Ivanov et al. has been shown to predict adequate values for K_a for water-soluble alkyl-based nonionic and ionic surfactants, both at water|air and water|oil [22,46], and also for the effective adsorption constant from water to a monolayer in the liquid expanded phase [21,46].

Ivanov's model predicts deviation from the simple exponential (Traube-like) dependence $K_a(n)$ for very short surfactants—there, a more complex dependence is expected:

$$K_a = \delta_a \left[1 - \exp\left(-n\frac{\Delta\mu_{CF_2}}{RT}\right)\right] \exp\left(-\frac{E_a(n)}{RT}\right) - nl_{CF_2}. \tag{5}$$

This is a variant of Equation (2.15) from ref. [46]. For long-chained surfactants, Equation (5) simplifies to Davies' expression: $K_a = \delta_a \exp(-E_a(n)/RT)$. However, for n equal to 1 or 2 (which are of interest here), the more complicated dependence on the size of the chain n has to be used, i.e. Equation (5).

Ivanov's model has also been generalized to the adsorption constant of diblock surfactants [25]: as in the original model, the molecule is represented as a stack of cylindrical segments of different lengths and different free energy penalties for transfer from oil to water (see Figure 1). However, the alpha carbon next to the –OH group has been found to behave as part of the polar group that remains immersed in the aqueous phase [15,22].

Therefore, for the fluorotelomer alcohols studied here, the hydrophobic tail is a fluorocarbon, and Equation (5) is sufficient. In this case, the adsorption length is:

$$\delta_a = \frac{RTl_{CF_2}}{2\Delta\mu_{CF_2}}, \quad (6)$$

and the adsorption free energy from water to water|oil is:

$$E_a = -\frac{2\sqrt{3}}{\pi}\alpha\gamma_0 - (n-1)\Delta\mu_{CF_2} - \Delta\mu_{CF_3}, \quad (7)$$

where $\Delta\mu_{CF_2}$ and $\Delta\mu_{CF_3}$ are the free energy changes of transfer from oil to water of the $-CF_2-$ and $-CF_3$ moieties, respectively. The interfacial tension of neat water|hexane is a linear function of the temperature: $\gamma_0/[\text{mN/m}] = 50.56 - 0.0876(T/[\text{K}] - 298.15)$ (an average from literature data [47–51]). The first term in Equation (7) is the interfacial energy gained by removing the $2\sqrt{3}/\pi \times \alpha$ contact area between hexane and water, when a surfactant molecule is adsorbed. The coefficient $2\sqrt{3}/\pi$ is the ratio between the hard-disk area and the partial area per surfactant molecule in the mixed monolayer of water and surfactant, assuming a quasi-hexagonal order. The second and third terms stand for the removal of the fluorocarbon chain|water contact area and the creation of a fluorocarbon chain|hexane contact area. For hydrocarbons, there is evidence that the $\Delta\mu_{CH_3}/\Delta\mu_{CH_2}$ is equal to the ratio of the contact areas with water for both moieties, as it follows from the hydrophobic/entropic origin of the transfer energies $\Delta\mu_{CH_2}$ and $\Delta\mu_{CH_3}$, i.e., $\Delta\mu_{CH_3}/\Delta\mu_{CH_2} \approx 2$ [22]. For fluorocarbons, even though the fluorine atom is bigger, the contact area of the CF_3- moiety with water is also approximately twice as large as that of the $-CF_2-$ moiety. Therefore, in the present report, we use the approximation $\Delta\mu_{CF_3} = 2\Delta\mu_{CF_2}$. This leaves only a single unknown in the adsorption model: the sought value of the transfer energy, $\Delta\mu_{CF_2}$.

3. Results and Discussion

The experimental results for the interfacial pressure as a function of the alcohol concentrations are presented in Figure 2. The increase in temperature results in an increase of the interfacial pressure, i.e., the adsorption of aqueous perfluoroalkylated surfactants to the water|oil interface is an endothermic process. This is in contrast to the adsorption of longer oil-dissolved fluorotelomer alcohols to water|oil, e.g., $CF_3(CF_2)_6CH_2OH$ studied by Takiue et al. [32], which is exothermic.

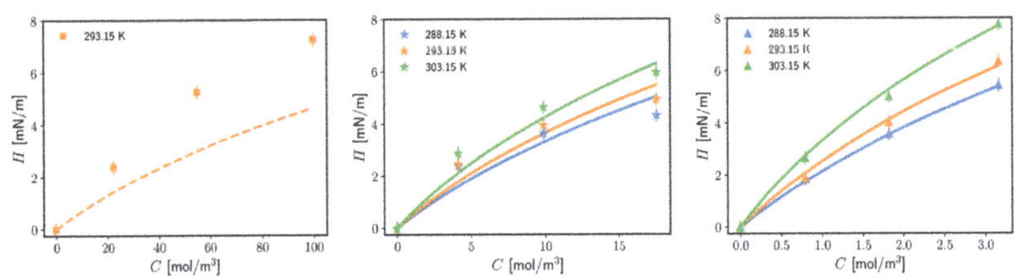

(a) CF_3CH_2OH (b) $CF_3CF_2CH_2OH$ (c) $CF_3(CF_2)_2CH_2OH$

Figure 2. The interfacial pressure as a function of surfactant concentration in the aqueous phase at different temperatures for the three alcohols. Points are experimental data; lines are the SD model combined with the molecular thermodynamic Equations (1)–(5) for the adsorption parameters, with a single fitting parameter $\Delta\mu_{CF_2}$. The fitting is done on the data for $CF_3CF_2CH_2OH$ and $CF_3(CF_2)_2CH_2OH$ at each temperature, while CF_3CH_2OH is omitted, since Ivanov's K_a model deviates from the data for the shortest homologue (see the text).

As a preliminary analysis, we calculated a $\Delta\mu_{CF_2}$ value from each experimental surface pressure Π. This is done by calculating from Equation (1) the value of the adsorption Γ, using Equation (2) to calculate the respective activity coefficient f_a, employing Equation (4) to calculate K_a, and finally solving Equations (5) and (7) for $\Delta\mu_{CF_2}$. The averaged results are presented in Figure 3a. From the graph, it would appear that there are two populations of data: CF_3CH_2OH produces values scattered around $2.15 \times RT_0$, while the $CF_3CF_2CH_2OH$ and $CF_3(CF_2)_2CH_2OH$ data are scattered around $1.8 \times RT_0$. The confidence intervals of the latter two overlap, while the distance of the fluorinated ethanol cluster from the fluorinated propanol and butanol cluster is more than twice the deviation within a dataset for a single surfactant. The average value of $\Delta\mu_{CF_2}$ obtained for CF_3CH_2OH is not only higher than that obtained from $CF_3CF_2CH_2OH$ and $CF_3(CF_2)_2CH_2OH$, but also rather high compared to previous estimates of $\Delta\mu_{CF_2}$ [25]. Therefore, we can conclude that the results for CF_3CH_2OH are qualitatively different than the other two alcohols. In the literature, there are other examples where the first member of a homologous series behaves differently from the rest, e.g., the closely related solvation free energy of hydrocarbons in water [52]. It is reasonable to assume that the K_a model (Equation (5)) is inadequate for $n = 1$. A likely reason for this is the interaction between the polar $-CF_3$ group and the aqueous phase, which may produce a more horizontal orientation of the adsorbed molecules. This violates the assumptions of Ivanov's model. Therefore, for CF_3CH_2OH, (i) the 1/2 factor in Equation (6) for the adsorption length should be reevaluated, and (ii) the area per molecule should be larger than α_{CF_2}. Hence, we exclude the fluorinated ethanol from the following analysis.

To calculate the transfer energy $\Delta\mu_{CF_2}$ with higher accuracy, we fit simultaneously all experimental points for a temperature with a single $\Delta\mu_{CF_2}$. The Levenberg–Marquardt algorithm is used to minimize the root-mean-square deviation between the model predictions and the experimental surface pressures. Since CF_3CH_2OH is excluded from the fitting procedure, each temperature has six experimental points coming from the propanol and butanol, i.e., a single parameter fit to six points from two homologues. The obtained optimal $\Delta\mu_{CF_2}$ values are presented in Table 2. The error of the fitted $\Delta\mu_{CF_2}$, as determined from the root of its variance, is very low: $0.02RT_0$ (ca. 50 J/mol).

Table 2. Results for the optimized transfer energy $\Delta\mu_{CF_2}$ and the values of the corresponding enthalpy Δh_{CF_2} and entropy Δs_{CF_2}.

T [K]	$\Delta\mu_{CF_2}/RT_0$ [-]
288.15	1.68 ± 0.02
293.15	1.75 ± 0.02
303.15	1.88 ± 0.02
Δh_{CF_2} [kJ/mol]	Δs_{CF_2} [J/mol K]
-10.0 ± 0.2	-48.7 ± 0.7

The optimal root-mean-square deviations between the experimental and the theoretical Π are 0.6 mN/m and 0.1 mN/m for $CF_3CF_2CH_2OH$ and $CF_3(CF_2)_2CH_2OH$, respectively (see Figure 2). For $CF_3(CF_2)_2CH_2OH$, the deviation 0.1 mN/m is less than the experimentally determined reproducibility. The higher deviation for $CF_3CF_2CH_2OH$ reflects the fact that the experimental points do not neatly fit the SD model (see Figure 2). This is assumed to be due to surface active impurities, up to 3% as reported by the vendor. On the other hand, since the butanol is about five times more surface active as compared to propanol, the impurities in $CF_3(CF_2)_2CH_2OH$ have a much smaller effect, thus, the lower deviation. Fortunately, the uncertainty in the tensiometric measurement propagates logarithmically to the transfer energy $\Delta\mu_{CF_2}$. Therefore, the effect of the impurities in $CF_3CF_2CH_2OH$ is small in terms of the uncertainty of the obtained transfer energy $\Delta\mu_{CF_2}$.

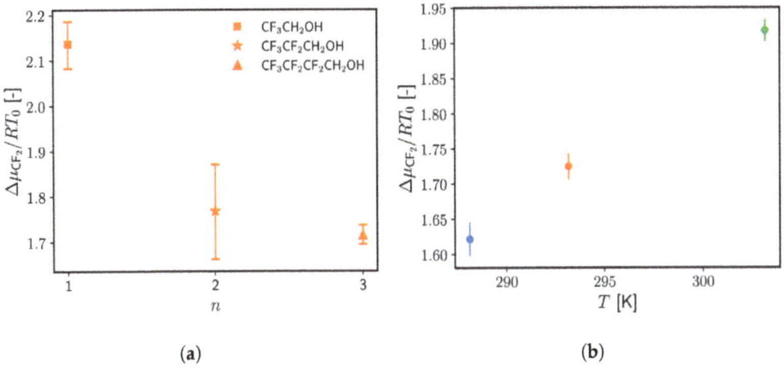

Figure 3. (**a**) Average free energy of transfer $\Delta\mu_{CF_2}$ calculated from each experimental surface tension at 293.15 K independently. The error bars specify the lowest and highest calculated $\Delta\mu_{CF_2}$. The data for $n = 1$ (CF_3CH_2OH) deviate significantly from the rest and are excluded from the following analysis. (**b**) Best fit value of the free energy of transfer $\Delta\mu_{CF_2}$ from simultaneous regression over the data for $CF_3CF_2CH_2OH$ and $CF_3(CF_2)_2CH_2OH$ as a function of the temperature. The different temperatures are color-coded in accordance with Figure 2.

A linear relationship between $\Delta\mu_{CF_2}$ and T is observed (Figure 3b). The free energy change is comprised of an entropic and an enthalpic term:

$$\Delta\mu_{CF_2} = \Delta h_{CF_2} - T\Delta s_{CF_2}. \tag{8}$$

Assuming that both are temperature independent, from the results in Figure 3, the enthalpy for transfer of a $-CF_2-$ group from hexane to water can be determined as $\Delta h_{CF_2} = -10.0$ kJ/mol and the respective entropy as $\Delta s_{CF_2} = -48.7$ J/mol K. The obtained value of Δs_{CF_2} is larger than expected (compare to methylene's Δs_{CH_2} on the order of -10 J/mol K [52]). At this length scale, the hydrophobic effect is approximately proportional to the surface area of the segment [53]. Therefore, Δs_{CF_2} is expected to be about 1.3 times higher than Δs_{CH_2}. However, we determine an entropy that is about 5 times higher than Δs_{CH_2}. This discrepancy might be due to the assumptions of our model, in particular about the temperature dependence of the adsorption parameters. For example, α, which appears in both Equation (7) for K_a and the equation of state (Equation (1)), is assumed to be independent of T, which would result in a compensation by an artificial variation of $\Delta\mu_{CF_2}$ with T and, respectively, an unrealistic value of Δs_{CF_2}. Nevertheless, the effect on the resultant $\Delta\mu_{CF_2}$ values appears to be relatively small: even if we used $\Delta s_{CF_2} = 1.3 \times \Delta s_{CH_2}$, for the current narrow temperature interval, the error introduced in $\Delta\mu_{CF_2}$ would be within 5% of the values in Figure 3b. The enthalpy Δh_{CF_2} is also higher than the corresponding Δh_{CH_2}, which is close to zero at room temperature [21]. This could be justified physically with the fact that, unlike $-CH_2-$ chains, for $-CF_2-$ chains, there is a significant Van der Waals contribution to $\Delta\mu_{CF_2}$ (the dispersion contribution to β as studied in ref. [25]).

4. Materials and Methods

All reagents—hexane (C_6H_{14}, $\geq$ 99%), 2,2,2-trifluoro-1-ethanol (CF_3CH_2OH, $\geq$ 99%), 2,2,3,3,3-pentafluoro-1-propanol ($CF_3CF_2CH_2OH$, $\geq$ 97%), and 2,2,3,3,4,4,4-heptafluoro-1-butanol ($CF_3CF_2CF_2CH_2OH$, $\geq$ 98%)—were purchased from Sigma-Aldrich and used without further purification. All water used was double distilled using a GFL 2001/2 distiller. Aqueous solutions close to the expected solubility limit were prepared volumetrically and subsequently diluted approximately two and four times (the exact concentrations are reported in Figure 2).

The interfacial tension of each solution was measured with a profile analysis tensiometer (PAT-1, Sinterface, Germany). A hexane drop with constant area, ca. 25 mm^2, was formed in a 25 cm^3 glass cuvette filled with the aqueous solution. The cuvette and the capillary were cleaned before each experiment by immersion in a dichromate solution for at least 24 h. The temperature of the measuring cell was controlled precisely with a thermostat (Ecoline E200, LAUDA DR. R. WOBSER GMBH & CO. KG, Lauda-Königshofen, Germany). Measurements were made at 288.15 K, 293.15 K, and 303.15 K. In a typical experiment, the interfacial tension was observed for ca. 6 h at a constant temperature so as to ensure that the value had settled. Within the studied concentration range, the solution density is assumed to be constant [54]. The average standard deviation of the measured interfacial tension determined from multiple repetitions is 0.26 mN/m.

5. Conclusions

We have determined experimentally the free energy of transfer of a –CF_2– group from oil to water $\Delta\mu_{CF_2}$ from data for the adsorption of water-soluble nonionic fluorosurfactants at water|hexane. As far as we are aware, this has been done for the first time, since only the shortest nonionic fluorinated surfactants (n = 1–3) are water-soluble. The values obtained are $1.68 \pm 0.02 \times RT_0$, $1.75 \pm 0.02 \times RT_0$, and $1.88 \pm 0.02 \times RT_0$ at 288.15 K, 293.15 K, and 303.15 K, respectively, and appear to be of lower uncertainty compared to previous reports [25].

The short length of the fluorocarbon chain results in complications—in particular, the deviations from Traube's rule must be taken into account. While we were successful in resolving this issue for $n \geq 2$ by using Ivanov's explicit model for $K_a(n)$, we found that the adsorption data for the first homologue CF_3CH_2OH suggest a very different behavior from the rest of the series.

The obtained value of $\Delta\mu_{CF_2}$ is within the expected range (see the discussion in Peychev and Slavchov [25]). As it can be deduced from the more hydrophobic behavior of fluorocarbons compared to hydrocarbons, the value of $\Delta\mu_{CF_2}$ is higher than the corresponding free energy of transfer of a –CH_2– group from oil to water ($\Delta\mu_{CH_2} = 1.39 \times RT_0$). However, the result at 293.15 K, $\Delta\mu_{CF_2} = 1.75 \pm 0.02 \times RT_0$, is lower than the value $2.05 \times RT_0$ reported by Mukerjee and Handa [24], based on similar tensiometric experiments but with fluorinated ionic surfactants. Due to the exponential relationship between K_a and $n\Delta\mu_{CF_2}$, this is a very large difference. For instance, for $n = 3$, the adsorption constant predicted from Equation (5) with Mukerjee and Handa's value $\Delta\mu_{CF_2} = 2.05 \times RT_0$ is twice as high as what we measured experimentally.

The obtained result for $\Delta\mu_{CF_2}$ should be useful for prediction of the partitioning of mixtures of short fluorinated amphiphiles between water and oil and between water and biomembranes [13], through the respective partition coefficient, $RT \ln K_p \propto n\Delta\mu_{CF_2}$ [25,55]. The transfer energy is also an essential parameter for predicting the incorporation of water-dissolved fluorinated amphiphiles into micelles and into adsorption monolayers made of hydrocarbon surfactants (compare to the effective adsorption constant of a liquid expanded monolayer [21,46]). These fundamental thermodynamic characteristics are essential for understanding the environmental fate of perfluoroalkyl substances and their health effects and for modeling separation processes that involve them.

Author Contributions: Conceptualization, R.I.S.; methodology, B.P., I.L.M. and R.I.S.; formal analysis, B.P.; investigation, D.A. and I.L.M.; resources, I.L.M. and E.M.; writing—original draft preparation, B.P.; writing—review and editing, D.A., I.L.M., E.M. and R.I.S.; visualization, B.P.; supervision and project administration, E.M. and R.I.S.; funding acquisition, I.L.M. and E.M. All authors have read and agreed to the published version of the manuscript.

Funding: This study is financed by the European Union-Next Generation EU, through the National Recovery and Resilience Plan of the Republic of Bulgaria, project № BG-RRP-2.004-0008-C01.

Institutional Review Board Statement: Not applicable.

Informed Consent Statement: Not applicable.

Data Availability Statement: Dataset available on request from the authors.

Acknowledgments: The investigation was performed in the Laboratory of Thin Liquid Films at the Institute of Physical Chemistry, Bulgarian Academy of Sciences, Sofia, Bulgaria.

Conflicts of Interest: The authors declare no conflicts of interest. The funders had no role in the design of the study; in the collection, analyses, or interpretation of data; in the writing of the manuscript; or in the decision to publish the results.

References

1. Glüge, J.; Scheringer, M.; Cousins, I.T.; DeWitt, J.C.; Goldenman, G.; Herzke, D.; Lohmann, R.; Ng, C.A.; Trier, X.; Wang, Z. An overview of the uses of per-and polyfluoroalkyl substances (PFAS). *Environ. Sci. Process. Impacts* **2020**, *22*, 2345–2373. [CrossRef]
2. Krafft, M.P.; Riess, J.G. Chemistry, physical chemistry, and uses of molecular fluorocarbon—Hydrocarbon diblocks, triblocks, and related compounds—Unique "apolar" components for self-assembled colloid and interface engineering. *Chem. Rev.* **2009**, *109*, 1714–1792. [CrossRef]
3. Kissa, E. *Fluorinated Surfactants and Repellents*, 2nd ed.; Marcel Dekker: New York, NY, USA, 2001.
4. Simcik, M.F. Aquatic processes and systems in perspective. Global transport and fate of perfluorochemicals. *J. Environ. Monit.* **2005**, *7*, 759–763.
5. Armitage, J.; Cousins, I.T.; Buck, R.C.; Prevedouros, K.; Russell, M.H.; MacLeod, M.; Korzeniowski, S.H. Modeling global-scale fate and transport of perfluorooctanoate emitted from direct sources. *Environ. Sci. Technol.* **2006**, *40*, 6969–6975. [CrossRef]
6. Johnson, G.R.; Brusseau, M.L.; Carroll, K.C.; Tick, G.R.; Duncan, C.M. Global distributions, source-type dependencies, and concentration ranges of per-and polyfluoroalkyl substances in groundwater. *Sci. Total Environ.* **2022**, *841*, 156602. [CrossRef]
7. Brunn, H.; Arnold, G.; Körner, W.; Rippen, G.; Steinhäuser, K.G.; Valentin, I. PFAS: Forever chemicals—persistent, bioaccumulative and mobile. Reviewing the status and the need for their phase out and remediation of contaminated sites. *Environ. Sci. Eur.* **2023**, *35*, 1–50.
8. Ehrlich, V.; Bil, W.; Vandebriel, R.; Granum, B.; Luijten, M.; Lindeman, B.; Grandjean, P.; Kaiser, A.; Hauzenberger, I.; Hartmann, C.; et al. Consideration of pathways for immunotoxicity of per- and polyfluoroalkyl substances (PFAS). *Environ. Health Glob. Access Sci. Source* **2023**, *22*, 19. [CrossRef] [PubMed]
9. ATSDR. Toxicological Profile for Perfluoroalkyls. 2021. Available online: https://doi.org/10.15620/cdc:59198 (accessed on 13 May 2023).
10. Brendel, S.; Fetter, É.; Staude, C.; Vierke, L.; Biegel-Engler, A. Short-chain perfluoroalkyl acids: Environmental concerns and a regulatory strategy under REACH. *Environ. Sci. Eur.* **2018**, *30*, 9. [CrossRef]
11. Wanninayake, D.M. Comparison of currently available PFAS remediation technologies in water: A review. *J. Environ. Manag.* **2021**, *283*, 111977. [CrossRef] [PubMed]
12. Garg, S.; Wang, J.; Kumar, P.; Mishra, V.; Arafat, H.; Sharma, R.S.; Dumée, L.F. Remediation of water from per-/poly-fluoroalkyl substances (PFAS)–Challenges and perspectives. *J. Environ. Chem. Eng.* **2021**, *9*, 105784. [CrossRef]
13. Zhang, M.; Peyear, T.; Patmanidis, I.; Greathouse, D.V.; Marrink, S.J.; Andersen, O.S.; Ingólfsson, H.I. Fluorinated alcohols' effects on lipid bilayer properties. *Biophys. J.* **2018**, *115*, 679–689. [CrossRef]
14. Davies, J.; Rideal, E. *Interfacial Phenomena*; Academic Press: London, UK, 1961.
15. Tanford, C. *The Hydrophobic Effect: Formation of Micelles and Biological Membranes*, 2nd ed.; J. Wiley.: New York, NY, USA, 1980.
16. Huckel, W. *Theoretical Principles of Organic Chemistry*; Elsevier Publishing: Amsterdam, The Netherlands, 1958.
17. Kitaigorodskii, A.I. *Organic Chemical Crystallography*; Consultants Bureau: New York, NY, USA, 1961.
18. Israelachvili, J.N. *Intermolecular and Surface Forces*, 3rd ed.; Academic Press: Burlington, MA, USA, 2011.
19. Möbius, D.; Miller, R.; Fainerman, V.B. *Surfactants: Chemistry, Interfacial Properties, Applications*; Elsevier: Amsterdam, The Netherlands, 2001.
20. Adam, N.K. *The Physics and Chemistry of Surfaces*, 3rd ed.; Milford, H., Ed.; Oxford University Press: London, UK, 1941.
21. Slavchov, R.; Dimitrova, I.; Ivanov, I. Cohesive and non-cohesive adsorption of surfactants at liquid interfaces. In *Without Bounds: A Scientific Canvas of Nonlinearity and Complex Dynamics*; Springer: Berlin/Heidelberg, Germany, 2013; pp. 199–225.
22. Slavchov, R.I.; Ivanov, I.B. Adsorption parameters and phase behaviour of non-ionic surfactants at liquid interfaces. *Soft Matter* **2017**, *13*, 8829–8848. [CrossRef]
23. Arp, H.P.H.; Niederer, C.; Goss, K.U. Predicting the partitioning behavior of various highly fluorinated compounds. *Environ. Sci. Technol.* **2006**, *40*, 7298–7304. [CrossRef]
24. Mukerjee, P.; Handa, T. Adsorption of fluorocarbon and hydrocarbon surfactants to air-water, hexane-water and perfluorohexane-water interfaces. Relative affinities and fluorocarbon-hydrocarbon nonideality effects. *J. Phys. Chem.* **1981**, *85*, 2298–2303. [CrossRef]
25. Peychev, B.; Slavchov, R.I. Adsorption model and phase transitions of diblock perfluoroalkylated surfactants at the water/alkane interface. *J. Colloid Interface Sci.* **2021**, *594*, 372–388. [CrossRef]
26. Wang, Z.; Ud-Daula, A.; Fiedler, S.; Schramm, K.W. Impact of fluorotelomer alcohols (FTOH) on the molecular and macroscopic phenotype of Tetrahymena thermophila. *Environ. Sci. Pollut. Res.* **2010**, *17*, 154–164. [CrossRef] [PubMed]
27. Schwickert, H.; Strobl, G.; Kimmig, M. Molecular dynamics in perfluoro-n-eicosane. I. Solid phase behavior and crystal structures. *J. Chem. Phys.* **1991**, *95*, 2800–2806. [CrossRef]

28. Albrecht, T.; Elben, H.; Jaeger, R.; Kimmig, M.; Steiner, R.; Strobl, G.; Stühn, B.; Schwickert, H.; Ritter, C. Molecular dynamics in perfluoro-n-eicosane. II. Components of disorder. *J. Chem. Phys.* **1991**, *95*, 2807–2816. [CrossRef]
29. Gang, O.; Ellmann, J.; Möller, M.; Kraack, H.; Sirota, E.; Ocko, B.; Deutsch, M. Surface phases of semi-fluorinated alkane melts. *Europhys. Lett.* **2000**, *49*, 761. [CrossRef]
30. Wang, J.; Ober, C.K. Solid state crystalline and liquid crystalline structure of semifluorinated 1-bromoalkane compounds. *Liq. Cryst.* **1999**, *26*, 637–648. [CrossRef]
31. Traube, J. Ueber die Capillaritätsconstanten organischer Stoffe in wässerigen Lösungen. *J. Liebigs Ann. Chem.* **1891**, *265*, 27–55. [CrossRef]
32. Takiue, T.; Sugino, K.; Higashi, T.; Toyomasu, T.; Hayami, Y.; Ikeda, N.; Aratono, M. Temperature effect on the adsorption of fluorooctanols at the hexane/water interface. *Langmuir* **2001**, *17*, 8098–8103. [CrossRef]
33. Ohta, A.; Murakami, R.; Urata, A.; Asakawa, T.; Miyagishi, S.; Aratono, M. Aggregation behavior of fluorooctanols in hydrocarbon solvents. *J. Phys. Chem. B* **2003**, *107*, 11502–11509. [CrossRef] [PubMed]
34. Murakami, D.; Fukuta, T.; Matsubara, H.; Aratono, M.; Takiue, T. Effect of the partial hydrogenation of hydrophobic chains on the mixing of fluoroalkanols in an adsorbed film at the hexane/water interface. *J. Phys. Chem. C* **2008**, *112*, 4564–4568. [CrossRef]
35. Murakami, D.; Fukuda, T.; Matsubara, H.; Aratono, M.; Takiue, T. Molecular orientation and miscibility of fluorinated α, ω-alkanediol and alcohol at the hexane/water interface. *Colloids Surf. A Physicochem. Eng. Asp.* **2010**, *354*, 205–209. [CrossRef]
36. Takiue, T.; Tottori, T.; Tatsuta, K.; Matsubara, H.; Tanida, H.; Nitta, K.; Uruga, T.; Aratono, M. Multilayer Formation of the Fluoroalkanol- ω-Hydrogenated Fluorocarbon Mixture at the Hexane/Water Interface Studied by Interfacial Tensiometry and X-ray Reflection. *J. Phys. Chem. B* **2012**, *116*, 13739–13748. [CrossRef]
37. Fukuhara, R.; Tanida, H.; Nitta, K.; Ina, T.; Uruga, T.; Matsubara, H.; Aratono, M.; Takiue, T. Effect of molecular orientation on monolayer and multilayer formations of fluorocarbon alcohol and fluorocarbon-α, ω-diol mixture at the hexane/water interface. *J. Phys. Chem. B* **2014**, *118*, 12451–12461. [CrossRef]
38. Mitani, K.; Imai, Y.; Ina, T.; Nitta, K.; Tanida, H.; Uruga, T.; Matsubara, H.; Aratono, M.; Takiue, T. Effect of hydrophobic chain structure on phase transition and domain formation of hybrid alcohol films adsorbed at the hexane/water interface. *J. Phys. Chem. B* **2015**, *119*, 12436–12445. [CrossRef]
39. Burrows, S.A.; Shon, J.W.; Peychev, B.; Slavchov, R.I.; Smoukov, S.K. Phase transitions of fluorotelomer alcohols at the water|alkane interface studied via molecular dynamics simulation. *Soft Matter* **2024**, *20*, 2243–2257. [CrossRef]
40. Gurkov, T.; Ivanov, I.B. Layers of non-ionic surfactants on fluid interfaces adsorption and interactions in the frames of a statistical model. In Proceedings of the 4th World Congress on Emulsions, Lyon, France, 3–6 October 2006; p. 509.
41. Ivanov, I.B.; Danov, K.D.; Dimitrova, D.; Boyanov, M.; Ananthapadmanabhan, K.P.; Lips, A. Equations of state and adsorption isotherms of low molecular non-ionic surfactants. *Colloids Surf. A Physicochem. Eng. Asp.* **2010**, *354*, 118–133. [CrossRef]
42. Helfand, E.; Frisch, H.; Lebowitz, J. Theory of the Two-and One-Dimensional Rigid Sphere Fluids. *J. Chem. Phys.* **1961**, *34*, 1037–1042. [CrossRef]
43. Slavchov, R.I.; Ivanov, I.B. Effective osmotic cohesion due to the solvent molecules in a delocalized adsorbed monolayer. *J. Colloid Interface Sci.* **2018**, *532*, 746–757. [CrossRef]
44. Rusanov, A. The essence of the new approach to the equation of the monolayer state. *Colloid J.* **2007**, *69*, 131–143. [CrossRef]
45. Tournilhac, F.; Bassoul, P.; Cortes, R. Structure of the smectic B phase formed by linear and branched perfluoroalkyl-alkanes. *Mol. Cryst. Liq. Cryst.* **2001**, *362*, 45–65. [CrossRef]
46. Slavchov, R.I.; Karakashev, S.I.; Ivanov, I.B. Ionic surfactants and ion-specific effects: Adsorption, micellization, thin liquid films. In *Surfactant Science and Technology: Retrospects and Prospects*; CRC Press: Boca Raton, FL, USA, 2014; p. 593.
47. Aveyard, R.; Haydon, D. Thermodynamic properties of aliphatic hydrocarbon/water interfaces. *Trans. Faraday Soc.* **1965**, *61*, 2255–2261. [CrossRef]
48. Rehfeld, S.J. Adsorption of sodium dodecyl sulfate at various hydrocarbon-water interfaces. *J. Phys. Chem.* **1967**, *71*, 738–745. [CrossRef]
49. Zeppieri, S.; Rodríguez, J.; López de Ramos, A. Interfacial tension of alkane + water systems. *J. Chem. Eng. Data* **2001**, *46*, 1086–1088. [CrossRef]
50. Goebel, A.; Lunkenheimer, K. Interfacial tension of the water/n-alkane interface. *Langmuir* **1997**, *13*, 369–372. [CrossRef]
51. Demond, A.H.; Lindner, A.S. Estimation of interfacial tension between organic liquids and water. *Environ. Sci. Technol.* **1993**, *27*, 2318–2331. [CrossRef]
52. Ben-Naim, A.Y. *Solvation Thermodynamics*; Springer Science & Business Media: New York, NY, USA, 2013.
53. Lum, K.; Chandler, D.; Weeks, J.D. Hydrophobicity at small and large length scales. *J. Phys. Chem. B* **1999**, *103*, 4570–4577. [CrossRef]
54. Rochester, C.; Symonds, J. Densities of solutions of four fluoroalcohols in water. *J. Fluor. Chem.* **1974**, *4*, 141–148. [CrossRef]
55. Leo, A.; Hansch, C.; Elkins, D. Partition coefficients and their uses. *Chem. Rev.* **1971**, *71*, 525–616. [CrossRef]

molecules

MDPI

Review

Law and Order of Colloidal Tectonics: From Molecules to Self-Assembled Colloids

Loïc Leclercq

Univ. Lille, CNRS, Centrale Lille, Univ. Artois, UMR 8181-UCCS, Unité de Catalyse et Chimie du Solide, Lille 59000, France; loic.leclercq@univ-lille.fr

Abstract: Since biochemists and biologists have progressed in understanding the mechanisms involved in living organisms, biological systems have become a source of inspiration for chemists. In this context, the concept of colloidal tectonics, describing the spontaneous formation of colloidal particles or supracolloidal structures in which the building blocks are called "tectons", has emerged. Therefore, a bottom-up edification of tectons towards (supra) colloidal structures is allowed. Each (supra) colloidal system has at least one of the following properties: amphiphilicity, predictability, versatility, commutability, and reversibility. However, for these systems to perform even more interesting functions, it is necessary for tectons to have very precise chemical and physical properties so that new properties emerge in (supra) colloidal systems. In this way, colloidal tectonics enables engineering at the nano- and micrometric level and contributes to the development of smart bioinspired systems with applications in catalysis, drug delivery, etc. In this review, an overview of the concept of colloidal tectonics is illustrated by some biotic systems. The design of abiotic (supra) colloidal systems and their applications in various fields are also addressed (notably Pickering emulsions for catalysis or drug delivery). Finally, theoretical directions for the design of novel self-assembled (supra) colloidal systems are discussed.

Keywords: colloidal tectonics; self-assembly; particles; biologic systems; abiotic systems; catalysis; pharmaceutical applications

1. Introduction

The physiological environment consists of "biological particles" such as proteins, organelles, viruses, cells, and a myriad of structures that move or are capable of self-assembling into tissues and organs [1]. These incredibly complex biological particles possess many very interesting properties such as surface chemistry, transport phenomena, catalytic activities, etc. [2]. A large majority of these "biological" particles result from the self-assembly of molecules into particles via a hierarchical sequential bottom-up construction in a given environment and conditions [3]. These particles are capable of self-assembling again into increasingly complex structures. For example, proteins have primary structures (defined as the sequence of amino acids linked together to form the polypeptide chain), secondary structures (defined as the local structures stabilized by H-bonds between certain amino acids of the polypeptide backbone and leading to α-helixes, β-sheets, and turns), and tertiary structures (defined as the overall 3D shape of a single protein molecule stabilized by nonlocal interactions such as hydrophobic forces, van der Waals interactions, salt bridges, and/or H-bonds) [4–6]. In addition to these three distinct aspects of protein structures, some of them have a quaternary structure, that is, the self-assembly of several polypeptide chains (subunits) to form a complex [7]. It is noteworthy that these four structures are relevant for proteins isolated under dilute conditions, but a quinary structure arises in vivo where protein surfaces are shaped by evolutionary adaptation to physiological conditions [8]. Most proteins self-assemble naturally and unassisted to form their native 3D ordered structures through non-covalent interactions. Generally speaking, self-assembly refers to the association of organic, inorganic, or hybrid entities, e.g., molecules, macromolecules, or nano-objects, into organized structures with nanometer to micrometer dimensions [9]. More specifically, in supramolecular chemistry, it refers to the process by which molecules spontaneously organize among themselves to form multimolecular structures whose stability relies on weak non-covalent interactions [10]. Self-assembly processes are essential in living organisms (e.g., the basic functions of the proteins are controlled by the tertiary structure) [11].

On the other hand, since van der Waals' postulate on intermolecular forces in 1873, Fischer's suggestion on enzyme–substrate interactions in 1894, and the description of the H-bond by Latimer and Rodebush in 1920, supramolecular chemistry has developed. With the elucidation of the structure of DNA and other fascinating biological entities, chemists began to emphasize the importance of non-covalent interactions. For example, the work of Pedersen, Cram, and Lehn (Nobel Prize in Chemistry in 1987) on crown ethers and cryptands allowed the design and synthesis of new highly selective molecular receptors. Following this work, research in the field has taken off rapidly with the work of Feringa, Stoddart, and Sauvage who received the Nobel Prize in Chemistry in 2016 for their work on molecular machines. In parallel, molecular self-assembly processes have found many applications, particularly in the development of new materials. Indeed, large structures are easily accessible using bottom-up synthesis because they are composed of small molecules requiring fewer synthesis steps. Currently, many smart materials are based on a bottom-up approach by molecular recognition. For example, in 1991, Simard et al. first introduced the concept of molecular tectonics, which offers a route for the bottom-up fabrication of predictable crystalline molecular networks from small molecules (tectons) through various interactions (van der Waals forces, electrostatic attraction, π-stacking, H-bonding, and coordination bonding) between the nearest neighbors [12]. The field of crystal engineering was born.

From a general perspective, there are two major types of self-assembled entities in living organisms: lipid assembly into dynamic, flexible, and fluid structures and protein assembly into rigid crystalline arrangements. The former is primarily driven by hydrophobic interaction while the latter is driven by a combination of hydrophobic, H-bonding, and electrostatic interactions. Lipid assembly is successfully mimicked by synthetic amphiphilic small molecules and polymers to form micellar, vesicular, and lamellar structures. In contrast, protein assembly that produces lamellae, tubules, and polyhedrons is largely unmatched by synthetic or nonpeptide molecules. The imbalance in lipid and protein mimicry research can be resolved using the colloidal tectonics approach. Indeed, since biological systems are a source of inspiration for chemists, the artificial construction by the self-assembly processes of molecules (including polymers), producing new abiotic particles with colloidal properties, refers to the concept of colloidal tectonics. This name was chosen by analogy with the pioneering work of Simard et al., which opened the field of crystal engineering [12]. In 1995, the same group proposed to define molecular tectonics as "*the art and science of supramolecular construction using tectonic subunits*" [13]. This concept has been extended to liquid crystals [14]. In 2018, I proposed the name colloidal tectonics, the self-assembly of complementary molecules leading to a variety of particles or rigid objects with colloidal properties (i.e., producing dispersions) [15]. This concept also includes interactions between objects giving rise to supracolloidal assemblies (e.g., Pickering emulsions, colloidosomes, etc.) since they emerge from the packing of particles under appropriate conditions [15]. However, all these tectons (molecules or particles) have complementary attractive sites allowing a bottom-up construction of large colloidal systems (down to the microscale) through non-covalent interactions driven by internal (e.g., hydrophilic/hydrophobic ratio, charge, H-bond donor and/or acceptors sites, and molecular conformation) and external factors (i.e., environmental conditions surrounding the nanostructure such as pH, concentration, temperature, and solvent) [15]. The assembling process is based on a recognition operating at the level of complementary tectons (molecules or particles) providing an infinite variety of (supra) colloidal systems [15]. In the strict sense of the term self-assembly, the resulting systems are intrinsically predictable, versatile, switchable, and can even be dissociated in their constitutive elements. Thus, over the last decade, new smart colloidal tectonics systems have been developed [15]. This approach is operational and versatile since it relies on self-assembly processes resulting from recognition events between programmed building blocks (molecules and/or particles), known as tectonic subunits or simply tectons. These events enable the design and construction, under mild conditions, of a variety of complex colloidal systems that are highly compatible with green chemistry. Indeed, all these systems are energy efficient, minimizing synthesis steps and harmful organic solvents [15]. As these abiotic particles can have the same properties as their biotic cousins (i.e., surface chemistry, transport phenomena, catalytic activities, etc.), all applications requiring the colloidal engineering of nanometric and micrometric systems can be envisaged, such as catalysis, cosmetics, drug delivery, vaccination, etc. [16]. All these systems, in which a bottom-up construction of large colloidal systems is allowed from tectons by simple molecular recognition, result in nanostructured systems and constitute a new and versatile field with promising green assets in the face of environmental problems (Figure 1) [15].

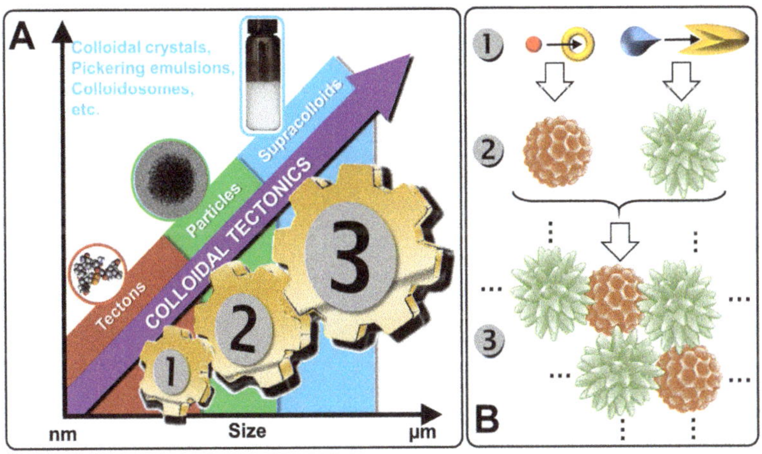

Figure 1. From tectons (molecules) to supracolloidal systems (**A**) and schematic illustration of the self-assembled process (**B**).

In this review, I focus on the fundamentals of colloidal tectonics by illustrating some self-assembled biological colloidal systems. The design of smart and switchable abiotic (supra) colloidal systems and their applications in various fields are also discussed (with a particular focus on Pickering emulsions for catalysis or drug delivery). Finally, theoretical directions for the design of novel self-assembled (supra) colloidal systems are discussed. I hope that this review can offer new insights and provide theoretical guidance in the design of novel self-assembled (supra) colloidal systems.

2. Colloidal Tectonics: A Bioinspired Concept

As is often the case, nature already uses the concept of colloidal tectonics to obtain supramolecular colloidal structures. As these ingenious biological systems can serve as inspiration for chemists, a few biotic self-assembled colloidal systems are presented in this review. It should be noted, however, that, for the sake of clarity, only a few examples are reported in this section. Therefore, the examples cited are not intended to constitute an exhaustive list.

2.1. Milk Casein Particles

Milk is the lactic secretion of mammals and contains valuable nutrients and immune components for the optimal growth of newborns [17]. Although the composition of milk varies depending on the stage of lactation and the growth rate of the newborn, the essential components include water, lipids (fatty acids, phospholipids, and cholesterol), vitamins, carbohydrates (lactose), minerals (calcium), and proteins (phosphoproteins and enzymes), thus contributing to the energy needs of newborn growth [18]. The most important proteins found in mammalian milk are caseins (phosphoproteins), which represent approximately 20 and 80% of the proteins in human and cow milk, respectively. Regardless of the type of mammal, there are four types of caseins (α_{S1}, α_{S2}, β, and κ) [19]. The most important amino acids present in casein isolates are, in order of importance, glutamic acid, proline, leucine, lysine, valine, and aspartic acid [20]. Caseins are non-crystallizable proteins, as the high number of proline acts as a structural disruptor preventing local compaction (i.e., the formation of secondary structural motifs common to proteins) and inhibiting the formation of ordered 3D structures (i.e., tertiary structure) [21]. The presence of hydrophobic (proline, leucine, and valine) and hydrophilic (lysine, glutamic acid, and aspartic acid) amino acids gives caseins an amphiphilic nature but the absence of a tertiary structure makes them poorly soluble in water due to the inevitable exposure to many hydrophobic residues. Therefore, milk is a physically stable aqueous suspension of caseins due to their strong self-association to form supramolecular particles called casein micelles [22]. Classically, the term "micelle" describes a supramolecular assembly of amphiphilic molecules (surfactants) leading to a colloidal suspension in which the hydrophilic "heads" of the surfactants reside on the surface in contact with the surrounding water, sequestering the hydrophobic "tails" inside the micellar core [23]. However, the core of surfactant micelles is not hydrated, unlike casein micelles (see below) [24]. The term "casein micelle" is therefore erroneous because the only similarity with surfactant micelles is that the hydrophilic parts are on the surface. Indeed, casein micelles should rather be considered as colloidal spherical particles of 20 to 600 nm in diameter (the average diameter is about 120 nm) composed of several thousand associated caseins [25]. It should be noted that κ-caseins are more hydrophilic than the other three and are therefore predominant on the surface of casein micelles while the more hydrophobic ones are found inside [26].

Although caseins have been known for nearly 70 years as unfolded proteins, the implications for the structure of casein micelles are still debated. In 1965, Waugh and Noble proposed a core–shell model consisting of spherical hydrophobic casein particles surrounded by κ-casein [27,28]. In 1966, Payens hypothesized that κ-casein formed the shell of the casein particles and that the compactly folded α_{S1}- and α_{S2}- caseins were attached to the free β-caseins (core), with calcium ions interacting with the phosphate or carboxylic acid groups of the proteins [29]. From the model of Slattery and Evard (1973), which excluded the function of colloidal calcium phosphate [30], Schmidt and later Walstra proposed, in 1980 and 1990, that the casein micelle was built by 10 to 100 submicelles (4 to 10 nm in diameter) [31,32]. Each submicelle is composed of a hydrophobic core containing mainly α_{S1}-, α_{S2}-, and β-casein, whereas the shell contains hydrophilic regions composed of the phosphate groups of α_{S1}-, α_{S2}-, β-caseins, and a variable proportion of κ-casein. The submicelles low in κ-casein are located on the surface to ensure solubility in whey while the others are at the core of the casein micelle. Since casein proteins are able to interact with colloidal calcium phosphate, Schmidt assumed that colloidal calcium phosphate filled the space between the submicelles. In 1999, this model was refined by Walstra, who proposed that colloidal calcium phosphate was located within the submicelles and not only between them [33]. In 2012, Dalgleish and Corredig hypothesized that the casein micelle is not based on submicelles but rather on a matrix structure with colloidal calcium phosphate nanoclusters surrounded by calcium-sensitive caseins (α_{S1}-, α_{S2}-, β-caseins) while the κ-caseins protect the casein micelle surface to ensure solubility in the aqueous environment and inhibit the growth of casein micelles by steric repulsions [34]. Interestingly, there are several other models that account for the spatial conformation of casein in micelles [35].

One of them proposes a double link between caseins for gelation to occur [36]. Fortunately, all models consider casein micelles as colloidal particles formed by casein aggregates wrapped up in hydrosoluble κ-casein molecules (Figure 2).

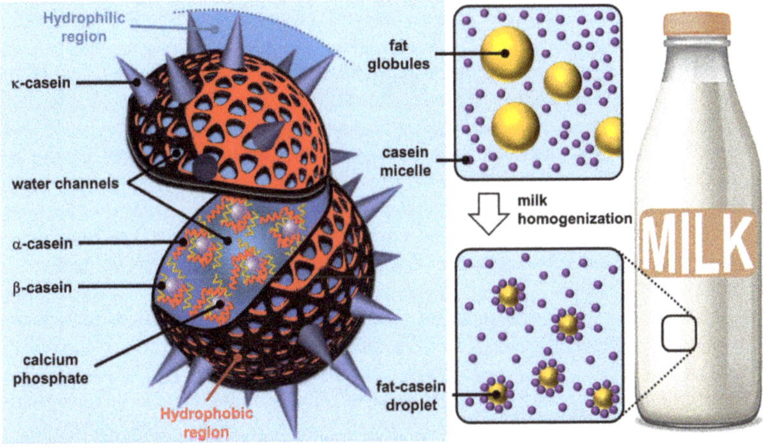

Figure 2. Structure of casein micelles (**left**) and effect of homogenization on fat and casein fractions in milk (**right**).

Despite the debate over its exact structure, casein micelle is built around strong interactions with calcium phosphate nanoclusters, while weaker, multivalent casein–casein interactions make casein micelles highly dynamic structures that complicate their modeling in the complex milk environment. Regardless of their exact structure, casein micelles are supramolecular colloidal assemblies with a lattice structure in which casein polymers and casein–calcium phosphate aggregates act together to maintain structural integrity through hydrophobic and electrostatic forces (i.e., calcium phosphate bridges), preventing spatial changes and the dissociation of the assembled structure unless the chemical environment is altered [37]. Unlike surfactant micelles, the core of casein particles remains highly hydrated: they are open sponge-like particles that are generally resistant to spatial changes due to their interlocked structure and the presence of multiple interactions [38]. Nevertheless, the structural arrangement within the casein particle can be altered by environmental factors that modify hydrophobic interactions and calcium phosphate solubility (e.g., pH, temperature) [39]. The spontaneous bottom-up edification of casein micelles using the interactions between casein proteins and calcium phosphate can be seen as tectonic subunits providing (supra) colloidal structures. The formation is driven by the interactions between the hydrophobic regions of the caseins and the phosphocalcic bridges. The self-assembly of casein micelle can be interpreted by calcium phosphate phase separation. In 2024, Antuma et al. supposed that the caseins would interact with the seed of calcium phosphate formed by pre-nucleation through their phosphoserine residues and subsequently self-assemble to form casein micelles, inhibiting the crystallization of calcium phosphate (i.e., the maturation into crystalline phases) [40]. Casein micelles give rise to a porous structure in which various other molecules or solvents can be inserted. These particles lead to Pickering-like emulsions (i.e., emulsions stabilized by particles absorbed at the water/oil interface) after milk homogenization (i.e., a process used to mix and disperse milk fat globules to avoid phase separation and achieve a homogeneous texture, hence the name of the process) when the casein micelles replace the milk fat globule membranes, which are damaged during homogenization (Figure 2) [41].

2.2. Plant Latex Particles

Another archetypal example of natural colloidal particles is found in plant latexes (i.e., wet rubber). Natural latex is a stable milky white, yellow, orange, brown, or even colorless colloidal dispersion of various particles (organic and inorganic) consisting of tiny droplets of organic matter dispersed in water [42]. Latex is present in at least 20,000 plants (with a high prevalence in flowering plants that form the angiosperm clade) and in several fungi such as *Lactarius deliciosus* [43]. The major botanical families producing latex in most of their species are *Euphorbiaceae* (which includes *Hevea*), *Moraceae* (especially *Ficus*), *Apocynaceae*, *Campanulaceae*, *Papaveraceae*, *Sapotaceae*, *Asclepiadaceae*, *Convolvulaceae*, *Asteraceae*, etc. [44] Latex is produced by extremely elongated secretory cells called laticifers constituting the laticiferous system [45]. Latexes arise from vacuoles occupying a very large part of the laticiferous cells. It should not be confused with plant sap because their functions are clearly different [45]. Indeed, after tissue injury, latex, which is the plant's first line of defense, coagulates upon exposure to air, serving mainly as a natural defense mechanism against herbivorous insects and other pathogens while sap ensures the distribution of water, mineral salts, or sugars [43]. It circulates in a distinct network of vessels. Indeed, latex forms a barrier against pathogen invasion due to its sticky and coagulating properties, limiting pathogen movements. Despite the variability of latex components, some chemical species can be distinguished, such as rubber, terpenoids, alkaloids, cardenolides, carotenoids, carbohydrates, phenolics, furanocoumarins, various metal ions, and proteins [46]. Some of them are toxic and/or dissuasive components depending on the plant species concerned.

In the following discussion, we focus on rubber latex, harvested from the rubber tree (*Hevea brasiliensis*) that grows in tropical regions, which is called fresh latex. Fresh latex is a stable colloidal dispersion of a polymeric substance (rubber) in a serum, which consists of phospholipids, proteins, inorganic salts, etc. In natural rubber, the polymeric substance is cis-1,4-polyisoprene. Fresh natural rubber latex is a whitish color fluid containing 30–33% natural rubber, 1–1.5% protein, 60–65% water, and some resinous substances [47,48]. After ultracentrifugation, fresh latex can be separated into four fractions (Figure 3) [49,50].

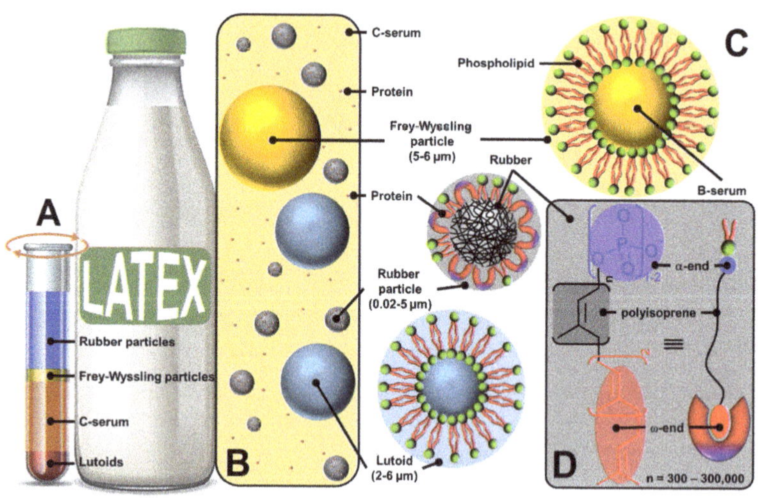

Figure 3. Ultracentrifuged fresh natural rubber latex fractions (**A**), schematic representation of liquid fresh rubber latex (**B**), structure of rubber and non-rubber particles (lutoid and Frey–Wyssling particles) (**C**), commonly accepted polyisoprene molecular structure (**D**) indicating the self-binding between the α-end and phospholipids and the ω-end and proteins through electrostatic and hydrophobic interactions (bottom right).

The upper layer (35%) is composed of 86% rubber particles dispersed in an aqueous medium, 3% lipids (sterol, sterol esters, waxes, and phospholipids), 1% proteins (α-globulin and hevien), and 0.05% metal ions such as Mg, K, and Cu. The shape of the rubber particles varies from spherical to oval depending on the age of the tree, and their size varies between 0.02 and 5 μm [42]. The second yellow or orange fraction (5%) contains Frey–Wyssling particles, which are composed of a lipid bilayer sequestering carotenoid pigments (giving them a yellow-orange color), oxidative enzymes, and proteins [42]. The third fraction (50%) corresponds to the centrifuged serum (C-serum) constituting the cytoplasm of laticifers and contains about 60% of all fresh latex proteins [42]. C-serum contains carbohydrates (methylinositol), water-soluble proteins (hevien), free nitrogenous bases, organic acids, inorganic anions (PO_4^{3-}, CO_3^{2-}), and metal ions (K, Mg, Fe, Na, and Cu). It is noteworthy that C-serum contains proteins that contribute to the colloidal stability of latex and rubber particles (see below). The lower fraction (10%) contains vacuole-like organelles called lutoids. Structurally, lutoid particles are composed of a lipid bilayer sequestering a fluid known as B-serum, which contains hydrolytic enzymes and proteins, and latex agglutination factors involved in the agglutination processes of latex particles [42]. Thus, fresh latex of *Hevea brasiliensis* contains three types of particles: rubber particles, present in the upper fraction; Frey–Wyssling particles, present mainly in the small second yellow or orange fraction; and Homans and van Gils lutoid particles in the lower fraction (Figure 3C). Since these rubber and non-rubber particles (lutoid and Frey–Wyssling particles) are dispersed in a continuous aqueous phase (i.e., serum C), fresh latex is a stable colloidal system [51].

On the other hand, the microstructure of polyisoprene molecules is known to have two terminal chain ends: α- and ω-ends (Figure 3D) [52,53]. Indeed, chain elongation in natural rubber latex biosynthesis caps one end of the rubber molecule with a mono- or diphosphate group (α-end), while the other end is capped with two *trans*-isoprene units (ω-end). The ω-end is linked to proteins resulting from the biosynthetic process probably via hydrophobic interactions. Indeed, both proteins (i.e., REF and SRPP) present a group of hydrophobic amino acids creating pockets and/or nonpolar surfaces, which allow hydrophobic interactions with the ω-end of polyisoprene molecules [54,55]. The association between the α-end and phospholipids occurs through direct electrostatic interactions or is mediated by Mg^{2+} ions or H-bonds with water molecules (see Figure 4).

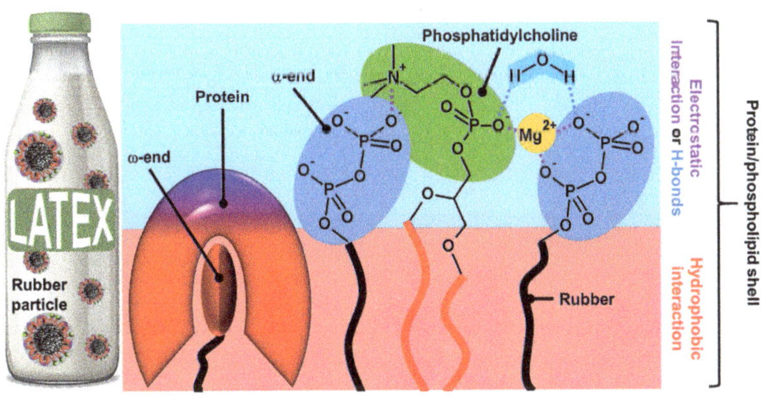

Figure 4. Presumed structure of rubber particles shell and non-covalent interactions between phosphatidylcholines and proteins (electrostatic and hydrophobic interactions).

Indeed, the phosphate groups are bonded to water via H-bonds within the surface, leading to a cooperative network [56]. The polyisoprene chain with the protein–phospholipid linkage can be moved and connected together as a network according to the reptation theory by de Gennes, leading to core–shell particles [57]. The phospholipid-protein layers are held on the surface of rubber particles by specific interactions. Indeed, proteins and phospholipids interact in the shell through hydrophobic contacts due to the clustering of hydrophobic amino acids in REF and SRPP (see above). It should be noted that electrostatic interactions also stabilize the shell. If there is no clear clustering of positive and negative residues in REF and SRPP, folding may occur in the presence of phospholipids to create charged surfaces and/or pockets that favor contacts with the phospholipid headgroup. In addition to the formation of direct ionic bonds between the negative charges of phospholipids, the polar head group of phospholipids participates in H-bonds with the polar solvent (i.e., water molecules) that restrict the mobility of the phosphate group, also stabilizing the mixed shell [56,58]. All the colloidal particles of fresh latexes allow the formation of amphiphilic regions that can be used as emulsifiers, leading to Pickering-type emulsions. Kumar and Basavaraj reported that plant latexes offer a sustainable route to reduce oil/water interfacial tension due to the spontaneous adsorption of surface-active species present in the latexes [59]. The surface-active species include molecular emulsifiers and particles. Indeed, under a cross-polarized microscope, the droplet surface is birefringent due to the interfacial adsorption of solid particles present in the latex. This mixed particle/surfactant emulsifier system of biological origin avoids the use of molecular surfactants involving complex synthesis routes, high production costs, and the generation of chemical waste.

2.3. Viral Particles

The last archetypal example of natural colloidal particles is found in viruses. Indeed, viruses are infectious agents, generally nanometric and stable in air or water, which use host cells to multiply. After contact and internalization in a target cell, the virus disassembles to transfer its genome, which is then translated into new viral components without their knowledge. Then, the viral components spontaneously self-assemble into new viruses that leave the cell to infect others, and the viral cycle begins again. The assembly and disassembly mechanisms are therefore closely linked to the proliferation of viruses and probably represent the most advanced case of biological supramolecular structures [60]. Viruses are composed of (i) the genome consisting of a nucleic acid chain (DNA or RNA); (ii) the capsid which surrounds and protects the genome from the external environment by a semi-rigid protein shell; and, sometimes, (iii) the envelope that covers the capsid and contains lipids and proteins essential for their binding to target cells (Figure 5) [61].

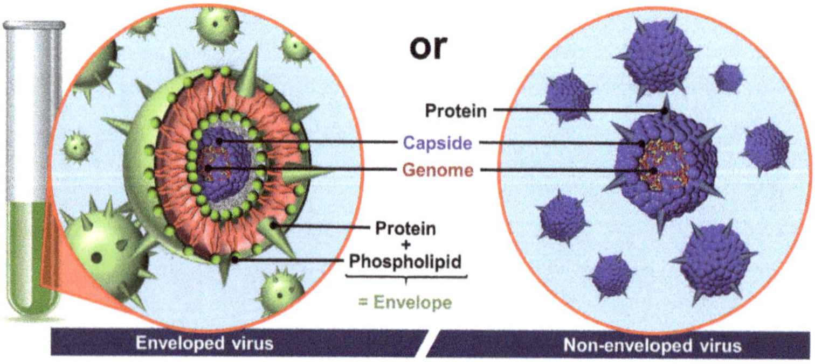

Figure 5. Schematic drawing of two basic types of viruses. In non-enveloped viruses, the genome is condensed in a capsid (coat protein), whereas enveloped viruses have a capsid or nucleocapsid wrapped in a phospholipid bilayer with protein (spike).

The most fascinating part of viruses is probably the capsid. Many capsids are icosahedral or nearly spherical with icosahedral symmetry as anticipated by Watson and Crick in 1956 and, more elaborately, by Caspar and Klug in 1962 [62,63]. The exterior of an icosahedral capsid is composed of repeating protein subunits (capsomeres). The regular icosahedron, characterized by 5:3:2 rotational symmetry, is the best way to form a closed shell from identical subunits: the proteins constituting the capsid are arranged on the faces, edges, or vertices of an icosahedron [64]. In more detail, the regular icosahedron is composed of 20 triangular faces and 12 vertices. Viruses with icosahedral symmetry contain 60 T proteins (where T is the number of distinct protein configurations), forming 12-vertex pentons (pentameric capsomers) and 10(T − 1) hexons (hexameric capsomers). For T = 1, the capsid contains only 12 pentagonal motifs (Figure 6A). In Figure 6B, the proteins have an identical chemical nature but with three different conformations (T = 3) identified by three colors leading to a capsid with 12 pentons centered on the vertices of the icosahedron (yellow proteins) and 20 hexagonal motifs (alternating dark and light green proteins). Thus, for larger viruses, the additional capsomers are arranged in a regular hexagonal lattice on the faces of the icosahedra. For instance, the cauliflower mosaic virus has an icosahedral capsid built from 420 proteins (12 pentons and 60 hexons): it is a T = 7 virus that obeys the rules of Caspar and Klug [65]. Icosahedral symmetry can be understood by simple geometric arguments: a structure consisting only of hexagons would have a maximum surface density, but it could not be closed because it would be flat. To introduce the curvature required for a closed structure, one must have exactly twelve pentagons, with the rest of the structure being composed of hexagons. If these twelve pentagons are distributed equidistantly, the structure exhibits icosahedral symmetry. However, when the twelve pentagons imposed by the topology are distributed non-uniformly, new geometric shapes are observed, and icosahedral symmetry is lost [65]. This is the case, for example, of the human immunodeficiency virus type 1 (HIV-1) or the tobacco mosaic virus (TMV), whose capsids are, respectively, conical or helical in shape [66,67].

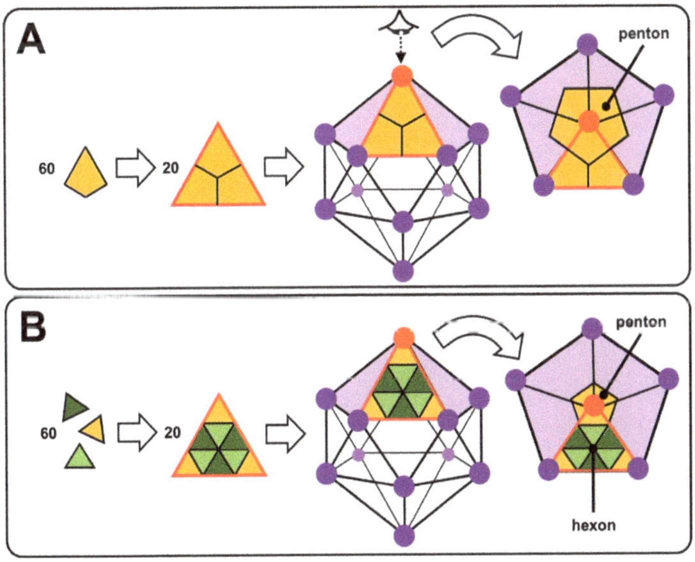

Figure 6. Self-assembly of 60 T asymmetric units (proteins) giving 20 triangular facets, leading to icosahedral nucleocapsid with T = 1 (**A**) and T = 3 (**B**).

Although our knowledge of genome packaging in a capsid is still embryonic, most capsids are formed by a self-assembly process that can be reproduced in vitro by mixing the genome with the subunits in an aqueous medium. These are often protein dimers because they are more stable than the proteins alone. This process is favorable (i.e., spontaneous) because the energy gain linked to the association of the subunits is greater than the entropy loss that is also associated with it. There is, however, a threshold concentration of subunits for self-assembly to be initiated. The energy gain is associated with non-covalent interactions between subunits (van der Waals and/or hydrophobic forces). After a transient regime, the association between the proteins eventually reaches a state of chemical equilibrium governed by a mass action law [68]. The genome, however, helps guide assembly. Protein–genome interactions are primarily electrostatic in origin because proteins often contain positively charged chemical groups, whereas the genome invariably carries negative charges. Indeed, there is a linear relationship between the total electrostatic charge of the genome and that of the protein subunits [69]. It is noteworthy that empty capsids (i.e., without genome) can also be formed under appropriate ionic conditions.

The self-assembly of an empty capsid is often interpreted in terms of nucleation/growth [70]. It should be noted that the term nucleation refers to a process by which atoms or molecules aggregate into a small unstable seed, from which a larger assembly will grow. Originally observed for crystal formation, nucleation appears to be a generic pathway, including many molecular assemblies [71]. The subunits begin by forming a seed under the action of a local concentration fluctuation; this is the nucleation phase. Other subunits that remain free then bind sequentially to this seed to make it grow rapidly until the formation of a complete capsid. For example, the seed is a trimer of subunits for the capsid of the hepatitis B virus [72]. Unfortunately, the seed varies according to the virus considered and remains difficult to identify due to its transient nature. The lower the free energy of association, the greater the number of subunits constituting the seed will increase. In addition, assembly and disassembly exhibit significant hysteresis that allows the assembled capsids to remain metastable even at high dilution, which is a crucial property for their survival outside the intracellular environment. In the presence of a genome, two scenarios are possible, depending on the interaction energies between the components: (i) the capsid assembles according to the nucleation/growth mechanism (see above) while simultaneously packaging the genome, or (ii) the subunits rapidly bind to the genome and then cooperatively reorganize into a capsid [73].

Some virus particles allow the formation of emulsions. For instance, the icosahedral capsids of cowpea mosaic virus (CPMV) and turnip yellow mosaic virus (TYMV) have the ability to stabilize Pickering emulsions [74,75]. However, viral envelope proteins can also be used to obtain long-term stable Pickering emulsions. For example, the amphiphilic envelope proteins of TMV can self-assemble at Pickering emulsion interfaces [76]. In all these cases, robust capsules can be fabricated by cross-linking between the coat proteins. By taking advantage of the amphiphilicity of coat proteins separately or directly on the capsid, the construction of robust Pickering membranes can have various applications, including drug delivery or virus recognition.

2.4. Nature and Colloidal Tectonics

All living organisms exhibit self-assembled entities. The three examples, presented in this section, show a core–shell structure with at least one layer of proteins (i.e., building blocks or tectons) surrounding the core. For casein micelles, the core is composed of hydrophobic proteins (α- and β-caseins) while the coat is composed of the more hydrophilic κ-caseins. For rubber particles, a mixed layer of proteins and phospholipids surrounds the hydrophobic polyisoprene core. For non-enveloped viruses, the nucleic acid is surrounded by capsid proteins. In all three systems, the main driving force of particle formation is hydrophobic forces supported by H-bonds and electrostatic interactions (relatively strong interactions). It should be noted that other biomolecules form self-assembled entities. For instance, lipids can self-assemble into membranes through primarily hydrophobic interactions, a relatively weak and non-directional interaction [77]. However, these systems are typically soft, fluid, and less-ordered (despite the coexistence of soft fluid and rigid gel domains in plasma membrane), whereas proteins self-assemble into rigid crystalline structures [78]. In general, chemists can easily reproduce lipid assemblies using synthetic amphiphilic small molecules and/or polymers to form micelles, vesicles, and/or lamellar structures [79]. In contrast, protein assemblies that produce assemblies with structural rigidity and crystallinity (e.g., lamellae, tubules, and/or polyhedrons) are more difficult to mimic by synthetic molecules [80]. Colloidal tectonics deals with the formation of artificial rigid and crystalline colloidal structures, similar to protein assemblies, which is a fundamental challenge in materials science [15].

In the three previous examples, the self-assembly of tectons into particles can be interpreted in terms of nucleation/growth. Indeed, some tectons start by forming a seed under the action of a local concentration fluctuation (nucleation) and then other tectons that remain free sequentially bind to this seed to make it grow rapidly until the formation of a complete particle. This process results in colloidal particles constituting hydrophilic and hydrophobic regions in which solvent(s) can be accommodated or at least interacted. The spontaneous formation of colloidal structures from tectonic subunits depends on the total interaction energy (i.e., the sum of attractive and repulsive interactions) [15]. The structure of the tectons influences the physicochemical properties of the particles and thus the possibility of obtaining supracolloidal structures such as dispersion, Pickering emulsions, colloidal crystals, etc. [15]. There is therefore a continuous transition between the properties of the tectons and those of (supra) colloidal systems [15]. In the first publication on the concept of colloidal tectonics, the following guidelines were proposed to obtain a bottom-up colloidal construction: two or more complementary tectons (i.e., with suitable binding sites to form supramolecular seeds essential for the growth of the colloidal particles) of opposite polarities, at least one of which has a rigid structure [15]. However, more recent papers have supplemented and refined these guidelines (see below). Indeed, the new formulation is as follows: colloidal systems can be generated using either a self-complementary tecton (single-component system) or two or more complementary tectons (multicomponent system). In addition to intermolecular interactions, structural flexibility/rigidity and fluidity/crystallinity are also essential for the spontaneous bottom-up formation of colloidal structures.

Let us focus here on the single-component system, supposing that a capsid-like structure forms. The tecton presented in Figure 7A has both polar and nonpolar binding sites. Due to this self-complementarity, tectons lead, under appropriate conditions (temperature, concentration, solvent, etc.), to the formation of pentamers (i.e., pentagonal pyramid). The mutual interconnection of pentamers is achieved by complementary interactions leading to a regular icosahedral envelope. The resulting polyhedron has 20 equilateral triangles as its faces, 30 edges, and 12 vertices. Therefore, two levels may be used to describe the growth of icosahedron colloidal arrangements: the formation of pentamers and the self-association of pentamers. The second possibility (Figure 7B) concerns the formation of icosahedrons using a two-component system based on two tectons of opposite polarity. However, the two tectons can be paired due to complementary attractive interactions forming discrete supramolecular structures. These structures can form pentamers by mutual bridging. For reasons of compaction, pentamers are grouped in regular icosahedrons (see above). In this case, three levels may be used to describe this system: the establishment of discrete supramolecular structures, the formation of pentamers, and the lateral association of pentamers leading to regular hollow icosahedrons. From a general point of view, intermolecular interactions, structural flexibility/rigidity, and fluidity/crystallinity are key parameters, as well as the morphology of tectons. Indeed, tectons of identical chemical nature can adopt different conformations allowing them to obtain pentagonal patterns and hexagonal patterns, leading to icosahedral symmetries. The self-assembly process is favorable and therefore spontaneous because the energy gain of association of the subunits is greater than the entropy loss that is also associated with it. However, there is a threshold concentration of subunits for self-assembly to be initiated. The energy gain is associated with short-range attractive interactions between the subunits (hydrophobic and/or van der Waals forces). After a transient regime, the association between the subunits finally reaches a state of chemical equilibrium governed by a mass action law. This self-assembly is often interpreted in terms of nucleation/growth (see above).

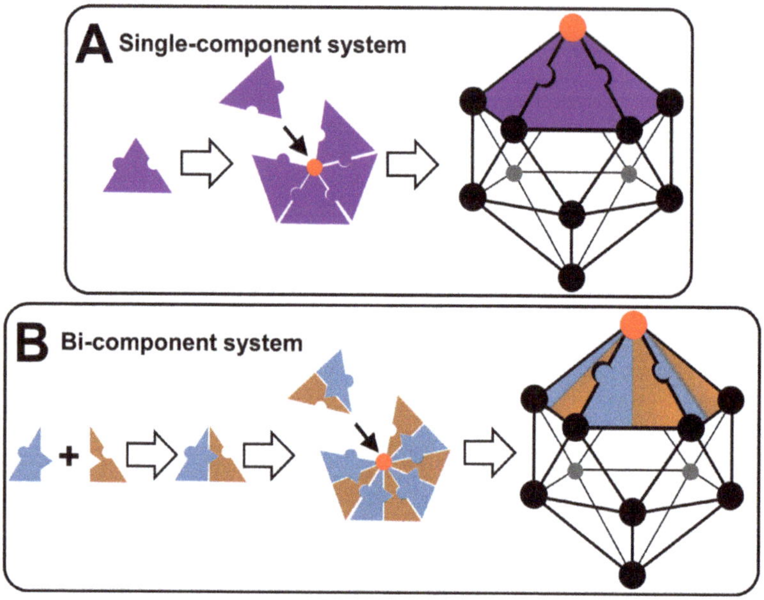

Figure 7. Schematic representation of colloidal arrangements in the form of regular icosahedrons (capsid-like structures) made with a single- or bi-component system (**A**) or (**B**).

Colloidal tectonics deals with the design and generation of colloidal systems based on tectons, which are active building units carrying recognition information and thus capable of recognizing each other. Colloidal arrangements are formed under self-assembly conditions, leading to reversible and adaptable systems with self-repair capacity. In other words, complementary tectons find their own way to the most stable situation under the given conditions (temperature, concentration, solvent, etc.). The main recognition events are solvophobic interactions, which may be complemented by a variety of other reversible attractive intermolecular interactions such as van der Waals, electrostatic forces, and H-bonds [15]. However, structural flexibility/rigidity and fluidity/crystallinity are also essential parameters (see above). However, unlike the concept of molecular tectonics [12,13], which deals with the design and preparation of infinite periodic molecular networks formed under self-assembly conditions between self-complementary or complementary tectons, the growth of colloidal tectonics assemblies is inhibited by various repulsions such as steric and electrostatic repulsions [15]. Therefore, a simple design rule for colloidal tectonics-based systems is the presence of (i) attractive interactions conferring rigidity and/or crystallinity, and (ii) repulsive forces to avoid the long-range order characteristic of a crystal. Since colloidal systems play a crucial role in nature and in many scientific and industrial applications, some recent examples of strategies to obtain such systems using colloidal tectonics are presented in the following section, together with their uses.

3. Design and Applications of Artificial Architectures Based on Colloidal Tectonics

In this section, it is noteworthy that artificial architectures based on colloidal tectonics are not presented from a historical perspective. Indeed, I have discarded this view to use a classification according to the number of components involved in the colloidal edification. Indeed, architectures can be generated using either a self-complementary tecton (single-component system) or two or more complementary tectons (multicomponent system). In each case, subdivisions are made according to the chemical nature of the tectons. It should be noted that, for the sake of clarity, only a few examples are reported in this section, with particular emphasis on Pickering emulsions for catalysis or drug delivery.

3.1. Single-Component Colloidal Systems

Colloidal tectonics can be used to generate colloids by self-assembly processes involving a single self-complementary tecton. Such a single-component system is the most attractive situation in terms of atoms and synthetic economy. Since the most important requirement is to have solvophobic (i.e., hydrophobic and organophobic) sections, solvophobic block copolymers can be good candidates. Indeed, the low solubility of one of the blocks provides a sufficient driving force for their self-assembly in a given solvent. In such aggregates, the solvophobic blocks associate into dense central domains surrounded by solvated coronas formed by solvophilic blocks. This driving force is counterbalanced by repulsive interactions between solvophilic blocks that ensure the formation of stable self-assemblies at the nanoscale. For example, in aqueous solutions, the assembly is clearly driven by the hydrophobic attraction between the associated blocks and counterbalanced by electrostatic or steric repulsions between ionic or neutral organophobic blocks. However, to avoid the formation of dynamic polymeric micelles similar to classical molecular surfactants, one has to keep in mind that the polymer must have structural flexibility/rigidity and fluidity/crystallinity in addition to complementary intramolecular interactions (see above). To manage flexibility/rigidity and fluidity/crystallinity, additional interactions have to be used, such as H-bonds, aromatic stacking, van der Waals interactions, etc. In practice, homopolymers and surfactant-based systems can also be used. In this section, some typical examples are given to illustrate the effectiveness of these approaches.

3.1.1. Homopolymers

In 2022, Douyère and his collaborators reported a system based on linear polyethyleneimine (LPEI) as a unique self-complementary tecton to design self-assembled gelled emulsions (i.e., emulgels) [81]. The polymer has a high molecular weight to obtain an appropriate rigidity/flexibility balance. The hydrophilic-lipophilic balance is obtained by the protonation/deprotonation of the amine groups. Indeed, depending on the LPEI protonation, the polymer is in a free form at pH < 2.3 or fully aggregated at pH > 10, leading, respectively, to a solution or an LPEI dispersion. Between these two behaviors, LPEI acts as a buffering agent by the continuous protonation/deprotonation of the amine groups. Thus, in an aqueous solution, LPEI molecules give a switchable gel depending on the pH and temperature. The formation of thermo-reversible hydrogels is due to adequate and spontaneous distributions of "hydrophobic" and "hydrophilic" regions on the LPEI chains (Figure 8). Indeed, the "hydrophilic" regions are protonated amine groups while the "hydrophobic" regions are neutral -CH_2CH_2NH- units. The gel phase is obtained by a pseudo-crystallization of "hydrophobic" regions, under the action of hydrophobic forces and H-bonds, while the protonated regions remain hydrated (Figure 8). In an aqueous solution, competition between hydrated and associated regions results in the formation of a 3D network where the crystallized hydrophobic domains act as knots (Figure 8). An interesting comparison can be made with poly(4-vinylpyridine), P4VP. Indeed, linear P4VP, even partially protonated, is not able to provide crystallites or other self-assembled colloidal entities capable of stabilizing Pickering-like emulsions [82]. In contrast, acidic colloidal particles of P4VP (39 mmolH+/g) cross-linked with 2% divinylbenzene (Reillex® 402 ion-exchange resin) are able to give heptane-in-water Pickering emulsions with long-term stability [83]. Recalling these observations, a simple design rule for self-assembly can be put forward: the presence of strong attraction is mandatory (H-bonds for LPEI molecules and cross-linking, involving the formation of covalent bonding between adjacent polymer backbone for P4VP particles). However, it is noteworthy that acidic P4VP particles containing 25% divinylbenzene (Reillex® 425) do not lead to hydrogels or Pickering-type emulsions. A higher cross-link density is the result of a greater number of bonds per polymer chain length, which results in greater rigidity and hardness, proving the importance of structural flexibility/rigidity and fluidity/crystallinity compared to morphology [82]. Indeed, the 2% cross-linked protonated P4VP particles have a spongy structure, leading to solvent penetration capable of interacting with the hydrophobic regions of P4VP. Therefore, the cross-linked systems are less adaptable than LPEI-based systems that take advantage of reversible H-bonds to adapt to the external environment.

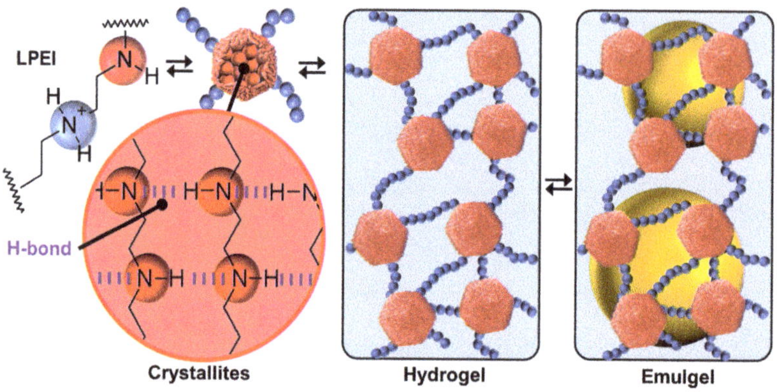

Figure 8. Structure of linear polyethyleneimine, LPEI, with protonated and unprotonated amine groups, leading to crystallites formed by H-bonds between unprotonated amine groups, thermoreversible hydrogels, and gelled emulsions (emulgels).

As LPEI provides a robust tecton with solvophilic and solvophobic regions, leading to structural flexibility/rigidity and fluidity/crystallinity, the spontaneous bottom-up self-assembly provides a good platform to obtain Pickering-like emulgels [81]. Since LPEI is not soluble in either conventional organic solvents (heptane, toluene, isopropyl myristate, and liquid paraffin) or vegetable oils (corn, sunflower, olive, and castor oil), the authors obtained long-term stable emulgels after the homogenization of biphasic oil/water systems (1:1 *w*/*w*) in the presence of LPEI [81]. Indeed, the stabilizing mechanism of the obtained systems is unusual: (i) crystallites are homogenously distributed on the surface of the droplets, leading to Pickering-like stabilization, and (ii) crystallites act as knots, contributing to the gelation of the aqueous continuous phase and the emulsion's stability (Figure 8). Moreover, due to their self-assembled nature, these emulgels provide reversible systems with a high degree of control. Indeed, sol/emulgel transitions are obtained by temperature or pH changes because the acidic environment and/or high temperature alter the self-assembly of the "hydrophilic" and "hydrophobic" regions, leading to destabilization by modifying the gelation of the aqueous continuous phase and the Pickering stabilization (alternating stabilization and phase separation up to 10 consecutive runs) [81].

3.1.2. Copolymers

Since it is possible to obtain tectons from homopolymers when the latter possess ionizable functions, the use of amphiphilic copolymers has been considered. For example, Bardoula and his collaborators reported the use of amphiphilic polymer particles based on copoly(2-methyl/phenyl-2-oxazoline)s, P(MeOx)-P(PhOx), where MeOx is the hyrophilic monomer and PhOx is the hydrophobic monomer, to obtain Pickering emulsions (Figure 9) [84]. Well-defined polymer particles are obtained by nanoprecipitation. The effects of chain length (degree of polymerization, DP = 50, 100, 150, or 200) and monomer distribution (block or gradient) on the properties of the polymer particles are also investigated. For a fixed MeOx/PhOx ratio of 25/75 (*w*/*w*), the intensity-weighted average hydrodynamic size of all particles measured by dynamic light scattering ranges from 69 $\pm$ 15 to 152 $\pm$ 10 nm for block copolymers and from 65 $\pm$ 3 to 114 $\pm$ 8 nm for gradient copolymers. In more detail, DP_{50}-based particles exhibit significant hydrophilic characteristics, close to water solubility, with measured sizes in the range of 115–140 μm and very low water contact angle (θ) values around 20°. For the other polymer particles, two distinct behaviors were observed depending on the monomer distribution. In the case of block-based particles, an increase in the DP leads to an increase in particle size (from 69 to 152 nm for DP_{100} and DP_{200}), while gradient particles exhibit a size around 65–72 nm, remarkably displaying lower polydispersity indices compared to their block counterparts. Interestingly, for polymer particles with DP_{100}, DP_{150}, and DP_{200}, an increase in chain length has little impact on their wettability, with θ values ranging from 53 to 65°. Furthermore, the particle glass transition temperature values follow the expected trend of increasing with chain length for both block and gradient particles. Recalling these results, it is possible to gain insight into the internal structure of polymer particles. Indeed, polymer nanoprecipitation occurs after the addition of a non-solvent to a polymer solution by a four-step mechanism: supersaturation, nucleation, condensation growth, and coagulation growth that leads to the formation of polymer particles [85]. Under such conditions, the essential driving force leading to the formation of polymer particles is clearly the hydrophobic effect between the PhOx monomers. To avoid considerable exposure of the hydrophobic residues in water, a strong self-association of the hydrophobic sections of the polymer chains occurs to form supramolecular assemblies. However, the presence of rigid aromatic phenyl rings leads to aromatic stacking of phenyl substituents with a shifted (or twisted) geometry relative to the face-to-face and/or T-shaped (or edge-to-face) orientation of the aromatics, thereby increasing the packing density of the polymer chains [86]. Therefore, we obtain core–shell polymer particles composed of a hydrophobic core containing PhOx monomers while the shell contains hydrophilic MeOx monomers. It is worth noting that this mechanism is valid for both block and gradient copolymers because for all copolymers, as the molar mass increases, a higher particle glass transition temperature value is observed due to the stronger π-stacking interactions [87]. Although these polymer particles, composed of a P(PhOx) core surrounded by P(MeOx) segments, are in dynamic equilibrium with their environment and the structural arrangement can be modified by environmental changes, the hydrophobic and aromatic stacking interactions prevent spatial changes and the dissociation of the supramolecular structure.

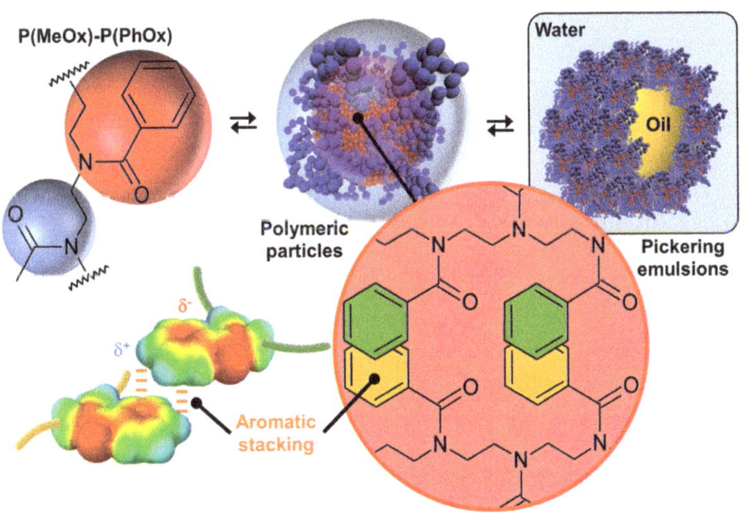

Figure 9. Structure of copoly(2-methyl/phenyl-2-oxazoline), P(MeOx)-P(PhOx), polymer particles formed by stacking interactions between phenyl side groups, and resulting Pickering emulsion.

Since the polymer particles are open sponge-like colloidal structures capable of accommodating solvent(s) in both hydrophilic and hydrophobic regions, they can be used for the long-term stabilization of Pickering emulsions with emulgel properties using paraffin oil and isopropyl myristate [84]. These polymer particles effectively stabilize Pickering emulsions starting from 0.9 wt. % and allow to easily obtain high internal phase emulsions entirely based on biocompatible, low-toxicity, and tunable P(MeOx)-P(PhOx), opening the way to wide applications. Finally, it is noted that the other MeOx/PhOx ratio of 50/50 or 75/25 (*w*/*w*), whatever the chain length (degree of polymerization, DP = 50, 100, 150 or 200) and the monomer distribution (block or gradient), leads to classical emulsions without the formation of polymer particles [88]. This observation is clearly a consequence of the minimization of hydrophobic and stacking interactions (see above), proving once again the importance of intermolecular interactions, structural flexibility/rigidity, and fluidity/crystallinity.

3.1.3. Surfactants

Surfactants are known to form micellar, lamellar, and vesicular structures. However, as previously discussed, these assemblies are soft and fluid because they are mainly driven by hydrophobic interactions. In contrast, the assembly of tectons using the concept of colloidal tectonics forms rigid and/or crystalline architectures driven by a combination of hydrophobic interactions, H-bonds, and electrostatic forces. Therefore, in order to obtain faceted structures, it is necessary to increase the rigidity and crystallinity of surfactant assemblies by additional strong and directional in-plane attractions (e.g., H-bond and/or electrostatic) and out-of-plane repulsions (e.g., electrostatic or steric).

Faceted vesicles (i.e., rigid non-spherical liposomes) can be obtained by enhancing the crystalline characteristic of membranes. One way to achieve this is to replace the ester bonds of natural phospholipids with artificial amides that stabilize the interfacial region by H-bonds. In this regard, Neuhaus and coworkers used an artificial amide (1,2-diamidophospholipid) homologue of 1,2-dipalmitoylphosphatidylcholine (DPPC) to form cuboidal vesicles without using a template (Figure 10) [89].

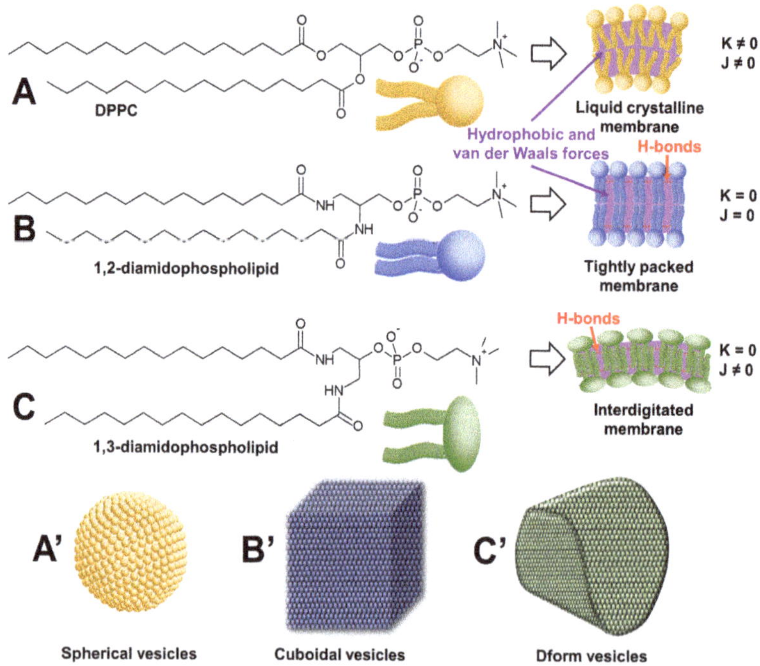

Figure 10. Structures of 1,2-dipalmitoylphosphatidylcholine (DPPC) (**A**), 1,2-diamidophospholipid (**B**), and 1,3-diamidophospholipid (**C**) and their self-assembly in spherical (**A'**), cuboidal (**B'**), and Dform (**C'**) vesicles. The figure highlights the dependence of the geometric shape of liposomes on the intrinsic Gaussian and extrinsic total curvature (K and J, respectively).

Interestingly, vesicles are typically spherical to minimize surface tension, e.g., DPPC forms spherical vesicles. The authors use complementary H-bonds to increase the interactions between phospholipids. Recalling that DPPC forms only slightly faceted vesicles in the gel phase and that the fishbone packing of the lipids is not found in DPPC vesicles, the authors argue that the H-bond network present between 1,2-diamidophospholipid molecules leads to a minimization of the intermolecular distance and thus a maximization of the chain–chain attractions (i.e., dispersion forces). Consequently, intermolecular H-bonds between amide groups in 1,2-diamidophospholipids lead to planar bilayers with exceptionally tight packing, where the molecules are packed in a fishbone pattern as suggested by wide-angle X-ray scattering measurements. In practice, these rigid bilayers must be heated above their melting temperature to obtain fluid membranes, leading to vesicles. When cooled below their melting temperature, phospholipid vesicles give rise to structures that maximize flat surfaces and minimize edges, i.e., cuboidal structures. The authors propose to use these cuboid vesicles as drug delivery devices.

However, other fascinating structures can be obtained. For instance, the same team reported that 1,3-diamidophospholipids form faceted vesicles [90]. In these vesicles, the 1,3-diamidophospholipid molecules take a rigid interdigitated arrangement inside the bilayer membrane. Indeed, the thickness of the bilayer is comparable to the length of a single acyl chain of the phospholipid [91]. This tight packing results in the absence of spontaneous curvature, leading to faceted tetrahedral vesicles called Dform because of their resemblance to the letter D (see Figure 10). Since these vesicles are stable in a salt-containing buffer and under static conditions, Zumbuehl et al. investigated their use as drug delivery devices [92]. This study suggests that such loaded nanocontainers could potentially be used to treat atherosclerotic patients, as the drug is preferentially released in constricted vessels where shear stress is high [93].

The two previous examples are very instructive because they highlight the importance of molecular morphology on the shape of the supramolecular object obtained. Indeed, 1,2- and 1,3-diamidophospholipids are positional isomers capable of forming intermolecular H-bonds between the amide groups, leading to a maximization of chain–chain attractions (i.e., dispersion forces). The maximization of chain–chain attractions is optimal between 1,2-diamidophospholipid molecules due to the proximity of the two alkyl chains, which are spatially close. This effect is less for 1,3-diamidophospholipids, where the two alkyl chains are further apart, hence the need for chain–chain interdigitation within the bilayer membrane, suggesting that interdigitation is a major contributor to the formation of D-form vesicles. Therefore, in both cases, as the vesicles were formed from a single type of phospholipid, all vesicles are spherical above the melting temperature of the phospholipids because the liquid crystalline phase imposes no constraints on the geometry of the vesicles. However, upon cooling below the melting temperature, the intrinsic Gaussian curvatures and the extrinsic total curvatures of the bilayers (K and J, respectively) force the spheres to adopt alternative shapes with a clear dependence on the geometric shape of the liposomes (see Figure 10). Therefore, 1,3-diamidophospholipid forms rigid interdigitated bilayers and thus D-shaped vesicles ($K = 0$ and $J \neq 0$). However, changing the phospholipid substitution pattern from 1,3 to 1,2 does not allow membrane interdigitation but preserves the H-bond network between phospholipid molecules, leading to cuboidal structures ($K = 0$ and $J = 0$). For DPPC, in the absence of an H-bond network, only spherical liposomes are observed ($K \neq 0$ and $J \neq 0$).

Polyhedral capsid-like cationic vesicles have also been observed. For instance, Pardin and his collaborators obtained this arrangement by the self-assembly of *N*,*N*′-dialkylmethylenediimidazolium ditriflate (Figure 11) [94]. Although there is no explanation for this extreme faceting of the vesicles in the original publication, it is now possible to propose an explanation based on current knowledge.

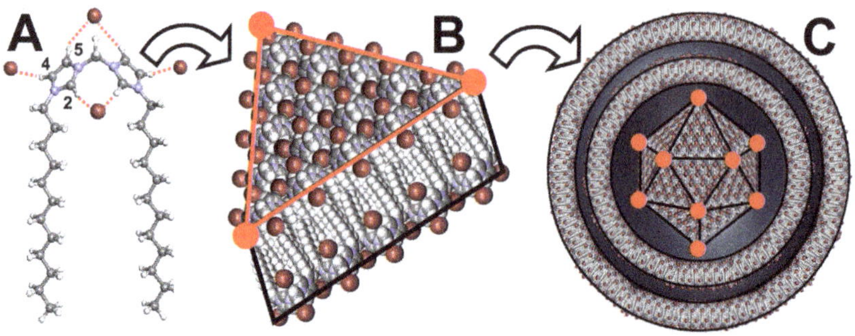

Figure 11. Structure of *N*,*N*′-didodecylmethylenediimidazolium ditriflate, schematic representation of H-bonds between imidazolium cations and anions (brown sphere (**A**)), planar bilayer arrangement with interdigitation of the akyl chains (**B**), and resulting vesicle with polyhedron core and spherical multilayer coat (**C**).

Indeed, it is known that imidazolium cations and anions are connected to each other to form H-bonded networks. In fact, the imidazolium ring serves as an H-bond donor through the H2 position and to a lesser extent through the H4 and H5 positions [95]. These large H-bond networks contribute to forming rigid membranes. In addition, the alkyl chains are spaced apart, leading to the intercalation of membrane sheets. Under such conditions, self-assembly into a closed 3D structure leads to a minimization of membrane intersections (edges) and a maximization of planar membrane faces, leading to a polyhedral vesicle. Recalling that the polyhedron core is surrounded by spherical multilayers, it is argued that spheres are produced to minimize the surface tension around the polyhedron core. Therefore, this arrangement resembles viruses: a polyhedron core surrounded by spherical layers. It is noteworthy that the polyhedron core depends on the length of the alkyl chain and the anion used. For instance, the dodecyl chain requires triflate anions to form polyhedrons while the hexadecyl chain requires bromides. These vesicles are used to entrap linear double-stranded (ds) DNA to protect it from enzymatic cleavage [95].

3.1.4. Other Systems

In 2004, Valéry and his collaborators reported that octapeptides (acetate salts of cyclic Laureotide of sequence NH_2-(D)Naph-Cys-Tyr-(D)Trp-Lys-Val-Cys-Thr-$CONH_2$ and its cyclic derivative of sequence NH_2-(D)Naph-Cys-Tyr-(D)Phe-Lys-Val-Cys-Thr-$CONH_2$) self-assemble into nanotubes in water. The nanotubes are arranged in hexagons (with a packing parameter of 365 Å) and are highly monodisperse [96]. The tube diameter and wall thickness are 244 and 18 Å, respectively. Furthermore, the tube diameter is tunable by modifications of the molecular structure. The self-assembly of the nanotubes is due to the association of amphiphilic β-sheets and a systematic segregation of aromatic/aliphatic side chains. The same year, Hill et al. reported that amphiphilic hexa-*peri*-hexabenzocoronene molecules self-assemble to form nanotubular objects [97]. These objects are uniform with a 14 nm wide open hollow space and a 3 nm thick wall. The wall consists of helical arrays of 13 fused benzene rings stacking via π-interactions. The inner and outer surfaces are covered with hydrophilic triethylene glycol chains since the hexa-*peri*-hexabenzocoronene molecules form interdigitated bilayers across the alkyl chains. In a water–THF mixture, a helical coil is formed by the loose winding of the bilayer band, while in THF, the nanotubes are formed by the tight winding of the bilayer band.

3.2. Multi-Component Colloidal Systems

To achieve the formation of self-assembled (supra) colloidal structures from several building blocks, tectons must meet both structural and energy criteria. Indeed, it must be kept in mind that complementary tectons must, on the one hand, recognize each other and thus generate the seed, and on the other hand, allow the growth that leads to the formation of a 3D particle. This assembly process is mainly driven by solvophobic interactions and also by other complementary interactions such as H-bonds, aromatic stacking, electrostatic, or van der Waals forces, etc. (see above). These latter interactions are crucial to managing the appropriate structural flexibility/rigidity and fluidity/crystallinity needed to limit the formation of dynamic systems characterized by the exchange of molecules leaving and joining the aggregate (i.e., to avoid the reversible change in the number of tectons and the overall morphology in the resulting colloidal structure). Indeed, the tecton must fulfill both structural (recognition) and energetic criteria (tecton–tecton interactions must be higher than tecton–solvent interactions) [15]. In the search for synthetic systems reproducing the morphology, rigidity, and crystallinity of protein assemblies of all living organisms (see above), several strategies have been developed over the last two decades. Some of them are described in detail in the following sections depending on the origin of the architectures (i.e., organic or hybrid) and their applications if available.

3.2.1. Organic Architectures

(i) Cyclodextrin-based architectures

Native cyclodextrins (CDs) are macrocyclic oligosaccharides, typically consisting of 6 (α-CD), 7 (β-CD), or 8 (γ-CD) glucose units linked by α-1,4 glycosidic bonds (Figure 12).

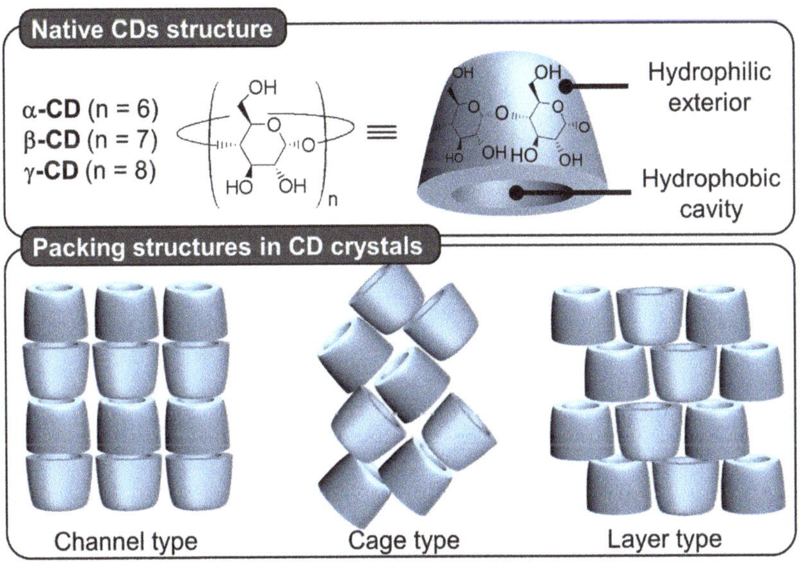

Figure 12. Native cyclodextrin (CD) chemical structure and schematic representation of packing structures of channel, cage, and layer type CD crystals.

Morphologically, CDs have a truncated cone shape (i.e., bowl-shaped without a bottom) and are water-soluble but not in typical organic solvents [98]. However, their cavities are considerably less hydrophilic than the aqueous medium and are therefore capable of forming inclusion complexes with hydrophobic molecules. Since the rims are covered with primary and secondary hydroxyl groups, H-bonds are formed with water molecules, explaining their solubility in water [98]. In the crystalline state, CD inclusion complexes adopt three types of assembly modes: channel, cage, and layer types. Channel-type assembly, commonly observed with polymers, forms CDs stacked like coins in a roll [99]. In this mode, CDs can be arranged head-to-head or head-to-tail by H-bonding between the hydroxyl groups of neighboring CDs. On the other hand, cage architecture is observed when CDs are packed in a herringbone pattern [99]. In the latter mode, CDs are arranged side-by-side, forming layers offset by about half a CD [99]. With the smallest native CD (i.e., α-CD), cage-like structures are formed with small guests, whereas long or ionic guests induce channel-like structures. Channel types are generally preferred with the β- and γ-CDs. This diversity of assembly can be very useful in producing various colloidal systems, including nanoparticles, crystallites, lamellae, helical tubes, and polyhedrons (see below).

Particles or crystallites can be easily obtained using native CDs and various non-charged guests. Indeed, mixing CDs in biphasic oil/water systems gives oil-in-water Pickering-like emulsions stabilized by partial wettable insoluble CD/oil inclusion complexes [100]. As the colloidal structure depends on the experimental conditions, the nature of the precipitated fraction can be crystals, crystallites, or spherical nanoparticles [15,101–103]. In detail, these systems result from the formation of inclusion complexes between CDs and oil molecules, leading to the formation of small nuclei in the dispersion; the nuclei then grow as insoluble inclusion complexes from the liquid attached to them. These steps allow the formation of a 3D structure by self-assembly or self-organization within the biphasic mixture, leading to crystallites or nanoparticles depending on the experimental conditions and the guest structure (Figure 13) [15].

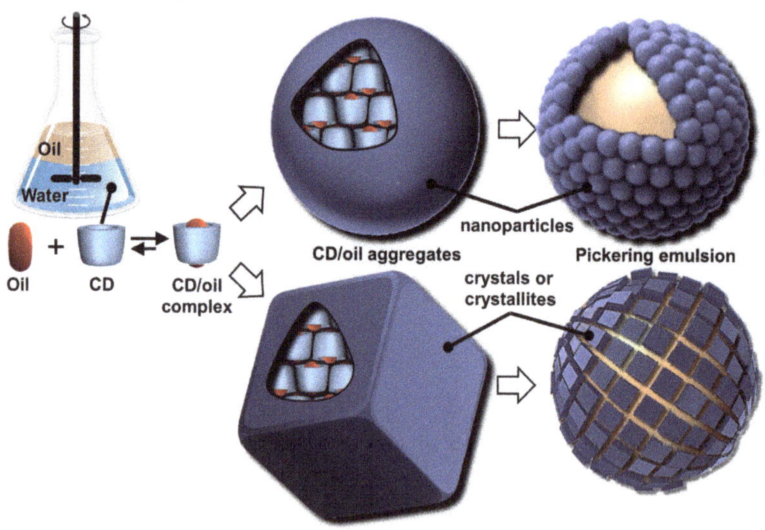

Figure 13. Schematic representation of sequential self-assembly of complementary tectons (CDs and oil molecules), leading to Pickering emulsions.

In 2024, Hou and Xu revealed the role of oils in the formation of CD/oil inclusion complex crystallites using different oils (e.g., linear alkanes, oleic acid, glycerol trioleate, and soybean oil). The authors observed that the inclusion complexes tend to grow in clusters and terminate at a certain finite size as long columns or lamella plates with well-defined facets, depending on subtle changes in the molecular architecture of the oil guests used [104]. Similar to the crystal structures of host–guest complexes, these assemblies are expected to contain channel, cage, and/or layer-like packing structures through the formation of a network of H-bonds between CDs (see above). The replacement of native CDs with chemically modified CDs alters the solubility of the inclusion complexes and renders them unable to assemble, highlighting the importance of H-bonds between CDs within the edifices. However, the growth of "particles" is limited due to the increase in interfacial rigidity with particle emergence, resulting in slower CD or oil transfer rates across the L/S/L interface [15]. Furthermore, the growth of these colloidal assemblies is also inhibited by steric repulsions.

Due to the virtually infinite number of systems that can be obtained by simple guest exchange, many applications have been considered. First, Pickering emulsions based on the extemporaneous formation and adsorption of insoluble CD/oil complexes at the water/oil interface are very interesting for performing catalytic reactions. One such system uses $[Na]_3[PW_{12}O_{40}]$ as a water-soluble catalyst to perform the oxidation of olefins, organosulfides, and alcohols in the presence of hydrogen peroxide as an oxidant [103]. These heptane-in-water Pickering catalytic emulsions are very efficient reaction media, e.g., the epoxidation of cyclooctene proceeds at a competitive rate (370 h^{-1}) with a good yield (>99% in 30 min) and a high selectivity (>99%). These good results can be attributed to the privileged interfacial contact between the substrate and the catalyst. Furthermore, phase separation is achieved simply by centrifugation or heating, and a catalytic system based on colloidal tectonics is highly compatible with some "green chemistry" concepts because these systems can be used without any organic hazardous solvents for liquid substrates [103]. Secondly, since biocidal phytochemicals (e.g., carvacrol and terpinen-4-ol) can be used as oil phases, the development of surfactant-free and silica-free phytochemical- and β-CD-based self-assembled Pickering emulsions was considered to potentiate the antimicrobial and antibiofilm activity of miconazoctylium bromide [102]. The results clearly show that the emulsion containing carvacrol and miconazoctylium bromide exhibits synergistic effects against fungi, additive responses against bacteria, and very high activity against methicillin-resistant *S. aureus* biofilms. In order to obtain fully bio-based antimicrobial Pickering emulsions, petroleum-based miconazoctylium bromide is replaced by undecylenic acid [105]. However, this castor oil derivative, already used as a bio-based drug to treat fungal infections, is less effective than petroleum-based drugs. The carvacrol-based emulsion is +390% and +165% more active against methicillin-resistant S. aureus compared to commercial undecylenic acid and azole emulsions. In addition, this emulsion is highly effective against *C. albicans* (up to +480% more potent than commercial undecylenic acid ointment). This eco-friendly emulsion also shows remarkable activity against *E. coli* and methicillin-resistant *S. aureus* biofilms.

In addition, CD-based Pickering emulsions allow the production of other derivative systems such as cyclodextrinosomes or beads (i.e., microcapsules) [106,107]. For instance, Mathapa and Paunov used oil-in-water Pickering emulsions (oil = *n*-tetradecane, tricaprylin, isopropyl myristate, sunflower, or silicone) stabilized by microrods and microplatelets of CD/oil inclusion complexes [106]. As these microcrystals remain irreversibly anchored at the droplet interface, these emulsions can serve as a model for the preparation of cyclodextrinosomes obtained solely by the assembly of CD/oil inclusion complexes, leading to a crystalline phase on the surface of the droplets that retains its stability after removal of the solvents. As the authors point out, cyclodextrinosomes can be used in cosmetics, personal and household care products, and pharmaceutical formulations. A very close system relies on the interactions between CDs and vegetable oils to produce CD-based beads [107]. By adding vegetable oil to an aqueous solution of CD, it is possible to obtain biphasic systems separated by an interfacial film consisting of triglyceride/CD inclusion complexes. By stirring, the authors obtained oil-in-water emulsions, leading to the crystallization of inclusion complexes. After a few days of continuous stirring, an aqueous suspension of CD-based beads is obtained [108]. X-ray diffraction studies reveal that the beads exhibit a crystalline organization, and microscopic analyses show that their internal structure consists of a matrix containing numerous oil compartments [107]. Indeed, unlike cyclodextrinosomes, beads are obtained after a few days of continuous agitation, with oil droplets interconnected via the fusion of crystalline zones, leading to droplets dispersed in a spherical crystalline matrix of CD/oil inclusion complexes [109–111]. Thus, regardless of their preparation, cyclodextrinosomes and beads refer to similar structures emerging from the packing of nanoparticles via a bottom-up construction.

If guest polymers are used, CD inclusion complexes form channels containing the guest polymers (see above). For instance, the formation of inclusion complexes between high-molecular-weight polyethylene glycol (PEG 20,000 or 35,000 g/mol) and native α-CDs, leading to the formation of supramolecular hydrogels after heating/cooling cycles. Indeed, the columnar CD domains (nanocrystallites), formed by nucleation and growth processes, act as physical cross-links alongside the non-included polymer chains (i.e., polypseudorotaxanes) [112]. However, crystallite growth is limited by the diffusion rate at which α-CDs are transferred from the aqueous bulk to the growing crystallites. After the addition of oil and stirring, these hydrogels allow the formation of oil-in-water Pickering emulsions. These emulsions are used to perform the biphasic catalytic hydroformylation of higher olefins. Under these conditions, the mass transfer is drastically enhanced in comparison to pure biphasic conditions or in the presence of α-CD or PEG alone due to the enhanced contact between the organic substrate and the rhodium catalyst associated with trisulfonated triphenylphosphine as a water-soluble ligand. The catalytic performance of these Pickering emulsions can be improved by adding randomly methylated β-CD (RAME-β-CD) [113]. Indeed, the catalytic activity is increased because this CD acts as a supramolecular carrier capable of transporting hydrophobic olefins. Unfortunately, an excess of RAME-β-CD leads to lower catalytic performance due to the instability of Pickering emulsions. Finally, to improve the thermal stability of the catalytic system, poloxamines (Tetronics®, a class of polyethylene oxide (PEO)/polypropylene oxide (PPO)-based amphiphiles linked to a central ethylene diamine moiety by way of the nitrogen atoms) were used instead of PEG [114]. The results reveal that the catalytic systems using reverse sequential Tetronic® 90R4 are more efficient than those containing conventional sequential Tetronic® 701 (Figure 14). As in reverse poloxamines (Tetronics® 90R4), the PPO segments are placed at the periphery, the PPO blocks act as "stoppers". Indeed, the α-CDs are kinetically trapped since the ends of the system are larger than the internal PEOm which prevents the dissociation (scrolling) of the α-CDs. The opposite is true for classical sequential poloxamines (Tetronic® 701), where the α-CDs can be displaced upon heating (see Figure 14). Consequently, α-CD/Tetronics® 90R4 nanocrystallites exhibit superior thermal stability compared to α-CD/Tetronic® 701. At 80 °C, the stability of these α-CD/Tetronics® 90R4-based emulsions is superior to that of α-CD/Tetronics® 701, leading to olefin conversion at competitive reaction rates due to enhanced mass transfer.

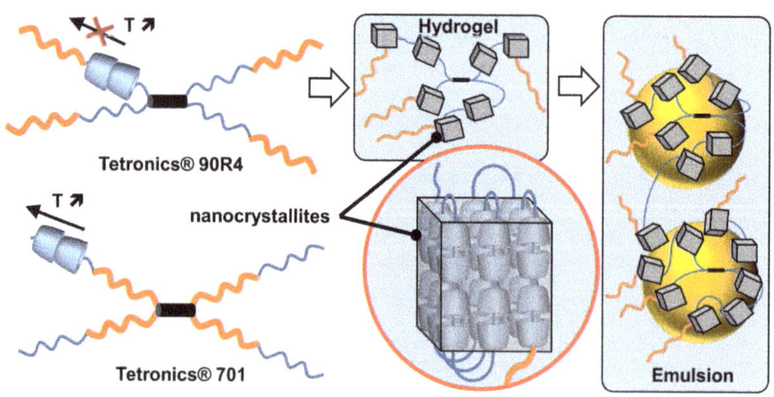

Figure 14. Schematic representation of Tetronics® 90R4 and 701 (orange = polypropylene oxide block and blue = polyethylene oxide block, black cylinder = central ethylene diamine), structure of α-CD/Tetronics® nanocrystallites and resulting Pickering emulsion.

The previously reported α-CD/PEG-based colloidal tectonics system can be used as a common base for formulation, leading to a range of dosage forms (i.e., hydrogels, Pickering emulsions, and cyclodextrinosomes) [16]. Stable ethanol-free hydrogels can be prepared by mixing, in water, α-CD, PEG, and antifungal drugs (miconazole, miconazole nitrate, or econazole nitrate) as inclusion complexes with 2-hydroxypropyl-β-CD to increase the solubility of the drugs in the hydrogels. As previously described, oil-in-water Pickering emulsions can be obtained from unloaded gels by adding paraffin oil containing antifungals in their native form. Once dried, the emulsions provide loaded cyclodextrinosomes. The properties are significantly improved compared to commercially available formulations in terms of stability, with even a marked improvement in antimicrobial activity (up to 1.6). These dosage forms avoid the use of petro-sourced surfactants or modified silica nanoparticles, providing solutions to the current trend of simplification of formulas in terms of ingredients.

If ionic guests are used, CD inclusion complexes preferentially form channels containing the guests (see above). When ionic guest molecules are surfactants, CDs disrupt the self-assembly properties of these last by sequestrating their hydrophobic regions. For instance, the complexation of β-CD with sodium dodecyl sulfate (SDS) leads to the formation of 1:1 and 2:1 inclusion complexes [115]. However, inclusion complexes, depending on the sample concentration and temperature, tend to organize into aggregates with an ordered structure. To mimic proteins that can readily assemble into rigid and crystalline structures such as viral capsids (see above), Jiang and coworkers reported the use of β-CD/SDS inclusion complexes (2:1 stoichiometry) capable of self-assembling into a variety of structures such as lamellar, spiral wound bilayers (multilamellar tubes), and hollow rhombic dodecahedral architectures (Figure 15) [116–118].

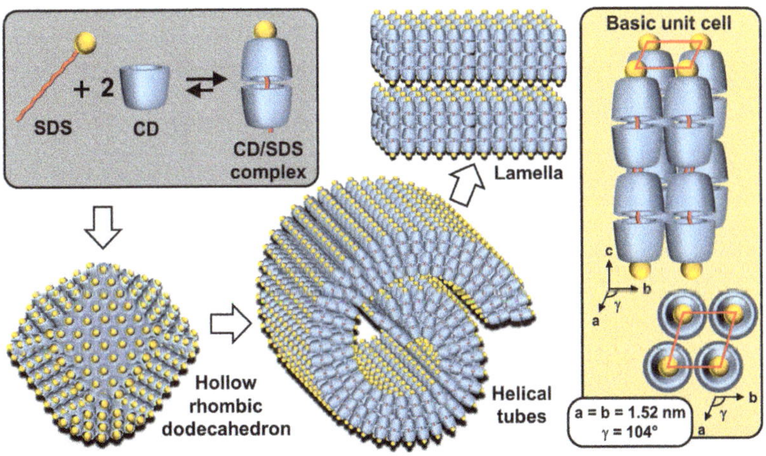

Figure 15. Schematic representation of self-assembled β-CD/SDS inclusion complexes (2:1 stoichiometry) into lamellar, helical tubular, and hollow rhombic dodecahedral architectures. The right inset shows the basic unit cell of columnar inclusion complexes in a 2D rhombic packing (quasi-monoclinic with the following parameters $a = b \neq c$ and $\alpha = \beta = 90°$ and $\gamma \neq 90°$).

The nature of architectures directly depends on the concentration of β-CD/SDS inclusion complexes, e.g., rhombic dodecahedra are obtained in the range of 4–6 wt. % while helical tubes and lamellae are observed, respectively, between 6–25 and 25–50 wt. %. In detail, the rhombic dodecahedral geometry (~1 μm) is hollow and has sharp edges as revealed by transmission electron microscope and atomic force microscope results. Unlike viral capsids where an icosahedral geometry is classically observed, the dodecahedral architecture is dominant in this case due to the 2D rhombic basic unit cell (see below and Figure 15).

Furthermore, it is worth noting that the size of the hollow rhombic dodecahedron is significantly larger than that of classical icosahedral viruses. Helical tubes consist of a bilayer wound with open ends. They are large (average length about 40 μm) and rigid. The diameter is monodisperse around 1 μm. The persistence length of the helical tubes is at least of the order of millimeters. The distance between the walls is large due to charge repulsion and depends on the condensation of cations. In addition, dilution causes swelling of the structure due to the penetration of water molecules into the interwall space. The bilayers are extremely rigid and planar due to their crystalline nature. The distance between two bilayers depends on the effective charge, with potential charge screening by counterions.

From a general point of view, the formation mechanism of these ordered architectures relies on CD/surfactant inclusion complexes as primitive building blocks. The self-assembly of these inclusion complexes is driven by strong direct and indirect (water-mediated) H-bonds, leading to channel-like arrangements, regardless of the size of the CD cavity. However, α-CD, β-CD, and γ-CD/surfactant form crystal structures with a hexagonal, rhombic, and square arrangement of channel-type inclusion complexes in the 2D lattice [119]. Therefore, β-CD/SDS inclusion complexes, with a 2:1 stoichiometry, form colloidal stable suspensions, presenting a crystalline substructure made from the primitive building blocks (i.e., the β-CD/SDS inclusion complexes). Since the basic unit consists of two inclusion complexes where the β-CDs are in a channel-type arrangement with a rhombic packing (see Figure 15), the basic unit grows into lamellae, spiral wound bilayers (multilamellar tubes), and/or folds into polyhedrons, depending on the concentration and temperature. This growth is driven by strong direct and indirect (water-mediated) H-bonds [116]. The importance of these H-bonds is confirmed by the fact that chemical substitutions on the hydroxyl groups of CDs and the use of chaotropic additives (disturbing the H-bonds) depress the assemblies. In addition, temperature is also a key parameter, as it weakens the H-bond network connecting the CDs, leading to structural changes in the observed structures. The bilayer-like assembly is determined by three factors: (i) the high rigidity of the membranes due to the H-bonds between CDs, (ii) the electrostatic repulsions between the faces due to the anionic nature of SDS, and (iii) the surface tension on the edges of the bilayer membrane that leads to the curvature of the edges and the formation of a closed 3D structure.

A clear similarity can be established between the colloidal architectures observed for β-CD/SDS inclusion complexes and protein self-assembly (see above). For example, capsid proteins are held together by specific directional interactions between the different protein units and by nonspecific, mainly electrostatic, interactions with the genetic material (see above). These supramolecular colloidal assemblies of β-CD and SDS exhibit characteristic sizes ranging from a few nanometers for a single primitive building block to several micrometers. Electrostatic interactions and H-bonds are identified as the driving forces for the self-assembly of β-CD/SDS inclusion complexes. The morphologies of the aggregates could be reversibly controlled using temperature or concentration changes.

In 2023, Liu et al. reported that polymer-like building blocks consisting of inclusion complexes between γ-CDs and *N,N′*-didodecyl-*N,N,N′,N′*-tetramethyl-*N,N′*-hexamethylenediamine surfactants exhibit various colloidal structures such as vesicles, nanotubes, and sheets in aqueous solutions with increasing concentrations (Figure 16) [120]. A single tetragonal unit with a 1:1 γ-CD/surfactant ratio is obtained by host–guest interactions (see Figure 16). This tetragonal unit enables the growth of bilayer sheets due to H-bonding and electrostatic interactions. However, the morphologies of the bilayer architectures are multi-responsive to stimuli such as temperature, solvent additives, and ions. Indeed, the decrease in H-bonds (i.e., increased in-plane fluidity) and the increase in out-of-plane electrostatic repulsions (i.e., increased repulsions between surfactant head groups) enable a transition from bilayers to nanotubes and then from tubes to spherical vesicles. As these transitions are reversible, this system has promising potential for drug delivery applications in the future. Due to the great interest in these systems, many studies have been conducted on them [121–124].

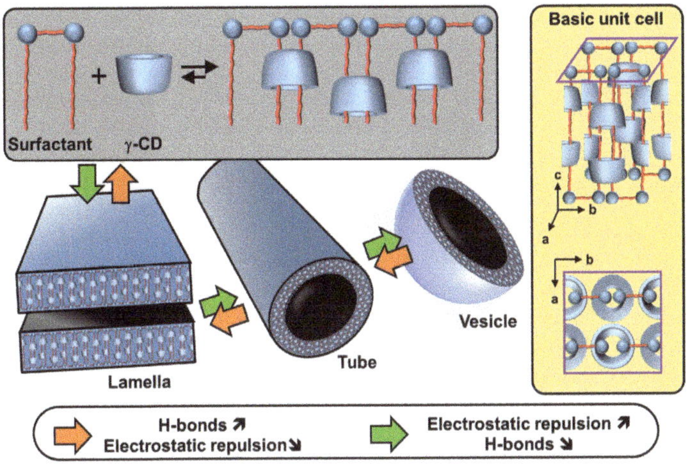

Figure 16. Schematic representation of self-assembled γ-CD/*N*,*N*′-didodecyl-*N*,*N*,*N*′,*N*′-tetramethyl-*N*,*N*′-hexamethylenediamines inclusion complexes into lamellar, tubular, and spherical vesicular architectures. The right inset shows the basic unit cell of columnar inclusion complexes in a tetragonal packing.

(*ii*) *Catanionic surfactants-based architectures*

Catanionic mixtures contain both anionic and cationic surfactants, where one ion acts as a counterion of the other. Under appropriate conditions, these catanionic mixtures have been shown to successfully mimic the morphologies of protein assemblies. These systems, which can self-assemble into regular hollow icosahedral shapes similar to those observed for viral capsid proteins (see above), have been extensively studied by Zemb and colleagues. The first report on close-packed icosahedra made from synthetic amphiphiles was published in 2001 [125]. The authors reported that salt-free mixtures of myristic acid and cetyltrimethylammonium hydroxide can self-assemble into hollow aggregates with a regular icosahedral shape for mole fractions between 0.5 and 0.75 and for a total surfactant volume fraction $<10^{-3}$ (Figure 17). However, some steps are required compared to previous systems. For instance, the solution of the two surfactants, where the anionic component (i.e., the myristic acid) is in excess, is first heated to 60 °C for a few minutes and then cooled to room temperature under constant stirring until micrometer-sized icosahedrons form in the fluid supernatant (phase separation leading to icosahedrons can be accelerated by gentle centrifugation at $3000\times g$).

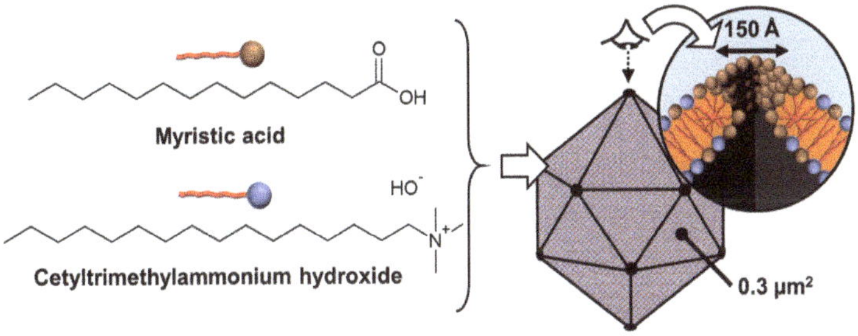

Figure 17. Schematic representation of self-assembled catanionic mixtures into icosahedral architectures made of about 10^6 ion pairs. The right inset shows one of the twelve pores produced by about 200 molecules due to the partial segregation of the anionic surfactant in excess.

Recalling that mild temperature cycling (i.e., heating to ensure redissolution and cooling to restore icosahedrons) and slow dissolution do not involve significant energy input other than the dispersion of colloidal objects, the authors argue that icosahedrons are maintained solely by molecular interactions. Indeed, these icosahedral aggregates are stabilized by the crystallization of surfactant alkyl chains. The chain melting transition is mole fraction-dependent and lies in the range of 50–65 °C according to differential scanning calorimetry. The icosahedral structures have a size of about 1 μm and a mass of ~1010 Da. As with β-CD/SDS systems, their sizes are larger than those of all known icosahedral viral capsids. Due to the excess of myristic acid, the electric charge of the icosahedrons is negative. In addition, pores, located at the vertices of the icosahedrons, are observed on freeze-fractured samples due to the excess of anionic surfactant molecules that cover the edges of the pores and form half-micelles (see Figure 17). Unfortunately, the icosahedrons are metastable, meaning that after several months, phase separation occurs. However, the addition of glycerol somewhat prolongs their stability. If the excess component is insufficient to form twelve pores per vesicle, this leads to the formation of nanodiscs or large perforated crystalline bilayers. In 2004, Zemb and his collaborators proposed a general mechanism explaining how the ratio of cationic and anionic surfactants controls the shape of the resulting crystallized colloidal structure [126]. The shapes of the aggregates are determined solely by the initial molar ratio. Indeed, molecular segregation occurs, leading to accumulation on the edges or pores of excess surfactant instead of being incorporated into the crystalline bilayers. As these icosahedral structures combine wall rigidity and vertex holes, they may be useful for the controlled release of drugs or DNA with probable functional superiority over soft vesicles obtained from classical surfactant systems.

Non-spherical structures can be generated from the self-assembly of unequal charged anionic and cationic amphiphiles, leading to small curved vesicles presenting flat ionic domains on the vesicle surface due to the interplay of electrostatic, hydrophobic, and steric forces. For instance, Greenfield et al. utilized the strong electrostatic interaction between the head groups of trivalent cationic and monovalent anionic surfactants to increase the cohesive energy of amphiphiles and promote the formation of faceted vesicles [127]. Faceted vesicles can also be obtained by in situ deprotonation of a surfactant. For instance, the mixture of perfluorononanoic acid (PFNA) or perfluorodecanoic acid (PFDA) with NaOH leads to polyhedral vesicles formed by PFNA/PFN^- or PFDA/PFD^- surfactants [128]. The PFDA/NaOH system is pH-sensitive as revealed by freeze fracture transmission electron microscopy images where a transition from the faceted vesicle phase to the sponge phase (i.e., L_3 phase) is observed. It is noteworthy that this transition is not observed for the PFDA/NaOH system. The mechanism leading to the formation of faceted vesicles and the mechanism of transition from vesicles to perforated lamellae and faceted vesicles have been widely discussed and modeled in the literature [129–131].

3.2.2. Inorganic–Organic Hybrid Architectures

The fascinating structures obtained using the tectonic colloidal approach can be used to obtain supramolecular hybrid functional architectures. In 2012, the colloidal engineering of novel hybrid nanoparticles was reported [132]. This system results from the ionic metathesis between negatively charged polyoxometalates (POMs) and cationic surfactants. These two "hydrophilic" and "hydrophobic" tectons lead to the formation of uncharged clusters. Indeed, in an aqueous solution, the neutralization of $H_3PW_{12}O_{40}$ (1 equiv) with dodecyltrimethylammonium hydroxide, $[C_{12}][OH]$ (3 equiv) leads to the formation of hybrid clusters $[C_{12}]_3[PW_{12}O_{40}]$. In order to decrease the hydrophobic/water contact, the clusters self-assemble to spontaneously form nanoparticles. Transmission electron microscopy and dynamic light scattering experiments reveal that the nanoparticles are monodisperse with a diameter of approximately 35 nm. The internal structure of the $[C_{12}]_3[PW_{12}O_{40}]$ nanoparticles, studied by small-angle X-ray scattering measurements, reveals a lamellar arrangement. Indeed, these nanoparticles consist of parallel inorganic planes of $[PW_{12}O_{40}]$ anions separated at a mesoscopic scale by organic bilayers of interdigitated cationic surfactant chains (Figure 18).

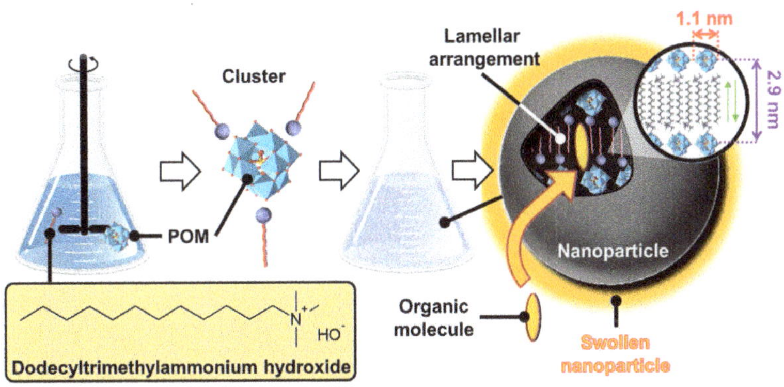

Figure 18. Schematic representation of self-assembled mixtures into inorganic–organic hybrid nanoparticle architectures made of $H_3PW_{12}O_{40}$ (1 equiv) and dodecyltrimethylammonium hydroxide (3 equiv). The lamellar internal arrangement allows the incorporation of small organic molecules, leading to swollen nanoparticles.

It should be noted that this arrangement is highly predictable as it is already observed in the literature in the solid state. For instance, surfactant/inorganic phases containing the adamantane thiogermanate anion $[Ge_4S_{10}]^{4-}$, in combination with a stoichiometric amount of alkyltrimethylammonium cations, crystallize in the triclinic space group *P*-1 and contain parallel planes of $[Ge_4S_{10}]^{4-}$, separated by interdigitated cationic surfactant bilayers [133]. As mentioned before, this self-assembly is governed by the nucleation and growth processes. However, during growth, some defects appear in the organization, leading to the appearance of charges [134]. The size or shape of the nanoparticles depends on the choice of POM. Indeed, various colloidal structures such as spherical, pseudospherical, and tubular structures can be obtained for $[PW_{12}O_{40}]^{3-}$, $[PW_{11}VO_{40}]^{4-}$, and $[PW_9V_3O_{40}]^{6-}$, respectively [135]. From TEM images, the size is estimated to be about 35, 60, and 392 nm for $[C_{12}]_3[PW_{12}O_{40}]$, $[C_{12}]_4[PW_{11}VO_{40}]$, and $[C_{12}]_6[PW_9V_3O_{40}]$, respectively [135]. Interestingly, however, internal lamellar packing is observed regardless of the size or shape of the POMs. Indeed, the lamellar organization results from the combined effects of hydrophobic, van der Waals, electrostatic, and steric interactions that allow the tectons to organize [135]. Like casein micelles, nanoparticles are porous due to their internal lamellar packing in which small organic molecules can be accommodated, leading to swollen particles (see Figure 18) [134]. Therefore, in the presence of water and aromatic oil, these nanoparticles can be used to stabilize water-in-oil Pickering emulsions consisting of shell-like architectures around the water droplets. These emulsions exhibit very high stability due to the strong cohesion between the particles located in the interfacial layer. Moreover, the nanoparticles are observed to interlock, which increases the interfacial elasticity due to the penetration of oil molecules into the nanoparticles [134]. Consequently, these emulsions offer a general route for the construction of colloidosomes [134].

As these Pickering emulsions, stabilized by inorganic–organic nanoparticles, exhibit high stability, a very large water/oil interface, and potential catalytic activity, their oxidation capabilities involving hydrogen peroxide as an oxidant were studied. Moreover, it is noted that although the emulsions exhibit long-term stability, the droplets break up and the final coalescence can be induced on demand by gentle centrifugation at $900\times\ g$, leading to three well-separated phases: water, oil, and hybrid nanoparticles. Two catalytic systems were used (Figure 19). The first, published in 2012, uses the $[C_{12}]_3[PW_{12}O_{40}]$ hybrid nanoparticles to perform the epoxidation of olefins in a biphasic water/toluene system in the presence of hydrogen peroxide as an oxidant [132]. As expected, the quantitative epoxidation of various olefins (oct-1-ene, cyclohexene, cyclooctene, and limonene) is achieved with easy separation of the product and catalyst. For instance, the epoxidation of cyclooctene proceeds at a competitive rate (32.3 h^{-1}), good yield (98% in 3 h), and high selectivity (>99%). This highly efficient reaction medium is due to the larger surface area of the water/oil interface where the catalytic nanoparticles are located. It is relevant to note that the catalytic performances are obtained under mild conditions (65 °C) and without stirring, indicating that the process is not limited by mass transfer. In other words, the reaction is solely driven by the catalytic cycle. The second catalytic system, published in 2014, uses $[C_{12}]_3[PW_{12}O_{40}]$ nanoparticles to perform the epoxidation of olefins directly and only in eco-compatible solvents (e.g., cyclopentyl methyl ether, 2-methyl tetrahydrofuran, methyl acetate, and glycerol triacetate) [136]. Indeed, hybrid particles are able to form stable dispersions due to their amphiphilic properties. As for the previous system, cyclooctene epoxidation proceeds at competitive rates (initial turnover frequency, $TOF_0 > 260\ h^{-1}$), with good yield (>95%) and high selectivity (>99%). It is worth noting that $[C_{12}]_3[PW_{12}O_{40}]$ nanoparticles exhibit a catalytic activity 10 times higher than that obtained with the native catalyst ($[Na]_3[PW_{12}O_{40}]$). This effect is attributed to the stable dispersion and the accommodation of cyclooctene molecules inside the porous nanoparticles. These two catalytic systems combine the advantages of heterogeneous and homogeneous catalysis, namely high activity and selectivity, easy phase separation, and catalyst reuse (after filtration and/or distillation).

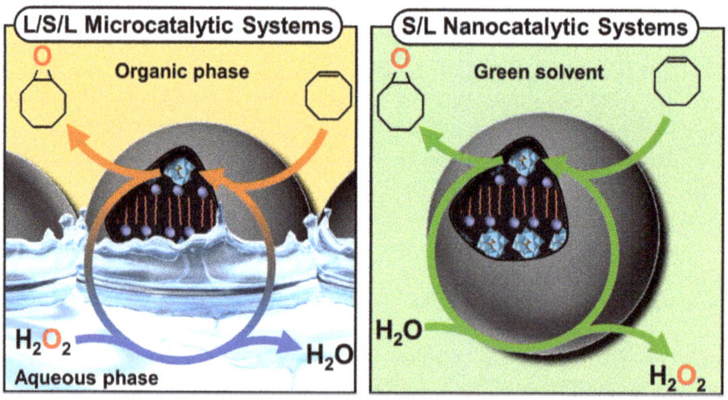

Figure 19. Schematic representation of the two catalytic systems using inorganic–organic hybrid nanoparticle architectures: Pickering emulsions (**left**) and dispersion (**right**).

Another inorganic–organic hybrid system can be easily obtained by reusing the self-assembled β-CD/guest inclusion complexes in nanoparticle architectures (see above). Indeed, POM anions have a strong propensity to adsorb on electrically neutral surfaces due to their large, sticky, polarizable, and salting-in nature. For instance, Stoddart and co-workers demonstrated that complexation between CDs and POMs leads to the formation of inorganic/organic hybrid sandwich complexes where a POM anion is encapsulated by the primary faces of two CDs [137]. Similar observations were made by Bauduin et al., who reported that the adsorption of POMs on the surface of nonionic micelles was mainly due to the entropy gain caused by the release of several molecules of water of hydration in the bulk [138,139]. Indeed, the adsorption of POMs on the surface of micelles leads to a partial dehydration of the polar heads of surfactants and POMs. Similar effects can be invoked with CD particles (see above). Based on this behavior, Pacaud and his collaborators construct core–shell hybrid nanoparticles or nanofibers in a hierarchical manner by the sequential self-assembly of three complementary tectons: β-CD, 1-decanol, and POM (Figure 20) [140]. The following mechanism occurs: (i) β-CD/1-decanol inclusion complexes are formed, (ii) β-CD/1-decanol inclusion complexes self-assemble to yield nanoparticles, and (iii) the adsorption of POMs onto the polar neutral interface of CDs allows the formation of spherical core–shell hybrid nanoparticles (~40 nm) and nanofibers (15–25 nm in width and 7.5 μm in length).

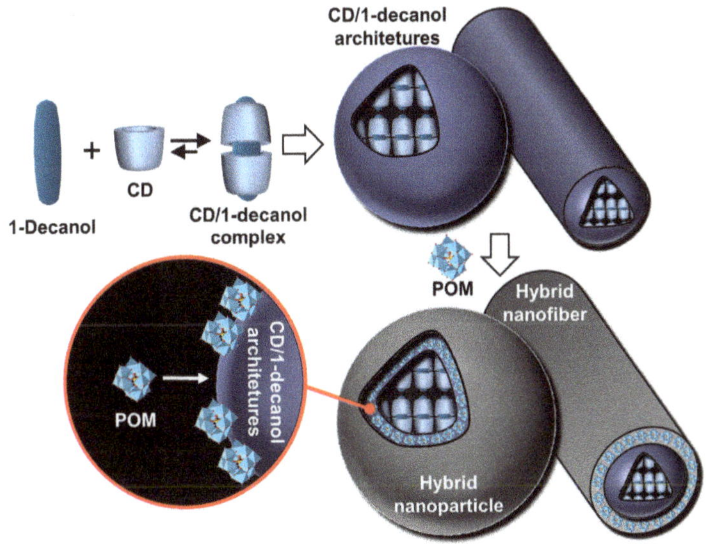

Figure 20. Schematic representation of the sequential synthesis of hybrid architectures by self-assembly of CD and 1-decanol followed by addition of POMs.

While it is possible to obtain hybrid nanoparticles using molecular tectons as in the two previous examples, it is also possible to directly influence the assembly of the particles. In this context, Yang and his collaborators used the colloidal tectonics approach to achieve tandem synergistic Pickering interfacial catalysis. This catalytic system is based on the use of two "surface-active" nanoparticles: $[C_{12}]_3[PW_{12}O_{40}]$ (see above) and silica nanoparticles functionalized by alkyl chain groups and sulfonic acids (C_n-SiO_2-SO_3H, see Figure 21).

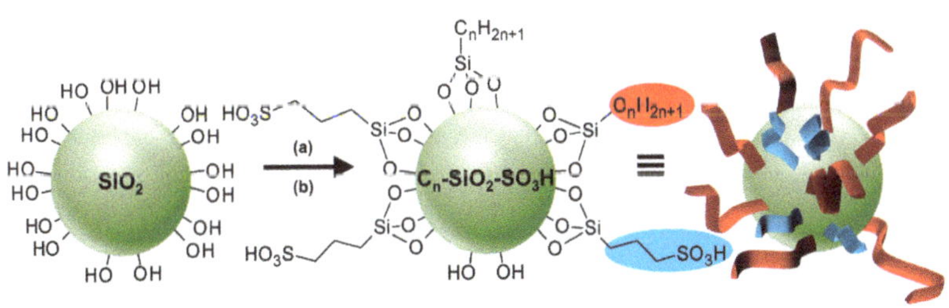

Figure 21. Synthesis of acidic/amphiphilic silica nanoparticles C_n-SiO_2-SO_3H (n = 3, 8 or 18) and their schematic representation: (a) 1 g of Aerosil® 200, 4 mmol of alkyltrimethoxysilane, 16 mmol of (3-mercaptopropyl)trimethoxysilane, H_2O/EtOH pH 9.6, reflux, 24 h; (b) 60 mL H_2O_2 (50%), CH_3CN, 40 °C, 24 h.

These nanoparticles, containing both recognition and catalytic sites, are used to control the formation and properties of Pickering emulsions, such as stability [141]. Due to the penetration of the alkyl chains of C_n-SiO_2-SO_3H nanoparticles into the self-assembled hybrid, in $[C_{12}]_3[PW_{12}O_{40}]$ nanoparticles that possess a porous structure composed of hydrophilic and lipophilic regions (see the internal arrangement of the particle, presented above), an interfacial self-assembly occurred, leading to the interlocking of the two nanoparticles. In a biphasic water/oil system, the two nanoparticles lead to elastic "springs" in the interfacial layer, giving rise to water-in-oil Pickering emulsions with long-term stabilities. These emulsions are used for the synthesis of adipic acid from the one-pot oxidative cleavage of epoxycyclohexane in the presence of aqueous hydrogen peroxide (Figure 22). It should be noted that, from an industrial point of view, adipic acid is the most important dicarboxylic acid (~2.5 billion kg/year) due to its use as a precursor for the synthesis of nylon. As expected, the catalytic performances were significantly enhanced due to the interfacial co-adsorption of the two nanoparticles catalyzing the hydrolysis and oxidation steps involved in this oxidative cleavage. For instance, under optimal conditions (80 °C, 500 rpm), the oxidative cleavage of epoxycyclohexane reached almost complete conversion (>99% after 12 h) and high selectivity (94%). To extend the scope of this system, the authors also studied the oxidative cleavage of cycloheptene, cyclooctene, 1-methylcyclohexene, and 4-methylcyclohexene oxides under the same reaction conditions. For all these substrates, high conversions and excellent selectivities are obtained. As with the previous Pickering catalytic system, the advantages of homogeneous (high activity and selectivity) and heterogeneous (simple phase separation and catalyst reuse) catalysis are combined. Moreover, the advantages of the colloidal tectonic approach, such as the flexibility and versatility of self-assembled systems, are also highlighted.

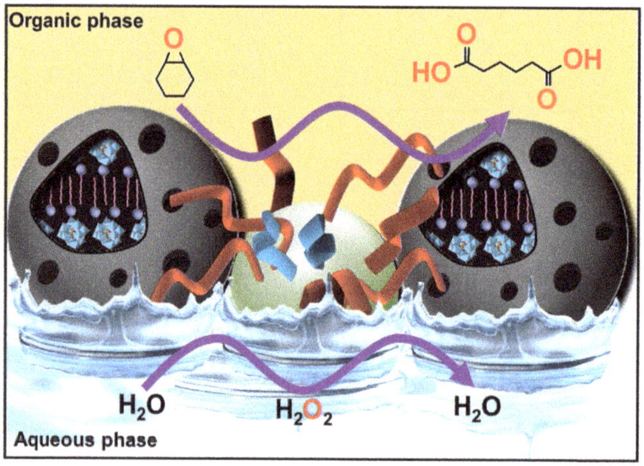

Figure 22. Schematic representation of oxidative cleavage of epoxycyclohexane in a water-in-toluene Pickering emulsion stabilized by $[C_{12}]_3[PW_{12}O_{40}]$ and C_n-SiO_2-SO_3H catalytic nanoparticles.

Using a similar approach, Feng et al. used a combination of plasmonic Au-loaded amphiphilic trimethoxy(propyl)- and aminopropyl-functionalized silica (Au/SiO_2-C_3) and $[C_{12}]_3[PW_{12}O_{40}]$ nanoparticles for the preparation of water-in-oil Pickering emulsions [142]. The Au/SiO_2-C_3 and $[C_{12}]_3[PW_{12}O_{40}]$ nanoparticles, adsorbed at the water/oil interface, act as an on-site photo-assisted heater/plasmon activator and a catalyst, respectively (Figure 23). After studying the physicochemical properties, stability, and interfacial plasmonic properties of the water-in-toluene emulsions stabilized by the two nanoparticles, the authors reported the catalytic performances of the system for the oxidation of cyclooctene using hydrogen peroxide as oxidant under UV irradiation. The results show that the turnover frequency (TOF) is 188 h^{-1} under light, while it is only 88 h^{-1} for the control experiment (i.e., $[C_{12}]_3[PW_{12}O_{40}]$ alone). Moreover, the epoxycyclohexane yield is almost 100% after 75 min under light (1000 mW cm^{-2}), but it is only 63% with conventional heating at 47 °C, corresponding to the temperature reached under light irradiation. The reaction proceeds without a loss of activity and selectivity after five consecutive recycles. Finally, the authors extended the scope of this system to a panel of olefins of industrial interest such as cyclooctene, (*R*)-limonene, and 1-octene. In each case, the results obtained under light irradiation are higher in terms of conversion and selectivity than after heating at 47 °C.

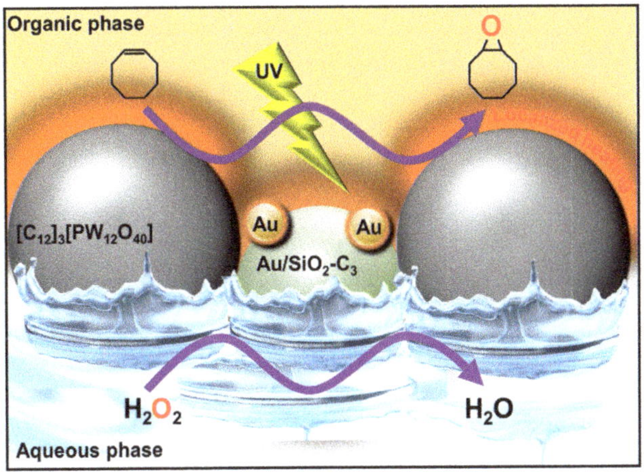

Figure 23. Schematic representation of epoxidation of cyclooctene in a water-in-toluene Pickering emulsion stabilized by $[C_{12}]_3[PW_{12}O_{40}]$ and Au/SiO_2-C_3 nanoparticles acting, respectively, as catalyst and as on-site heater/plasmon activators under UV irradiation.

4. Conclusions and Perspectives

This manuscript provides an overview of the emerging concept of colloidal tectonics illustrated by some biotic and abiotic (supra) colloidal systems and their applications in various fields. From a historical point of view, the formation of supramolecular colloidal structures using complementary tectons (molecular building blocks), called colloidal tectonics, appeared in 2018, but a few isolated examples without generalization have been observed for two decades [15]. Following this work, new research in the field has progressed at a rapid pace. From a theoretical point of view, the phenomenon of colloidal tectonics is a limiting case between the assembly of small amphiphilic molecules or polymers into soft and fluidic micellar, vesicular, or lamellar structures, mainly due to hydrophobic interactions, and the crystalline state characterized by a regular and time-invariant three-dimensional periodic arrangement of molecules in space. Indeed, crystallization depends on molecular interactions in a simple way: highly heterogeneous interactions between molecules promote crystallization. Between these two worlds, a new paradigm is possible, as revealed by the assembly of proteins into rigid crystalline structures driven by a combination of H-bonds and hydrophobic and electrostatic interactions, producing structures such as lamellae, tubules, and polyhedrons.

For the past two decades, synthetic molecules have been used to reproduce these astonishing supramolecular assemblies that can range from the nanometer to the micrometer scale (i.e., from discrete colloidal to supracolloidal systems). The concept of colloidal tectonics, defined as the art and science of the supramolecular formation of (supra) colloidal structures using the packing of complementary molecular building blocks (i.e., tectons), creates an exciting environment for research at the intersection of molecular, colloidal, and crystal sciences. Interestingly, the colloidal tectonic approach can be compared to the colloidal building block approach. Indeed, these two terms are different but, at the same time, related in some way and may have an interdependent relationship. While the colloidal tectonic approach uses the packing of complementary molecular building blocks (tectons) to form supramolecular colloidal structures, the colloidal building block approach uses the packing of particles to obtain superstructures. The aggregates of colloidal building block particles, which can be defined as specific geometric structures obtained under the effect of attractive forces and/or external environmental effects, leading to an optimal packing of building block particles, are called colloidal clusters (or colloidal molecules) [143–146]. Therefore, the major difference between these two approaches is related to the nature of the building blocks: molecules for colloidal tectonics and particles for the colloidal building block approach. Therefore, colloidal tectonics involves using molecules (called tectons) with or without colloidal properties to provide a colloidal structure solely by self-assembly. However, these self-assembled particles can be used to form supracolloidal structures, where they act as colloidal building blocks (see above).

The major advantage of the colloidal tectonic approach is that the preparation of the systems uses bottom-up self-assembly processes. However, the self-assembled structures must be stable in solvents for possible applications. Indeed, tecton–tecton interactions (or particle–particle for supracolloidal structures) must be higher than tecton–solvent interactions. Therefore, finding the right tectons is not so easy. However, in light of the scientific data illustrated in this review of the recent literature, the following empirical guidelines can be formulated: (i) colloidal architectures can be generated using either a self-complementary tecton (single-component system) or two or more complementary tectons (multi-component system); (ii) the existence of solvophobic interactions allows molecules to cluster and repel solvent molecules (e.g., for hydrophobic interactions, nonpolar molecules are held together by weak forces, such as van der Waals forces, exerted by molecular surfaces); (iii) high structural rigidity and/or well-defined crystallinity are required via strong and/or directional attraction (H-bonds, electrostatic interactions, etc.), leading to the growth of the assemblies; and (iv) steric and/or electrostatic repulsions are needed to inhibit growth at the colloidal scale. Thus, tectons can readily assemble into rigid and crystalline (or pseudo-crystalline) structures driven by a combination of hydrophobic, H-bonding, and electrostatic interactions, whereas the assembly of surfactants forms dynamic, soft, and fluidic objects mainly driven by hydrophobic interactions (Figure 24).

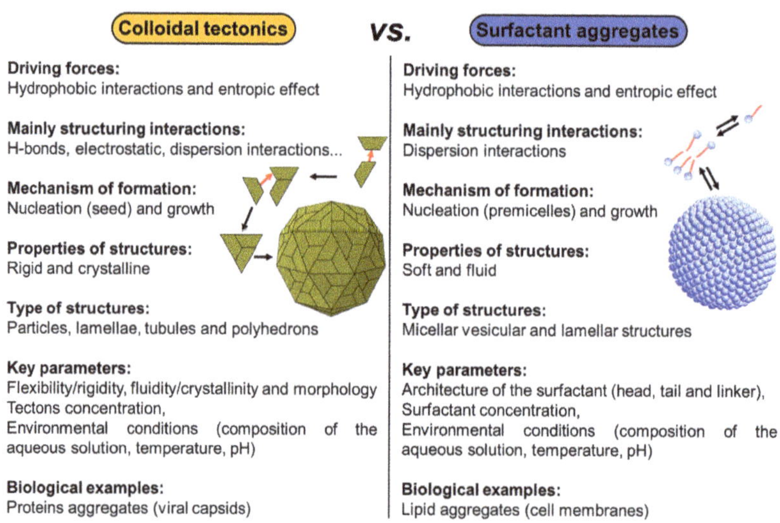

Figure 24. Self-assembled colloids from the colloidal tectonics approach versus surfactant aggregates.

These two opposing behaviors are due to the existence of complementary interactions (i.e., H-bonds and/or electrostatic interactions). The self-assembled structures obtained from the colloidal tectonics approach resemble the assembly of proteins such as viral capsids as opposed to the traditional self-assembly of amphiphilic molecules that are fluid and flexible. Consequently, colloidal tectonics opens the way to new protein-mimetic materials. However, while surfactant morphology is a key parameter to guide surfactant assembly (lipid mimicry), the control of intermolecular interactions, structural flexibility/rigidity, and fluidity/crystallinity are essential to successfully mimic the morphologies of protein assemblies. On the other hand, it is commonly accepted that micellization is a two-step process: (i) surfactant molecules in the solution start to associate into small aggregates at a concentration below the CMC, and (ii) they grow to form larger aggregates (e.g., micelles at the CMC) as the concentration increases. In a similar way, self-assembly by colloidal tectonics can be interpreted in terms of nucleation/growth: (i) the subunits begin by forming a seed under the action of a local concentration fluctuation (nucleation), and (ii) other subunits that remain free then bind sequentially to this seed to make it grow rapidly until the formation of the aggregate. However, the experimental monitoring of the assembly processes of these objects remains a challenging task, as it requires probing both nanometric spatial scales and time scales ranging from microseconds to hours. Investigations are therefore needed to explain the mechanisms. For example, static light scattering, time-resolved X-ray scattering, and numerical computation techniques (Monte Carlo and molecular dynamics simulations) will undoubtedly contribute to the understanding of the mechanisms at the molecular scale by providing the structure of the most stable transient species without presupposing their nature.

The common goal is to control molecular self-assembly to produce colloidal systems with predetermined properties and/or functions in order to apply them in various fields. Current or planned applications cover catalysis, the controlled release of drugs or nucleic acids, and cosmetics. However, other applications are possible, such as obtaining molecular machines, artificial viruses for medicine, intelligent and environmentally adaptable objects, etc. Indeed, the colloidal tectonics approach can be used for the development of new smart materials because large structures are easily accessible using a bottom-up approach as they are composed of small molecules, requiring fewer synthesis steps. This approach is therefore highly multi- and interdisciplinary because the scope of this research goes far beyond the traditional boundaries of chemistry. Furthermore, it should be noted that the colloidal tectonics approach is an environmentally friendly process since it uses "green" syntheses by supramolecular assemblies (e.g., energy efficiency, minimization of synthesis steps, minimization of hazardous solvents, and possible use of bio-sourced materials). However, from an economic point of view, the cost and economic viability of the colloidal tectonic approach depend on the system (nature of the tectons, their prices, and the required quantity) and also on the intended applications. For example, the use of native CDs as tectons seems to be very interesting because they come from renewable bio-based raw materials and are also highly biocompatible, environmentally friendly, and not harmful to human health. However, considering the cost, even if the industrial production of native CDs is a relatively low-cost production, the formation of stable emulsions by colloidal tectonics (see above) requires a non-negligible amount of CDs (10 wt. %), which remains a significant limiting factor compared to current formulations stabilized by molecular surfactants. However, the value of a product depends on its composition, formulation, and application, i.e., the finished products show considerable variability in their value. Indeed, high-value-added products are intended for personal care and pharmaceuticals. Medium-value products are intended for the food and animal, chemical, and materials industries, while low-value-added products are available in larger quantities and are used in the energy and bioremediation sectors. Since the three sectors where producers can obtain higher market prices are cosmetics, healthcare, and food additives, the use of the colloidal tectonic approach is currently limited to high- and medium-value products. However, the growing demand for environmentally friendly products and processes could change this.

With the number of supramolecular colloidal systems being limited only by our imagination, colloidal tectonics, through its power over the expression of matter, invites creativity and innovation to solve current and future scientific challenges. Chemistry has expanded from molecular chemistry to supramolecular chemistry and is now evolving towards adaptive chemistry through constitutional dynamic chemistry. Indeed, the adaptation of colloidal assemblies, in response to external agents, switching processes, and morphological or shape changes, opens the way to intelligent and evolving systems with multiple potential applications. From an academic and industrial point of view, colloidal tectonics constitutes an almost infinite playground for the systems of tomorrow, thus contributing to limitless inventiveness. Like a construction game for children, the motto of colloidal tectonics is without a doubt Build and Play!

Funding: This research received no external funding.

Acknowledgments: This review is dedicated to Michel Simard (retired from the University of Montreal, Canada) to honor his contributions to chemistry, including his participation in the development of molecular tectonics with James D. Wuest (University of Montreal, Canada), and for teaching me X-ray diffraction for crystal structure determination and for the numerous discussions on crystallization processes.

Conflicts of Interest: The author declares no conflicts of interest.

References

1. Wedlich-Söldner, R.; Betz, T. Self-organization: The fundament of cell biology. *Philos. Trans. R. Soc. Lond. B Biol. Sci.* **2018**, *373*, 20170103. [CrossRef]
2. Ruiz-Mirazo, K.; Briones, C.; de la Escosura, A. Prebiotic systems chemistry: New perspectives for the origins of life. *Chem. Rev.* **2014**, *114*, 285–366. [CrossRef] [PubMed]
3. Hamley, I.W. Nanotechnology with soft materials. *Angew. Chem. Int. Ed.* **2003**, *42*, 1692–1712. [CrossRef] [PubMed]
4. Sanger, F.; Tuppy, H. The amino-acid sequence in the phenylalanyl chain of insulin. I. The identification of lower peptides from partial hydrolysates. *Biochem. J.* **1951**, *49*, 463–481. [CrossRef] [PubMed]
5. Pauling, L.; Corey, R.B.; Branson, H.R. The structure of proteins; two hydrogen-bonded helical configurations of the polypeptide chain. *Proc. Natl. Acad. Sci. USA* **1951**, *37*, 205–211. [CrossRef] [PubMed]
6. Wuyun, Q.; Chen, Y.; Shen, Y.; Cao, Y.; Hu, G.; Cui, W.; Gao, J.; Zheng, W. Recent progress of protein tertiary structure prediction. *Molecules* **2024**, *29*, 832. [CrossRef]
7. Marciano, S.; Dey, D.; Listov, D.; Fleishman, S.J.; Sonn-Segev, A.; Mertens, H.; Busch, F.; Kim, Y.; Harvey, S.R.; Wysockid, V.H.; et al. Protein quaternary structures in solution are a mixture of multiple forms. *Chem. Sci.* **2022**, *13*, 11680–11695. [CrossRef]
8. Cohen, R.D.; Pielak, G.J. Electrostatic Contributions to Protein Quinary Structure. *J. Am. Chem. Soc.* **2016**, *138*, 13139–13142. [CrossRef]
9. Service, R.F. How Far Can We Push Chemical Self-Assembly? *Science* **2005**, *309*, 95. [CrossRef]
10. Mattia, E.; Otto, S. Supramolecular systems chemistry. *Nat. Nanotechnol.* **2015**, *10*, 111–119. [CrossRef]
11. Wang, H.; Feng, Z.; Xu, B. Bioinspired assembly of small molecules in cell milieu. *Chem. Soc. Rev.* **2017**, *46*, 2421–2436. [CrossRef] [PubMed]
12. Simard, M.; Su, D.; Wuest, J.D. Use of hydrogen bonds to control molecular aggregation. Self-assembly of three-dimensional networks with large chambers. *J. Am. Chem. Soc.* **1991**, *113*, 4696–4698. [CrossRef]
13. Su, D.; Wang, X.; Simard, M.; Wuest, J.D. Molecular tectonics. *Supramol. Chem.* **1995**, *6*, 171–178. [CrossRef]
14. Leclercq, L.; Schmitzer, A.R. The liquid crystal state: An intermediate state to obtain crystal packing. *J. Mol. Liq.* **2014**, *200*, 283–288. [CrossRef]
15. Leclercq, L. Get beyond limits: From colloidal tectonics concept to the engineering of eco-friendly catalytic systems. *Front. Chem.* **2018**, *6*, 168. [CrossRef]
16. Leclercq, L.; Dechézelles, J.-F.; Rauwel, G.; Nardello-Rataj, V. In vitro study of versatile drug formulations based on α-cyclodextrin and polyethylene glycol using colloidal tectonics. *J. Drug Deliv. Sci. Technol.* **2020**, *59*, 101913. [CrossRef]
17. Chai, C.; Oh, S.; Imm, J.-Y. Roles of milk fat globule membrane on fat digestion and infant nutrition. *Food Sci. Anim. Resour.* **2022**, *42*, 351–371. [CrossRef]
18. Ballard, O.; Morrow, A.L. Human milk composition: Nutrients and bioactive factors. *Pediatr. Clin.* **2013**, *60*, 49–74. [CrossRef]
19. Sadiq, U.; Gill, H.; Chandrapala, J. Casein micelles as an emerging delivery system for bioactive food components. *Foods* **2021**, *10*, 1965. [CrossRef]
20. Gorissen, S.H.M.; Crombag, J.J.R.; Senden, J.M.G.; Huub Waterval, W.A.; Bierau, J.; Verdijk, L.B.; van Loon, L.J.C. Protein content and amino acid composition of commercially available plant-based protein isolates. *Amino Acids* **2018**, *50*, 1685–1695. [CrossRef]
21. Uversky, V.N.; Dunker, A.K. Understanding Protein Non-Folding. *Biochim. Biophys. Acta* **2010**, *1804*, 1231–1264. [CrossRef] [PubMed]
22. McMahon, D.J.; Oommen, B. Supramolecular Structure of the Casein Micelle. *J. Dairy Sci.* **2008**, *91*, 1709–1721. [CrossRef] [PubMed]
23. Perinelli, D.R.; Cespi, M.; Lorusso, N.; Palmieri, G.P.; Bonacucina, G.; Blasi, P. Surfactant self-assembling and critical micelle concentration: One approach fits all? *Langmuir* **2020**, *36*, 5745–5753. [CrossRef]
24. Huppertz, T.; Gazi, I.; Luyten, H.; Nieuwenhuijse, H.; Alting, A.; Schokker, E. Hydration of casein micelles and caseinates: Implications for casein micelle structure. *Int. Dairy J.* **2017**, *74*, 1–11. [CrossRef]

25. Kamigaki, T.; Ito, Y.; Nishino, Y.; Miyazawa, A. Microstructural observation of casein micelles in milk by cryo-electron microscopy of vitreous sections (CEMOVIS). *Microscopy* **2018**, *67*, 164–170. [CrossRef]
26. Møller, T.L.; Nielsen, S.B.; Pedersen, J.S.; Corredig, M. Structural and compositional characterization of Ca- and β-casein enriched casein micelles. *Food Hydrocoll.* **2024**, *151*, 109811. [CrossRef]
27. Noble, R.W., Jr.; Waugh, D.F. Casein micelles. Formation and structure. I. *J. Am. Chem. Soc.* **1965**, *87*, 2236–2245. [CrossRef]
28. Noble, R.W., Jr.; Waugh, D.F. Casein micelles. Formation and structure. II. *J. Am. Chem. Soc.* **1965**, *87*, 2246–2257. [CrossRef]
29. Payens, T.A.J.; Schmidt, D.G. Boundary spreading of rapidly polymerizing α_{S1}-casein B and C during sedimentation: Numerical solutions of the Lamm-Gilbert-Fujita equation. *Arch. Biochem. Biophys.* **1966**, *115*, 136–145. [CrossRef]
30. Slattery, C.W.; Evard, R. A model for the formation and structure of casein micelles from subunits of variable composition. *Biochim. Biophys. Acta* **1973**, *317*, 529–538. [CrossRef]
31. Schmidt, D.G. Colloidal aspects of casein. *Neth. Milk Dairy J.* **1980**, *34*, 42–64.
32. Walstra, P. On the stability of casein micelles. *J. Dairy Sci.* **1990**, *73*, 1965–1979. [CrossRef]
33. Walstra, P. Casein sub-micelles: Do they exist? *Int. Dairy J.* **1999**, *9*, 189–192. [CrossRef]
34. Dalgleish, D.G.; Corredig, M. The structure of the casein micelle of milk and its changes during processing. *Annu. Rev. Food Sci. Technol.* **2012**, *3*, 449–467. [CrossRef]
35. Horne, D.S. Casein micelle structure: Models and muddles. *Curr. Opin. Colloid Interface Sci.* **2006**, *11*, 148–153. [CrossRef]
36. Casein, D.S. Interactions: Casting light on the black boxes, the structure in dairy products. *Int. Dairy J.* **1998**, *8*, 171–177. [CrossRef]
37. Fox, P.F.; Brodkorb, A. The casein micelle: Historical aspects, current concepts and significance. *Int. Dairy J.* **2008**, *18*, 677–684. [CrossRef]
38. Bouchoux, A.; Gésan-Guiziou, G.; Pérez, J.; Cabane, B. How to squeeze a sponge: Casein micelles under osmotic stress, a SAXS study. *Biophys. J.* **2010**, *99*, 3754–3762. [CrossRef]
39. Fan, Z.; Fehér, B.; Hettinga, K.; Voets, I.K.; Bijl, E. Effect of temperature, pH and calcium phosphate concentration on the properties of reassembled casein micelles. *Food Hydrocoll.* **2024**, *149*, 109592. [CrossRef]
40. Antuma, L.J.; Stadler, M.; Garamus, V.M.; Boom, R.M.; Keppler, J.K. Casein micelle formation as a calcium phosphate phase separation process: Preparation of artificial casein micelles through vacuum evaporation and membrane processes. *Innov. Food Sci. Emerg. Technol.* **2024**, *92*, 103582. [CrossRef]
41. Zamora, A.; Ferragut, V.; Guamis, B.; Trujillo, A.J. Changes in the surface protein of the fat globules during ultra-high pressure homogenisation and conventional treatments of milk. *Food Hydrocoll.* **2012**, *29*, 135–143. [CrossRef]
42. Rodríguez Urbina, J.C.; Osswald, T.A.; Estela Garcia, J.E.; Román, A.J. Environmentally safe preservation and stabilization of natural rubber latex in an acidic environment. *SPE Polym.* **2023**, *4*, 93–104. [CrossRef]
43. Gracz-Bernaciak, J.; Mazur, O.; Nawrot, R. Functional studies of plant latex as a rich source of bioactive compounds: Focus on proteins and alkaloids. *Int. J. Mol. Sci.* **2021**, *22*, 12427. [CrossRef] [PubMed]
44. Bauer, G.; Gorb, S.N.; Klein, M.-C.; Nellesen, A.; von Tapavicza, M.; Speck, T. Comparative study on plant latex particles and latex coagulation in *Ficus benjamina*, *Campanula glomerata* and three *Euphorbia* species. *PLoS ONE* **2014**, *9*, e113336. [CrossRef]
45. Lahmar, I.; Nasri-Ayachi, M.B.; Belghith, K. Laticifer identification, rubber characterization, phenolic content, and antioxidant activity of *Pergularia tomentosa* latex extract. *BioMed Res. Int.* **2022**, *2022*, 7158905. [CrossRef]
46. Konno, K. Plant latex and other exudates as plant defense systems: Roles of various defense chemicals and proteins contained therein. *Phytochemistry* **2011**, *72*, 1510–1530. [CrossRef]
47. Nor, H.M.; Ebdon, J.R. Telechelic liquid natural rubber: A review. *Prog. Polym. Sci.* **1998**, *23*, 143–177. [CrossRef]
48. Sakdapipanich, J.T. Structural characterization of natural rubber based on recent evidence from selective enzymatic treatments. *J. Biosci. Bioeng.* **2007**, *103*, 287–292. [CrossRef]
49. Lam, K.L.; Yang, K.L.; Sunderasan, E.; Ong, M.T. Latex C-serum from Hevea brasiliensis induces non-apoptotic cell death in hepatocellular carcinoma cell line (HepG2). *Cell Prolif.* **2012**, *45*, 577–585. [CrossRef]
50. Rochette, L.; Ghibu, S.; Richard, C.; Zeller, M.; Cottin, Y.; Vergely, C. Direct and indirect antioxidant properties of α-lipoic acid and therapeutic potential. *Mol. Nutr. Food Res.* **2013**, *57*, 114–125. [CrossRef]
51. Rippel, M.M.; Leite, C.A.P.; Lee, L.T.; Galembeck, F. Formation of calcium crystallites in dry natural rubber particles. *J. Colloid Int. Sci.* **2005**, *288*, 449–456. [CrossRef]
52. Buaksuntear, K.; Limarun, P.; Suethao, S.; Smitthipong, W. Non-covalent interaction on the self-healing of mechanical properties in supramolecular polymers. *Int. J. Mol. Sci.* **2022**, *23*, 6902. [CrossRef] [PubMed]
53. Nawamawat, K.; Sakdapipanich, J.T.; Ho, C.C.; Ma, Y.; Song, J.; Vancso, J.G. Surface nanostructure of Hevea brasiliensis natural rubber latex particles. *Colloids Surf. A Physicochem. Eng. Asp.* **2011**, *390*, 157–166. [CrossRef]
54. Wadeesirisak, K.; Castano, S.; Vaysse, L.; Bonfils, F.; Peruch, F.; Rattanaporn, K.; Liengprayoon, S.; Lecomte, S.; Bottier, C. Interactions of REF1 and SRPP1 rubber particle proteins from Hevea brasiliensis with synthetic phospholipids: Effect of charge and size of lipid headgroup. *Biochem. Biophys. Res. Commun.* **2023**, *679*, 205–214. [CrossRef]
55. Dai, L.; Nie, Z.; Kang, G.; Li, Y.; Zeng, R. Identification and subcellular localization analysis of two rubber elongation factor isoforms on Hevea brasiliensis rubber particles. *Plant. Physiol. Biochem.* **2017**, *111*, 97–106. [CrossRef]
56. Garcia, A.; Zou, H.; Hossain, K.R.; Xu, Q.H.; Buda, A.; Clarke, R.J. Polar interactions play an important role in the energetics of the main phase transition of phosphatidylcholine membranes. *ACS Omega* **2019**, *4*, 518–527. [CrossRef]
57. De Gennes, P.G. Reptation of a polymer chain in the presence of fixed obstacles. *J. Chem. Phys.* **1971**, *55*, 572–579. [CrossRef]

58. Sansatsadeekul, J.; Sakdapipanich, J.; Rojruthai, P. Characterization of associated proteins and phospholipids in natural rubber latex. *J. Biosci. Bioeng.* **2011**, *111*, 628–634. [CrossRef]
59. Kumar, H.; Basavaraj, M.G. Plant latex as a versatile and sustainable emulsifier. *Langmuir* **2022**, *38*, 13217–13225. [CrossRef]
60. Bruinsma, R.F.; Wuite, G.J.L.; Roos, W.H. Physics of viral dynamics. *Nat. Rev. Phys.* **2021**, *3*, 76–91. [CrossRef]
61. Mateu, M.G. Assembly, stability and dynamics of virus capsids. *Arch. Biochem. Biophys.* **2013**, *531*, 65–79. [CrossRef] [PubMed]
62. Crick, F.H.; Watson, J.D. Structure of small viruses. *Nature* **1956**, *177*, 473–475. [CrossRef] [PubMed]
63. Caspar, D.L.; Klug, A. Physical principles in the construction of regular viruses. *Cold Spring Harb. Symp. Quant. Biol.* **1962**, *27*, 1–24. [CrossRef]
64. Rossmann, M.G.; Rao, V.B. Principles of virus structural organization. *Viral Mol. Mach.* **2012**, *726*, 17–47. [CrossRef]
65. Hoh, F.; Uzest, M.; Drucker, M.; Plisson-Chastang, C.; Bron, P.; Blanc, S.; Dumas, C. Structural insights into the molecular mechanisms of cauliflower mosaic virus transmission by its insect vector. *J. Virol.* **2010**, *84*, 4706–4713. [CrossRef]
66. Campbell, E.M.; Hope, T.J. HIV-1 capsid: The multifaceted key player in HIV-1 infection. *Nat. Rev. Microbiol.* **2015**, *13*, 471–483. [CrossRef]
67. Lomonossoff, G.P.; Wege, C. TMV particles: The journey from fundamental studies to bionanotechnology applications. *Adv. Virus Res.* **2018**, *102*, 149–176. [CrossRef]
68. Morozov, A.Y.; Bruinsma, R.F.; Rudnick, J. Assembly of viruses and the pseudo-law of mass action. *J. Chem. Phys.* **2009**, *131*, 155101. [CrossRef]
69. Ting, C.L.; Wu, J.; Wang, Z.G. Thermodynamic basis for the genome to capsid charge relationship in viral encapsidation. *Proc. Natl. Acad. Sci. USA* **2011**, *108*, 16986–16991. [CrossRef]
70. Zandi, R.; van der Schoot, P.; Reguera, D.; Kegel, W.; Reiss, H. Classical nucleation theory of virus capsids. *Biophys. J.* **2006**, *90*, 1939–1948. [CrossRef]
71. Wilson, C.J.; Bommarius, A.S.; Champion, J.A.; Chernoff, Y.O.; Lynn, D.G.; Paravastu, A.K.; Liang, C.; Hsieh, M.-C.; Heemstra, J.M. Biomolecular assemblies: Moving from observation to predictive design. *Chem. Rev.* **2018**, *118*, 11519–11574. [CrossRef] [PubMed]
72. Perlmutter, J.D.; Hagan, M.F. Mechanisms of virus assembly. *Annu. Rev. Phys. Chem.* **2015**, *66*, 217–239. [CrossRef] [PubMed]
73. Niklasch, M.; Zimmermann, P.; Nassal, M. The Hepatitis B virus nucleocapsid—Dynamic compartment for infectious virus production and new antiviral target. *Biomedicines* **2021**, *9*, 1577. [CrossRef] [PubMed]
74. Russell, J.T.; Lin, Y.; Böker, A.; Su, L.; Carl, P.; Zettl, H.; He, J.; Sill, K.; Tangirala, R.; Emrick, T.; et al. Self-assembly and cross-linking of bionanoparticles at liquid-liquid interfaces. *Angew. Chem. Int. Ed.* **2005**, *44*, 2420–2426. [CrossRef]
75. Kaur, G.; He, J.; Xu, J.; Pingali, S.V.; Jutz, G.; Böker, A.; Niu, Z.; Li, T.; Rawlinson, D.; Emrick, T.; et al. Interfacial Assembly of Turnip Yellow Mosaic Virus Nanoparticles. *Langmuir* **2009**, *25*, 5168–5176. [CrossRef]
76. Wang, Z.; Gao, S.; Liu, X.; Tian, Y.; Wu, M.; Niu, Z. Programming self-assembly of tobacco mosaic virus coat proteins at Pickering emulsion interfaces for nanorod-constructed capsules. *ACS Appl. Mater. Interfaces* **2017**, *9*, 27383–27389. [CrossRef]
77. Lee, A.G. How lipids affect the activities of integral membrane proteins. *Biochim. Biophys. Acta-Biomembr.* **2004**, *1666*, 62–87. [CrossRef]
78. Pike, L.J. The challenge of lipid rafts. *J. Lipid Res.* **2008**, *50*, S323–S328. [CrossRef]
79. Lombardo, D.; Kiselev, M.A.; Magazù, S.; Calandra, P. Amphiphiles self-assembly: Basic concepts and future perspectives of supramolecular approaches. *Adv. Condens. Matter Phys.* **2015**, *2015*, 151683. [CrossRef]
80. Jagodzinski, F.; Clark, P.; Grant, J.; Liu, T.; Monastra, S.; Streinu, I. Rigidity analysis of protein biological assemblies and periodic crystal structures. *BMC Bioinform.* **2013**, *14*, S2. [CrossRef]
81. Douyère, G.; Leclercq, L.; Nardello-Rataj, V. From polyethyleneimine hydrogels to Pickering-like smart "On/Off" emulgels switched by pH and temperature. *J. Colloid Int. Sci.* **2022**, *628*, 807–819. [CrossRef] [PubMed]
82. Douyère, G.; Leclercq, L.; Nardello-Rataj, V. Université de Lille, Lille 59000, France. 2024; *unpublished work*.
83. Douyère, G.; Leclercq, L.; Nardello-Rataj, V. Cross-linked poly(4-vinylpyridine) particles for pH- and ionic strength-responsive "on–off" Pickering emulsions. *Colloids Surf. A Physicochem. Eng. Asp.* **2021**, *631*, 127705. [CrossRef]
84. Bardoula, V.; Leclercq, L.; Hoogenboom, R.; Nardello-Rataj, V. Amphiphilic polymeric particles based on gradient and block copoly(2-oxazoline)s: Interplay between structure and Pickering emulsion properties. *Colloids Surf. A Physicochem. Eng. Asp.* **2024**, *699*, 134634. [CrossRef]
85. Rivas, C.J.M.; Tarhini, M.; Badri, W.; Miladi, K.; Greige-Gerges, H.; Nazari, Q.A.; Galindo Rodríguez, S.A.; Román, R.A.; Fessi, H.; Elaissari, A. Nanoprecipitation process: From encapsulation to drug delivery. *Int. J. Pharm.* **2017**, *532*, 66–81. [CrossRef]
86. Martinez, C.R.; Iverson, B.L. Rethinking the term "pi-stacking". *Chem. Sci.* **2012**, *3*, 2191–2201. [CrossRef]
87. Kohlan, T.B.; Atespare, A.E.; Yildiz, M.; Menceloglu, Y.Z.; Unal, S.; Dizman, B. Synthesis and structure–property relationship of amphiphilic poly(2-ethyl-co-2-(alkyl/aryl)-2-oxazoline) copolymers. *ACS Omega* **2022**, *7*, 40067–40077. [CrossRef]
88. Valentin Bardoula, V.; Leclercq, L.; Hoogenboom, R.; Nardello-Rataj, V. Amphiphilic nonionic block and gradient copoly(2-oxazoline)s based on 2-methyl-2-oxazoline and 2-phenyl-2-oxazoline as efficient stabilizers for the formulation of tailor-made emulsions. *J. Colloid Int. Sci.* **2023**, *632*, 223–236. [CrossRef]
89. Neuhaus, F.; Mueller, D.; Tanasescu, R.; Balog, S.; Ishikawa, T.; Brezesinski, G.; Zumbuehl, A. Vesicle origami: Cuboid phospholipid vesicles formed by template-free self-assembly. *Angew. Chem. Int. Ed.* **2017**, *56*, 6515–6518. [CrossRef]

90. Tanasescu, R.; Lanz, M.A.; Mueller, D.; Tassler, S.; Ishikawa, T.; Reiter, R.; Brezesinski, G.; Zumbuehl, A. Vesicle origami and the influence of cholesterol on lipid packing. *Langmuir* **2016**, *32*, 4896–4903. [CrossRef]
91. Weinberger, A.; Tanasescu, R.; Stefaniu, C.; Fedotenko, L.A.; Favarger, F.; Ishikawa, T.; Brezesinski, G.; Marques, C.M.; Zumbuehl, A. Bilayer Properties of 1,3-diamidophospholipids. *Langmuir* **2015**, *31*, 1879–1884. [CrossRef]
92. Holme, M.N.; Fedotenko, I.A.; Abegg, D.; Althaus, J.; Babel, L.; Favarger, F.; Reiter, R.; Tanasescu, R.; Zaffalon, P.L.; Ziegler, A.; et al. Shear-stress sensitive lenticular vesicles for targeted drug delivery. *Nat. Nanotechnol.* **2012**, *7*, 536–543. [CrossRef] [PubMed]
93. Zumbuehl, A. Artificial phospholipids and their vesicles. *Langmuir* **2019**, *35*, 10223–10232. [CrossRef] [PubMed]
94. Pardin, C.; Leclercq, L.; Schmitzer, A.R. *N,N′*-Methylenediimidazolium salts: From self-assembly to an efficient DNAse protection system. *Chem. Eur. J.* **2010**, *16*, 4686–4692. [CrossRef]
95. Dong, K.; Zhang, S.; Wang, D.; Yao, X. Hydrogen bonds in imidazolium ionic liquids. *J. Phys. Chem. A* **2006**, *110*, 9775–9782. [CrossRef]
96. Valéry, C.; Paternostre, M.; Robert, B.; Gulik-Krzywicki, T.; Narayanan, T.; Dedieu, J.-C.; Keller, G.; Torres, M.-L.; Cherif-Cheikh, R.; Calvo, P.; et al. Biomimetic organization: Octapeptide self-assembly into nanotubes of viral capsid-like dimension. *Proc. Natl. Acad. Sci. USA* **2004**, *100*, 10258–10262. [CrossRef]
97. Hill, J.P.; Jin, W.; Kosaka, A.; Fukushima, T.; Ichihara, H.; Shimomura, T.; Ito, K.; Hashizume, T.; Ishii, N.; Aida, T. Self-assembled hexa-*peri*-hexabenzocoronene graphitic nanotube. *Science* **2004**, *304*, 1481–1483. [CrossRef]
98. Leclercq, L. Interactions between cyclodextrins and cellular components: Towards greener medical applications? *Beilstein J. Org. Chem.* **2016**, *12*, 2644–2662. [CrossRef]
99. Saenger, W. Crystal packing patterns of cyclodextrin inclusion complexes. *J. Incl. Phenom.* **1984**, *2*, 445–454. [CrossRef]
100. Hashizaki, K.; Kageyama, T.; Inoue, M.; Taguchi, H.; Ueda, H.; Saito, Y. Study on preparation and formation mechanism of *n*-alkanol/water emulsion using α-cyclodextrin. *Chem. Pharm. Bull.* **2007**, *55*, 1620–1625. [CrossRef]
101. Davarpanah, L.; Vahabzadeh, F. Formation of oil-in-water (O/W) pickering emulsions via complexation between β-cyclodextrin and selected organic solvents. *Starch-Särke* **2021**, *64*, 898–913. [CrossRef]
102. Leclercq, L.; Tessier, J.; Douyère, G.; Nardello-Rataj, V.; Schmitzer, A.R. Phytochemical- and cyclodextrin-based Pickering emulsions: Natural potentiators of antibacterial, antifungal, and antibiofilm activity. *Langmuir* **2020**, *36*, 4317–4323. [CrossRef] [PubMed]
103. Leclercq, L.; Company, R.; Mühlbauer, A.; Mouret, A.; Aubry, J.-M.; Nardello-Rataj, V. Versatile eco-friendly Pickering emulsions based on substrate/native cyclodextrin complexes: A winning approach for solvent-free oxidations. *ChemSusChem* **2013**, *6*, 1533–1540. [CrossRef] [PubMed]
104. Hou, J.; Xu, H.-N. Faceted crystal growth of cyclodextrin-oil inclusion complexes. *Carbohydr. Polym.* **2024**, *343*, 122446. [CrossRef]
105. Leclercq, L.; Tessier, J.; Nardello-Rataj, V.; Schmitzer, A.R. Highly active, entirely biobased antimicrobial Pickering emulsions. *ChemMedChem* **2021**, *16*, 2223–2230. [CrossRef]
106. Mathapa, B.G.; Paunov, V.N. Cyclodextrin stabilised emulsions and cyclodextrinosomes. *Phys. Chem. Chem. Phys.* **2013**, *15*, 17903–17914. [CrossRef]
107. Bochot, A.; Trichard, L.; Le Bas, G.; Alphandary, H.; Grossiord, J.-L.; Duchêne, D.; Fattal, E. α-Cyclodextrin/oil beads: An innovative self-assembling system. *Int. J. Pharm.* **2007**, *339*, 121–129. [CrossRef]
108. Hamoudi, C.; Bochot, A. Oil-cyclodextrin based beads for oral delivery of poorly-soluble drugs. *Curr. Top. Med. Chem.* **2014**, *14*, 510–517. [CrossRef]
109. Trichard, L.; Fatta, E.; Besnard, M.; Bochot, A. α-Cyclodextrin/oil beads as a new carrier for improving the oral bioavailability of lipophilic drugs. *J. Control. Release* **2007**, *122*, 47–53. [CrossRef]
110. Trichard, L.; Delgado-Charro, M.B.; Guy, R.H.; Fattal, E.; Bochot, A. Novel beads made of α-cyclodextrin and oil for topical delivery of a lipophilic drug. *Pharm. Res.* **2008**, *25*, 435–440. [CrossRef]
111. Trichard, L.; Chaminade, P.; Grossior, J.-L.; Le Bas, G.; Huang, N.; Durand, D.; Fattal, E.; Bochot, A. Beads made of α-cyclodextrin and vegetable oils: Oil composition and physicochemical properties influence bead feasibility and properties. *J. Drug Deliv. Sci. Technol.* **2011**, *21*, 189–194. [CrossRef]
112. Potier, J.; Menuel, S.; Chambrier, M.-H.; Burylo, L.; Blach, J.-F.; Woisel, P.; Monflier, E.; Hapiot, F. Pickering emulsions based on supramolecular hydrogels: Application to higher olefins' hydroformylation. *ACS Catal.* **2013**, *3*, 1618–1621. [CrossRef]
113. Potier, J.; Menuel, S.; Monflier, E.; Hapiot, F. Synergetic Effect of randomly methylated β-cyclodextrin and a supramolecular hydrogel in Rh-catalyzed hydroformylation of higher olefins. *ACS Catal.* **2014**, *4*, 2342–2346. [CrossRef]
114. Chevry, M.; Vanbésien, T.; Menuel, S.; Monflier, E.; Hapiot, F. Tetronics/cyclodextrin-based hydrogels as catalyst-containing media for the hydroformylation of higher olefins. *Catal. Sci. Technol.* **2017**, *7*, 114–123. [CrossRef]
115. Lin, C.E.; Huang, H.C.; Chen, H.W. A capillary electrophoresis study on the influence of β-cyclodextrin on the critical micelle concentration of sodium dodecyl sulfate. *J. Chromatogr. A* **2001**, *917*, 297–310. [CrossRef]
116. Jiang, L.; Peng, Y.; Yan, Y.; Deng, M.; Wang, Y.; Huang, J. "Annular ring" microtubes formed by SDS@2β-CD complexes in aqueous solution. *Soft Matter* **2010**, *6*, 1731–1736. [CrossRef]
117. Jiang, L.; Peng, Y.; Yan, Y.; Huang, J. Aqueous self-assembly of SDS@2β-CD complexes: Lamellae and vesicles. *Soft Matter* **2011**, *7*, 1726–1731. [CrossRef]
118. Yang, S.; Yan, Y.; Huang, J.; Petukhov, A.V.; Kroon-Batenburg, L.M.J.; Drechsler, M.; Zhou, C.; Tu, M.; Granick, S.; Jiang, L. Giant capsids from lattice self-assembly of cyclodextrin complexes. *Nat. Commun.* **2017**, *8*, 15856. [CrossRef]

119. dos Santos Silva Araújo, L.; Lazzara, G.; Chiappisi, L. Cyclodextrin/surfactant inclusion complexes: An integrated view of their thermodynamic and structural properties. *Adv. Colloid Interface Sci.* **2021**, *289*, 102375. [CrossRef]
120. Liu, Z.; Zhao, Q.; Gao, S.; Yan, Y.; Xu, B.; Ma, C.; Huang, J. Multi-responsive self-assembly of polymeric-like building blocks nC12C6C12(Me)@nγ-CD: Vesicles, nanotubes and sheet crystal. *Colloids Surf. A Physicochem. Eng. Asp.* **2023**, *676*, 132269. [CrossRef]
121. Wu, C.; Xie, Q.; Xu, W.; Tu, M.; Jiang, L. Lattice self-assembly of cyclodextrin complexes and beyond. *Curr. Opin. Colloid Interface Sci.* **2019**, *39*, 76–85. [CrossRef]
122. Liu, K.; Ma, C.; Wu, T.; Qi, W.; Yan, Y.; Huang, J. Recent advances in assemblies of cyclodextrins and amphiphiles: Construction and regulation. *Curr. Opin. Colloid Interface Sci.* **2020**, *45*, 44–56. [CrossRef]
123. Gu, T.; Huang, J.; Yan, Y. New opportunities for cyclodextrins in supramolecular assembly: Metal organic frameworks, crystalline self-assembly, and catalyzed assembly. *Chem. Commun.* **2023**, *59*, 14759–14775. [CrossRef] [PubMed]
124. Landman, J.; Ouhajji, S.; Prévost, S.; Narayanan, T.; Groenewold, J.; Philipse, A.P.; Kegel, W.K.; Petukhov, A.V. Inward growth by nucleation: Multiscale self-assembly of ordered membranes. *Sci. Adv.* **2018**, *4*, eaat1817. [CrossRef] [PubMed]
125. Dubois, M.; Demé, B.; Gulik-Krzywicki, T.; Dedieu, J.-C.; Vautrin, C.; Désert, S.; Perez, E.; Zemb, T. Self-assembly of regular hollow icosahedra in salt-free catanionic solutions. *Nature* **2001**, *411*, 672–675. [CrossRef]
126. Dubois, M.; Lizunov, V.; Meister, A.; Gulik-Krzywicki, T.; Verbavatz, J.M.; Perez, E.; Zimmerberg, J.; Zemb, T. Shape control through molecular segregation in giant surfactant aggregates. *Proc. Natl. Acad. Sci. USA* **2004**, *101*, 15082. [CrossRef]
127. Greenfield, M.A.; Palmer, L.C.; Vernizzi, G.; Olvera de la Cruz, M.; Stupp, S.I. Buckled membranes in mixed-valence ionic amphiphile vesicles. *J. Am. Chem. Soc.* **2009**, *131*, 12030–12031. [CrossRef]
128. Zhang, J.; Xu, G.; Song, A.; Wang, L.; Lin, M.; Dong, Z.; Yang, Z. Faceted fatty acid vesicles formed from single-tailed perfluorinated surfactants. *Soft Matter* **2015**, *11*, 7143–7150. [CrossRef]
129. Bowick, M.J.; Sknepnek, R. Pathways to faceting of vesicles. *Soft Matter* **2013**, *9*, 8088–8095. [CrossRef]
130. Guttman, S.; Ocko, B.M.; Deutsch, M.; Sloutskin, E. From faceted vesicles to liquid icoshedra: Where topology and crystallography meet. *Curr. Opin. Colloid Interface Sci.* **2016**, *22*, 35–40. [CrossRef]
131. Shen, Y.; Ou-Yang, Z.; Hao, J.; Lin, H.; Jiang, L.; Liu, Z.; Gao, X. The mechanism for the transition from vesicles to punctured lamellae and faceted vesicles in cationic and anionic fluorinated surfactant mixture. *Colloids Surf. A Physicochem. Eng. Asp.* **2016**, *500*, 40–44. [CrossRef]
132. Leclercq, L.; Mouret, A.; Proust, A.; Schmitt, V.; Bauduin, P.; Aubry, J.-M.; Nardello-Rataj, V. Pickering emulsion stabilized by catalytic polyoxometalate nanoparticles: A new effective medium for oxidation reactions. *Chem. Eur. J.* **2012**, *8*, 14352–14358. [CrossRef] [PubMed]
133. Bonhomme, F.; Kanatzidis, M.G. Structurally characterized mesostructured hybrid surfactant–inorganic lamellar phases containing the adamantane $[Ge_4S_{10}]^{4-}$ anion: Synthesis and properties. *Chem. Mater.* **1998**, *10*, 1153–1159. [CrossRef]
134. Leclercq, L.; Mouret, A.; Bauduin, P.; Nardello-Rataj, V. Supramolecular colloidosomes based on tri(dodecyltrimethylammonium) phosphotungstate: A bottom-up approach. *Langmuir* **2014**, *30*, 5386–5393. [CrossRef] [PubMed]
135. Leclercq, L.; Mouret, A.; Renaudineau, S.; Schmitt, V.; Proust, A.; Nardello-Rataj, V. Self-assembled polyoxometalates nanoparticles as pickering emulsion stabilizers. *J. Phys. Chem. B* **2015**, *119*, 6326–6337. [CrossRef]
136. Mouret, A.; Leclercq, L.; Mühlbauer, A.; Nardello-Rataj, V. Eco-friendly solvents and amphiphilic catalytic polyoxometalate nanoparticles: A winning combination for olefin epoxidation. *Green Chem.* **2014**, *16*, 269–278. [CrossRef]
137. Wu, Y.; Shi, R.; Wu, Y.-L.; Holcroft, J.M.; Liu, Z.; Frasconi, M.; Wasielewski, M.R.; Li, H.; Stoddart, J.F. Complexation of polyoxometalates with cyclodextrins. *J. Am. Chem. Soc.* **2015**, *137*, 4111–4118. [CrossRef]
138. Naskar, B.; Diat, O.; Nardello-Rataj, V.; Bauduin, P. Nanometer-size polyoxometalate anions adsorb strongly on neutral soft surfaces. *J. Phys. Chem. C* **2015**, *119*, 20985–20992. [CrossRef]
139. Girard, L.; Naskar, B.; Dufrêche, J.-F.; Lai, J.; Diat, O.; Bauduin, P. A thermodynamic model of non-ionic surfactants' micellization in the presence of polyoxometalates. *J. Mol. Liq.* **2019**, *293*, 111280. [CrossRef]
140. Pacaud, B.; Leclercq, L.; Dechézelles, J.-F.; Nardello-Rataj, V. Hybrid core-shell nanoparticles by "plug and play" self-assembly. *Chem. Eur. J.* **2018**, *24*, 17672–17676. [CrossRef]
141. Yang, B.; Leclercq, L.; Schmitt, V.; Pera-Titus, M.; Nardello-Rataj, V. Colloidal tectonics for tandem synergistic Pickering interfacial catalysis: Oxidative cleavage of cyclohexene oxide into adipic acid. *Chem. Sci.* **2019**, *10*, 501–507. [CrossRef]
142. Feng, Y.; Dechezelles, J.-F.; D'Acremont, Q.; Courtade, E.; De Waele, V.; Pera-Titus, M.; Nardello-Rataj, V. Light-driven Pickering interfacial catalysis for the oxidation of alkenes at near-room temperature. *Green Chem.* **2023**, *25*, 1417–1423. [CrossRef]
143. van Blaaderen, A. Colloidal molecules and beyond. *Science* **2003**, *301*, 470–471. [CrossRef] [PubMed]
144. Manoharan, V.N.; Elsesser, M.T.; Pine, D.J. Dense packing and symmetry in small clusters of microspheres. *Science* **2003**, *301*, 483–487. [CrossRef]

145. Kim, Y.-J.; Moon, J.-B.; Hwang, H.; Kim, Y.S.; Yi, G.-R. Advances in colloidal building blocks: Toward patchy colloidal clusters. *Adv. Mater.* **2023**, *35*, 2203045. [CrossRef]
146. Cho, Y.-S.; Kim, S.-H.; Yi, G.-R.; Yang, S.-M. Self-organization of colloidal nanospheres inside emulsion droplets: Higher-order clusters, supraparticles, and supraballs. *Colloids Surf. A Physicochem. Eng. Asp.* **2009**, *345*, 237–245. [CrossRef]

MDPI AG
Grosspeteranlage 5
4052 Basel
Switzerland
Tel.: +41 61 683 77 34

Molecules Editorial Office
E-mail: molecules@mdpi.com
www.mdpi.com/journal/molecules

www.ingramcontent.com/pod-product-compliance
Lightning Source LLC
Chambersburg PA
CBHW061503210926
51850CB00008B/1800
9783725859597

MDPI AG
Grosspeteranlage 5
4052 Basel
Switzerland
Tel.: +41 61 683 77 34

Catalysts Editorial Office
E-mail: catalysts@mdpi.com
www.mdpi.com/journal/catalysts

www.ingramcontent.com/pod-product-compliance
Lightning Source LLC
Chambersburg PA
CBHW061507210926
51850CB00008B/1845
9783725867547